Optoelectronic Technology

光电子技术

"世界学术研究前沿丛书"编委会
THE EDITORIAL BOARD OF
WORLD ACADEMIC FRONTIERS

中国出版集团公司
世界图书出版公司
广州 · 上海 · 西安 · 北京

图书在版编目（CIP）数据

光电子技术：英文 / “世界学术研究前沿丛书”编委会编.—广州：世界图书出版广东有限公司，2017.8
ISBN 978-7-5192-2460-8

Ⅰ. ①光… Ⅱ. ①世… Ⅲ. ①光电子技术—文集—英文 Ⅳ. ①TN2-53

中国版本图书馆CIP数据核字(2017)第040979号

书　　名：光电子技术
Guangdianzi Jishu
编　　者：“世界学术研究前沿丛书”编委会
责任编辑：康琬娟
出版发行：世界图书出版广东有限公司
地　　址：广州市海珠区新港西路大江冲25号
邮　　编：510300
电　　话：（020）84460408
网　　址：http：//www.gdst.com.cn/
邮　　箱：wpc_gdst@163.com
经　　销：新华书店
印　　刷：广州市德佳彩色印刷有限公司
开　　本：787 mm×1092 mm　1/16
印　　张：28
插　　页：4
字　　数：620千
版　　次：2017年8月第1版　　2017年8月第1次印刷
国际书号：ISBN 978-7-5192-2460-8
定　　价：598.00元

Preface

Optoelectronic technology is the application of electronic devices that source, detect and control light, usually considered a sub-field of photonics. In this context, light often includes invisible forms of radiation such as gamma rays, X-rays, ultraviolet and infrared, in addition to visible light. Optoelectronics is based on the quantum mechanical effects of light on electronic materials, especially semiconductors, sometimes in the presence of electric fields. Optoelectronic devices are electrical-to-optical or optical-to-electrical transducers, or instruments that use such devices in their operation. Electro-optics is often erroneously used as a synonym, but is a wider branch of physics that concerns all interactions between light and electric fields, whether or not they form part of an electronic device.[1]

In the present book, thirty-five literatures about optoelectronic technology published on international authoritative journals were selected to introduce the worldwide newest progress, which contains reviews or original researches on Optoelectronic information technology, communications, optoelectronic technology, biological science and medical optoelectronic technology, military optoelectronic technology *ect*. We hope that this book can demonstrate advances in optoelectronic technology as well as give references to the researchers, students and other related people.

March 9, 2017

[1]From Wikipedia: https://en.wikipedia.org/wiki/Optoelectronics

Selected Authors

Yu Ye, State Key Lab for Mesoscopic Physics and School of Physics, Peking University, Beijing, China.

E. Cuesta, Dept. of Manufacturing Engineering, University of Oviedo, Asturias, Spain.

Andrzej Kolek, Department of Electronics Fundamentals, Rzeszow University of Technology, Al. Powstancow Warszawy 12, Rzeszow, Poland.

Zhiming M. Wang, State Key Laboratory of Electronic Thin Films and Integrated Devices, University of Electronic Science and Technology of China, Chengdu, China.

Ning Xu, Professor, Shanghai Engineering Research Center of Ultra-Precision Optical Manufacturing, Department of Optical Science and Engineering, Fudan University, Shanghai, China.

David Hernández-de la Luz, Science Institute-Center of Investigation for Semiconductors Devices (IC-CIDS) of Autonomous University of Puebla, Mexico.

J. Alberto Luna López, Professor, IC-CIDS Benemérita Universidad Autónoma de Puebla, Ed. 103 C o D, Col. San Manuel, C.P, Puebla, Pue , Mexico.

Yeong Hwan Ko, Department of Electronics and Radio Engineering, Kyung Hee University, Seocheon-dong, Giheung-gu, Yongin-si, Gyeonggi-do, Republic of Korea.

Koichi Baba, Department of Visual Regenerative Medicine, Osaka University Graduate School of Medicine, 2-2 Yamadaoka, Suita, Osaka, Japan.

Contents

Preface..........I

Selected Authors..........III

Chapter 1..........1
Low Temperature Optodic Bonding for Integration of Micro Optoelectronic Components in Polymer Optronic Systems
by Y. Wang and L. Overmeyer

Chapter 2..........17
Novel Optoelectronic Devices Based on Single Semiconductor Nanowires (Nanobelts)
by Yu Ye, Lun Dai, Lin Gan, et al.

Chapter 3..........33
Symmetric Periodic Solutions of Delay-Coupled Optoelectronic Oscillators
by Chunrui Zhang and Hongpeng Li

Chapter 4..........55
An Optoelectronic Monitoring System for Aviation Hydraulic Fluids
by Andreas Helwig, Konrad Maier, Gerhard Müller, et al.

Chapter 5..........63
Compositional and Optical Properties of SiO_x Films and (SiO_x/SiO_y) Junctions Deposited by HFCVD
by Diana E. Vázquez-Valerdi, Jose A. Luna-López, Jesús Carrillo-López, et al.

Chapter 6......81
Conformity Analysis in the Measurement of Machined Metal Surfaces with Optoelectronic Profilometer
by E. Cuesta, B. J. Álvarez, M. García-Diéguez, et al.

Chapter 7......95
Modeling of Optoelectronic Devices with One-Band Effective Mass Equation: Nonequilibrium Green's Function Approach
by Andrzej Kolek

Chapter 8......105
Effects of Rapid Thermal Annealing on the Optical Properties of Strain-Free Quantum Ring Solar Cells
by Jiang Wu, Zhiming M. Wang, Vitaliy G. Dorogan, et al.

Chapter 9......117
Electrical and Optical Properties of Binary CN_x Nanocone Arrays Synthesized by Plasma-Assisted Reaction Deposition
by Xujun Liu, Leilei Guan, Xiaoniu Fu, et al.

Chapter 10......131
Synthesis, Characterization and Interpretation of Screen-Printed Nanocrystalline CdO Thick Film for Optoelectronic Applications
by Rayees Ahmad Zargar, Santosh Chackrabarti, Manju Arora, et al.

Chapter 11......145
Microstructure and Optical Properties of Ag/ITO/Ag Multilayer Films
by Zhaoqi Sun, Maocui Zhang, Qiping Xia, et al.

Chapter 12......157
Modification of Optical and Electrical Properties of Zinc Oxide-Coated Porous Silicon Nanostructures Induced by Swift Heavy Ion
by Yogesh Kumar, Manuel Herrera-Zaldivar, Sion Federico Olive-Méndez, et al.

Chapter 13..173
Morphological, Compositional, Structural, and Optical Properties of Si-Nc Embedded In SiO_x Films
by J. Alberto Luna López, J. Carrillo López, D. E. Vázquez Valerdi, et al.

Chapter 14..195
One-Dimensional CuO Nanowire: Synthesis, Electrical, and Optoelectronic Devices Application
by Lin-Bao Luo, Xian-He Wang, Chao Xie, et al.

Chapter 15..213
Optical Properties of GaP/GaNP Core/Shell Nanowires: A Temperature-Dependent Study
by Alexander Dobrovolsky, Shula Chen, Yanjin Kuang, et al.

Chapter 16..223
Optical Properties of Ni and Cu Nanowire Arrays and Ni/Cu Superlattice Nanowire Arrays
by Yaya Zhang, Wen Xu, Shaohui Xu, et al.

Chapter 17..235
Optoelectronic Spin Memories of Electrons in Semiconductors
by M. Idrish Miah

Chapter 18..243
Optoelectronic System for the Determination of Blood Volume in Pneumatic Heart Assist Devices
by Grzegorz Konieczny, Tadeusz Pustelny, Maciej Setkiewicz, et al.

Chapter 19..265
Sol-Gel Synthesized Zinc Oxide Nanorods and Their Structural and Optical Investigation for Optoelectronic Application

by Kai Loong Foo, Uda Hashim, Kashif Muhammad, et al.

Chapter 20..........287
Effect of 2-Mercaptoethanol as Capping Agent on ZnS Nanoparticles: Structural and Optical Characterization
by Abbas Rahdar

Chapter 21..........299
Structural and Optical Properties of ZnO Nanorods by Electrochemical Growth Using Multi-Walled Carbon Nanotube-Composed Seed Layers
by Yeong Hwan Ko, Myung Sub Kim and Jae Su Yu

Chapter 22..........311
Synthesis of ZnSe Quantum Dots with Stoichiometric Ratio Difference and Study of Its Optoelectronic Property
by Uzma B. Memon, U. Chatterjee, M. N. Gandhi, et al.

Chapter 23..........323
Growth and Optical Properties of ZnO Nanorod Arrays on Al-Doped ZnO Transparent Conductive Film
by Suanzhi Lin, Hailong Hu, Weifeng Zheng, et al.

Chapter 24..........337
Preparation of 1,4-Bis(4-Methylstyryl)Benzene Nanocrystals by a Wet Process and Evaluation of Their Optical Properties
by Koichi Baba and Kohji Nishida

Chapter 25..........353
AlGaInP LED with Low-Speed Spin-Coating Silver Nanowires as Transparent Conductive Layer
by Xia Guo, Chun Wei Guo, Cheng Wang, et al.

Chapter 26..........365
Cu-Doped ZnO Nanorod Arrays: The Effects of Copper Precursor and

Concentration

by Musbah Babikier, Dunbo Wang, Jinzhong Wang, et al.

Chapter 27..383

Electromagnetic Enhancement of Graphene Raman Spectroscopy by Ordered and Size-Tunable Au Nanostructures

by Shuguang Zhang, Xingwang Zhang and Xin Liu

Chapter 28..399

Functionalized Silicon Quantum Dots by N-Vinylcarbazole: Synthesis and Spectroscopic Properties

by Jianwei Ji, Guan Wang, Xiaozeng You, et al.

Chapter 29..415

Improving the Photoelectric Characteristics of MoS_2 Thin Films by Doping Rare Earth Element Erbium

by Miaofei Meng and Xiying Ma

Chapter 30..427

Investigation of Optoelectronic Properties of N3 Dye-Sensitized TiO_2 Nano-Crystals by Hybrid Methods: ONIOM (QM/MM) Calculations

by Mohsen Oftadeh and Leila Tavakolizadeh

Chapter 1

Low Temperature Optodic Bonding for Integration of Micro Optoelectronic Components in Polymer Optronic Systems

Y. Wang*, L. Overmeyer

Institute of Transport- and Automation Technology, Leibniz Universität Hannover, An der Universität 2, 30823 Garbsen, Germany

Abstract: Large area, planar optronic systems based on flexible polymer substrates allow a reel-to-reel mass production, which is widely adopted in modern manufacturing. Polymer optronic systems are fully integrated with micro optical and optoelectronic components as light sources, detectors and sensors to establish highly functional sensor networks. To achieve economical production, low-cost polymer sheets are employed. Since they are mostly thermally sensitive, this requires a restricted thermal loading during processing. Furthermore, a short process time improves production efficiency, which plays a key role in manufacturing processes. Thus, in this contribution we introduce a new bare chip bonding technique using light instead of heat to meet both requirements. The technique is based on the conventional flip-chip die bonding process. Ultraviolet radiation curing adhesives are applied as bonding material, accordingly a sideway ultraviolet radiation source, a so-called optode, is designed. Before implementing the concept, the light distribution in the contact spot is simulated to examine the feasibility of the solution. Besides, we investigate two different UV lamps regarding induced thermal influence on polymer substrate to choose one to be employed in the optode. Process factors, irradiation intensity and irradiation time are studied. Based on these results, the mechanical and electrical reliability of the integrated components is finally evaluated.

Keywords: Flip-Chip, Die Bonding, UV-Curing Adhesives, Optoelectronic, Component Integration, Low-Temperature

1. Introduction

With rapid development of modern information technology the demands of sensor systems for more reliability of signal detection and conversion, higher speed of data transmission, more flexibility of carrier materials, and lower cost of mass production are increasingly growing. Fully integrated, large area optronic systems based on planar polymer substrates are the innovative solution. Optronics implies the employing of optical or optoelectronic components as light sources, detectors or sensors as well as optical waveguides as transmission medium to create a fully optical system that is well-established for sensing, processing of signals and transmission of data. Polymer films with a thickness of a hundred micro meters as carrier substrate offer high flexibility so that the optronic systems allow a reel-to-reel mass production, which facilitates a highly efficient manufacturing process. Besides, polymer films can mostly be purchased on a low budget, which makes the production of optronic systems more economical.

To realize such polymer based planar optronic systems, one key focus is on the integration of optical or optoelectronic components into the thin polymer sheets. We employ the optical or optoelectronic components in form of bare chips. Given the fact that most polymer products are thermally sensitive because of their low glass transition temperature Tg, e.g. PMMA with Tg 105°C, PET 70°C and PVC 80°C, the thermal loading on polymer sheets must be restricted during processing[1]. Another particular aspect is the requirement of the high positioning accuracy of the bare optoelectronic chips. As shown in **Figure 1** left, a small tilt of the bonded laser diode may cause a great loss of light transmission into the waveguide[2].

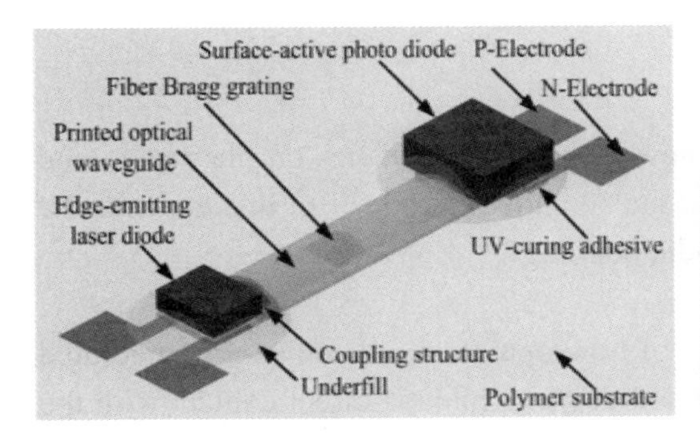

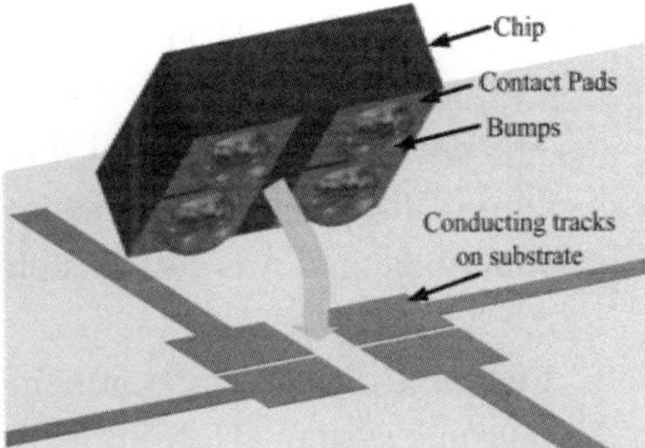

Figure 1. Left: Planar optronic system based on conventional semiconductor optoelectronic components; right: Flip-chip technology.

According to the state of the art in the field of chip mounting technology, there are two bonding methods available to accomplish the electric connections: wire bonding and flip-chip die bonding. In wire bonding technology, the chips are mounted face up, and the wires are used to build interconnections between the pads of chip and external[3]. Flip-chip die bonding is referred to as a mounting technology, in which the functional side of bare chip is faced down, and the pads on chip are aligned directly with the external circuit[4]. By this way, the electric connection is completed simultaneously with the mechanical mounting, which makes the whole bonding process much more efficient and cost-effective. Regarding the fact that optronic systems are built on a large area, flexible polymer substrate, wire bonding technology is excluded, since thin wire interconnections cannot withstand a variety of environmental disturbances. Thus, flip-chip die bonding technology is preferred.

In most applications of flip-chip technology, thermal effects are utilized, typically using solder bumps combined with hot air reflow or heat curing adhesives[5]. However, as already mentioned, thermal loading has to be avoided in polymer optronic systems to protect the polymer substrate from damage. With above considerations, we developed a new bonding method based on flip-chip technology using light instead of heat to realize the integration of bare optoelectronic chips into polymer optronic systems. Using light means applying ultraviolet curing adhesives (abbreviated as UV curing adhesive) as bonding material and getting it cured by UV irradiation. In order to enforce a stable UV irradiation during the curing process, a reliable UV radiation source is necessary.

2. Optodic Bonding

The development focus of this new bonding method lies on the design and realization of the UV radiation source, the so-called optode[6]. In this context, the new bonding method was called "optodic bonding".

Figure 2 illustrates the principle of how optoelectronic chips are integrated in the optronic system. The chip is flipped down to make a direct contact with the pads on the circuit while it is mounted on the polymer substrate. UV-curing adhesives serve hereon as bonding material mainly to assure the mechanical connection. Furthermore, some UV-curing adhesives contribute to electric contacting as well, e.g. isotropic conductive adhesive (ICA) and anisotropic conductive adhesive (ACA)[7]. Because of the danger leading to an electrical short of circuit, the ICA UV-curing adhesives are out of choice. ACA UV-curing adhesives are also excluded for the moment due to the shortage of the sorts on the market and the high cost of purchase. Therefore, in this work we employ non-conductive UV-curing adhesives. By triggering the optode, the UV rays are getting started to irradiate and will be guided in the direction of the spot of the adhesive. While getting the adhesive cured, the chip is getting mounted mechanically on the polymer films and electrically interconnected with the circuit.

2.1. Optode for Sideway Irradiation

On the basis of the concept shown in **Figure 2**, an optode serving as sideway irradiation was designed and is illustrated in **Figure 3**.

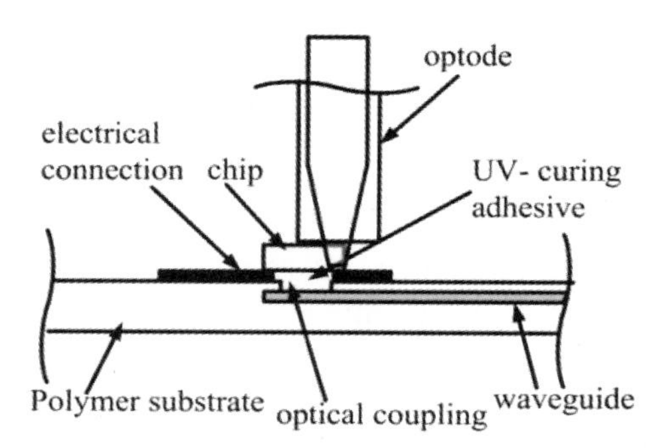

Figure 2. Schematic illustration of chip integration in optronic system using optodic bonding (Krühn T, 2012).

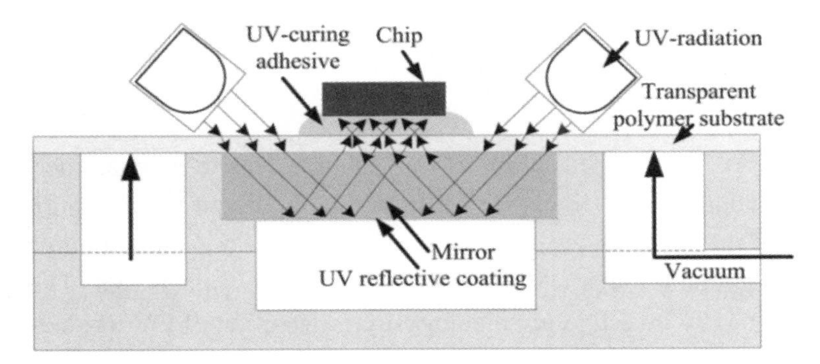

Figure 3. Schematic illustration of optode for sideway irradiation.

A manual flip-chip die bonding assembly system named *Fineplacer®* is used which is equipped with a heat plate as thermode for a conventional bonding process. To realize the optode illustrated in **Figure 3**, the heat plate has to be replaced by a newly constructed underplate. As a result, the vacuum configuration attached within the heat plate was eliminated as well, leaving the polymer films unfixed. This gives rise to unwished moving of the polymer films during processing. Even a tiny shifting can cause great positioning faults. Given this effect, we built a vacuum channel in form of a circle, which consists of many small holes around the bonding spot on the surface of the underplate.

The transparent polymer substrate lay on the underplate must have a sufficient transmission grade of UV irradiation. **Figure 4** shows the property of the transmission, reflection and absorption of the light in the UV range from 350nm to 400nm for the polymer substrates PET with a thickness of 200μm and the PMMA with a thickness of 175μm.

Polymer substrate PET-GAG has a sufficient UV transmission grade for the range from 350μm to 400μm. In contrast, for PMMA a UV transmission grade of about 80% starts only from the wavelength of 380μm. This optical property of polymer substrates has a significant effect on choosing UV-radiation source.

The chip is placed face-down, aligned with the conducting track on the polymer substrate after the UV-curing adhesive is dispensed on this contact spot. To accomplish the junction between chip and substrate, the key is the curing of the adhesive. However, because of the ultrathin thickness of the adhesive layer barely adequate UV rays can reach there to get the adhesive cured. To solve this problem, we place a mirror under the contact spot utilizing the principle of light reflection to collect sufficient UV irradiation.

As a first step, the feasibility of this solution was to be examined. For this purpose, we conducted a simulation of an optode with and without mirror. As simulation tool the software Zemax was applied. Aim of the simulation was to find out whether and how much UV irradiation reaches the contact spot. The question can be answered by investigating the intensity distribution of UV irradiation on the contact spot. The result presented in **Figure 5** shows a clear contrast between the simulation with and without mirror. The number in the figure has the unit mW/cm^2. In the left figure for the simulation without mirror, a blue circle can be seen that stands for an intensity of nearly 0mW/cm^2, *i.e.* no UV irradiation at all. In contrast, a red circle in the right figure that describes the simulation with mirror indicates an intensity of 7663.70mW/cm^2, *i.e.* sufficient UV irradiation. Therefore, the simulation

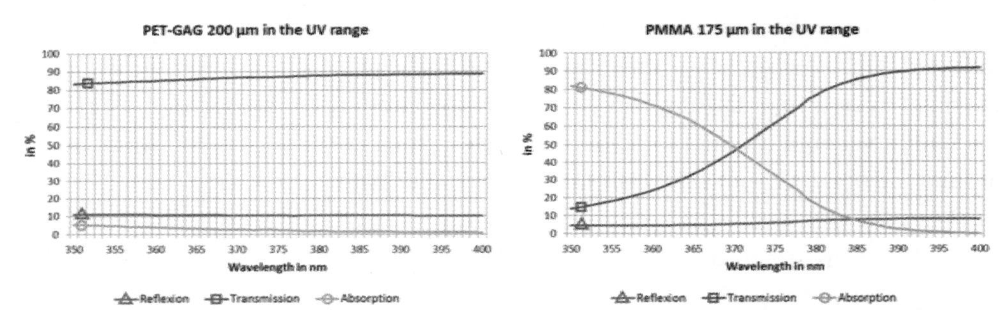

Figure 4. Light transmission, reflection and absorption in the UV range. left: PET 200μm; right: PMMA 175μm.

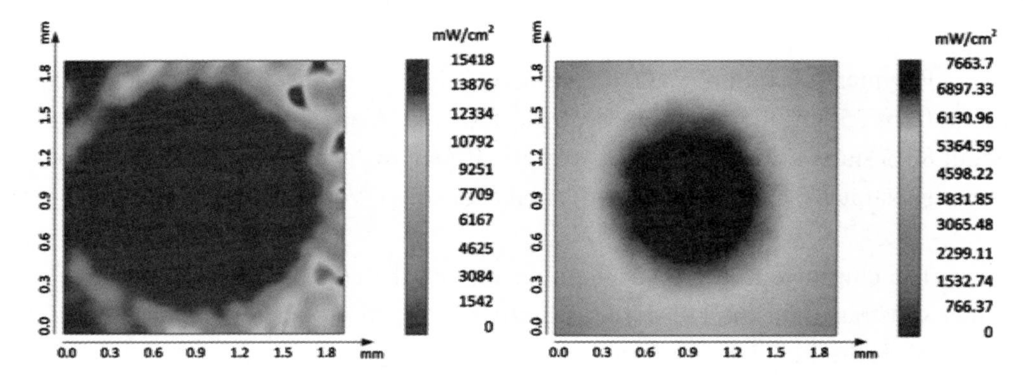

Figure 5. Simulation of the intensity distribution of UV irradiation on the contact spot from sideway optode. left: without mirror; right: with mirror.

with mirror verified the feasibility of a sideway optode.

As shown in **Figure 3**, the UV ray is emitted by a UV lamp, passes through the thin transparent polymer substrate, reaches the coated surface of the mirror and then is reflected to the contact spot, where the adhesive lies. The employed mirror has a thickness of 5mm. If the coated surface of the mirror faces upwards, the adhesive can hardly be irradiated. Due to the low thickness of the substrate, the propagation path of the UV rays is too short to create a sufficiently horizontal shifting distance between incident light and emerging light, so that the UV rays reach the adhesive. They will only be reflected once and then emitted away immediately. By flipping the mirror faced down the coated surface lies far away from the polymer substrate by the thickness of the mirror itself. Thus, the propagation path of the UV rays is extended. It results in a larger horizontal shifting distance between incident light and emerging light. In this way, the UV irradiation will reach the contact spot to get the adhesive cured.

In addition, the position of the UV lamp and the angle of incidence of UV rays play a large role in the curing process as well. Thus, when designing the optode, both factors were taken into account, trying to make them adjustable. By adjusting the angles and axes, an optimum position is to be found out, in which the adhesive is maximally irradiated and cured by UV rays. It must be subsequently fixed before starting the bonding process.

Due to the spatial restriction of *Fineplacer®*, only the left UV radiation can be realized.

2.1. Choice of UV Lamp

Figure 6 shows an irradiation head, which is the head of a LED UV lamp. An appropriate lamp can prevent the polymer substrate from thermal damage, if it induces no or very few heat loading. Furthermore, it can facilitate a shortening of the process time, which improves production efficiency. As investigated, there are mainly two types of UV lamps on the market: the conventional UV lamp based on plasma physics and the LED UV lamp based on semiconductor technology. The **Table 1** below shows a brief comparison of these two different lamps considering the items that are relevant to optodic bonding.

Figure 6. Photo of realized sideway optode (only the right side of UV irradiation).

Table 1. Comparison of the different types of UV lamps.

	conventional UV technology	UV-LED technology
Functional principle	Built on plasma physics and optics	Built on semiconductor technology and optics
Spectrum	Continuous UV spectrum between 200nm and 450nm	Quasi-monochromatic radiation at defined wavelength, e.g. 365nm, 385nm
Lifetime	1000h–5000h	>10000h
Advantages & disadvantages	Warm-up period necessary	No warm-up period, instant on and off
	Heat portion	No heat loading of substrate

In order to find out the different effects that these two UV lamps have on polymer substrate, we conducted a test, where the UV head was installed in the optode, enabling it radiating directly towards the polymer sheet without adhesives. According to the right graph from **Figure 4** we chose a head of the UV LED lamp with a wavelength of 385nm, at which the UV transmission grade of PMMA is around 85%. Both UV irradiations lasted for 40 seconds with a maximal intensity of 10100mW/cm^2. As test material we employed PMMA. **Figure 6** shows the different changes between the samples after irradiation by conventional UV lamp and LED UV lamp.

For the film that is irradiated by a conventional lamp, an obvious deformation is observed, which is caused by the heat portion of the irradiation. The maximum deformation is about 120μm. In contrast, the film irradiated by a LED lamp remains nearly unchanged and was not damaged. Only a tiny deformation of 9μm round the corner can be observed.

The explanations can be founded in **Table 1**. In UV LED technology, the common warm-up period for on and off is eliminated, thus improving the process efficiency. In addition, very barely or even no heat loading will be induced over the whole irradiation. This is exactly what was expected as the goal of this research.

3. Analysis of Process Parameters

In order to achieve a reliable integration of optoelectronic bare chips into polymer substrates to establish a high functional sensor network, it is necessary to investigate the process parameters in optodic bonding. Only by mastering the influences of these parameters, the process can be controlled and improvements made to meet all requirements. Based on the concept and on the realized optode, mainly two parameters are considered: irradiation intensity and irradiation time.

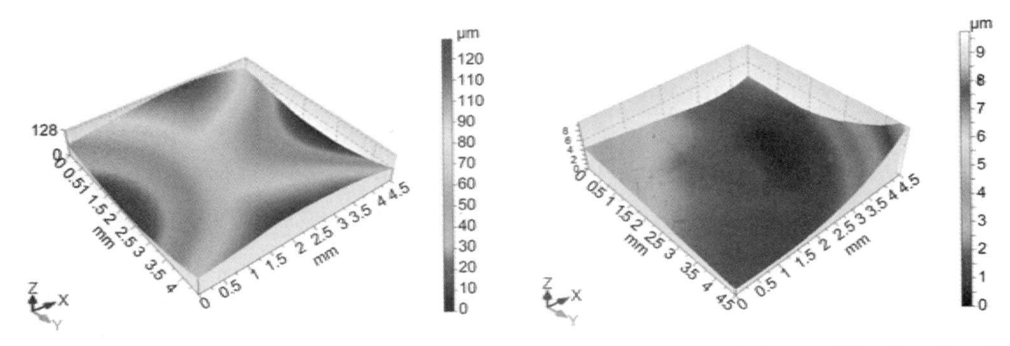

Figure 7. Photos of polymer films. left: irradiated by conventional lamp; right: irradiated by LED lamp.

To investigate the influence of both process parameters, bare chips of 1mm × 1mm size were used. The contact pads have an area of 80μm × 80μm, which is similar to most optoelectronic chips. Some pairs of adjacent pads on the chip are designed as two nested daisy chains, so that the electric conductivity can be measured. For the following tests, 4 daisy-chain pairs of each chip are applied[8].

3.1. Irradiation Time

Generally, process time is one of the most relevant parameters in manufacturing engineering. Here the irradiation time is part of the entire process time, taking up a great portion. Thus, shortening irradiation time means shortening process time, which is of great significance for improving production efficiency. Prepared as specimens, the chips were integrated on polymer substrate by optodic bonding under the same conditions but different irradiation time. The intensity was adjusted to 7070mW/cm^2. By checking the electric conductivity between adjacent contacts, we affirmed whether the chip had been bonded successfully. Meanwhile, we measured the electric resistance to conform with the conductivity quantitatively, when the chip was bonded successfully.

After starting cautiously at 70 seconds, time was reduced always by 20 seconds till 10 seconds. Shorter than 10 seconds, tests were made at 5 seconds and 3 seconds. For every time stage, 10 chips, altogether 40 pairs of adjacent pads, were connected using optodic bonding technology. We observe in **Figure 8**, the success rate from 70 seconds to 5 seconds remains approximately around 95%, which is nearly constant. At 3 seconds, it is strongly decreased. This means, 5 seconds UV irradiation time is sufficient to cure the adhesive successfully. Shorter than 5 seconds, here at 3 seconds, it can clearly be seen that the possibility to fail increases. A similar phenomenon can be observed in the electric resistance measuring. The arithmetic mean and standard deviation of the adjacent junctions stay around 0.7ohm ± 0.3ohm from 70 seconds to 5 seconds. At 3 seconds, the value is evidently greater than others, which stands for a worse electric conductivity.

It is noted that in this work the chips were always bonded manually using optodic bonding technique. Thus, a success rate of 95% can be a sufficient result. For the automatic process about 100% should be reachable.

Furthermore, we conducted a shear test on these specimens to investigate the mechanical strength of the connections that are joined by optodic bonding. The result is shown in the following **Figure 9**.

It can be recognized, that the mean shear forces for irradiation time from 70 seconds to 10 seconds are practically the same, staying around 2.25kgf, which is nearly 22.5N. According to the size of chips for $1mm^2$, the mean mechanical strength is calculated to about $22.5N/mm^2$. At 5 seconds, the shear force declines strongly to less than 1.4kgf, and it keeps falling to 0.9kgf at 3 seconds. The irradiation time has no significant effects on standard deviation.

Synthesizing the results above, an irradiation time of 5 seconds is sufficient for realizing the electric contacting between the optoelectronic chips and circuit in

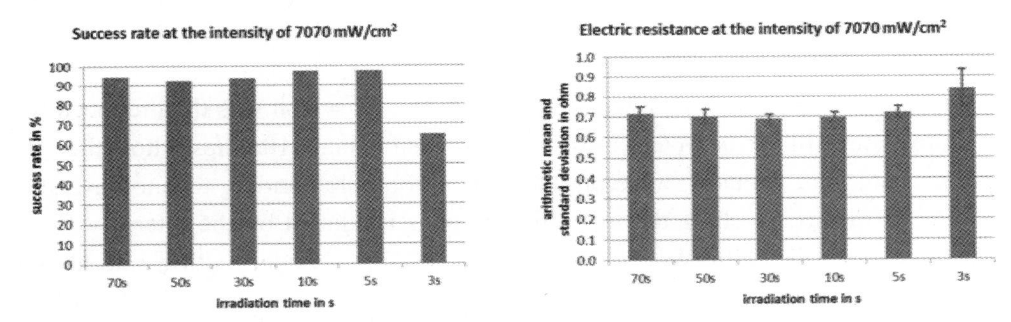

Figure 8. Success rate (left) of optodic bonded chips and their electric resistance (right) in dependence on irradiation time.

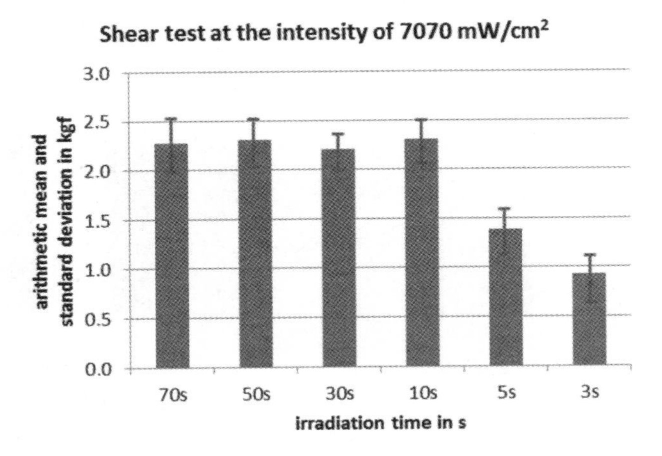

Figure 9. Shear force of optodic bonded chips in dependence on irradiation time.

optronic system. However, considering the aspect of mechanical strength, the irradiation time of 5 seconds does not achieve the best result the UV adhesive can reach. An irradiation time of 10 seconds is still a time that is short enough for the whole production process.

3.2. Irradiation Intensity

The intensity of irradiated UV rays affects the curing degree of the basis resin of NCA adhesives. The demanded minimum intensity for curing differs in dependence on adhesives. Analogously to investigating the irradiation time, we prepared the specimens in the same way. This time only the irradiation intensity was changed and all other conditions remained identical. The irradiation time was set to 5 seconds, with which a good success rate can be achieved. Afterwards we examined the electric conductivity and measured the electric resistance.

It was started with an intensity of 7070mW/cm^2 which was then reduced by 2020mW/cm^2 till 1010mW/cm^2. For every intensity stage, 10 chips, altogether 40 pairs of adjacent contacts, were produced using optodic bonding technology. The results in **Figure 10** show that the adjacent pads irradiated by the intensity from 7070mW/cm^2 till 3030mW/cm^2 are most successfully bonded with a rate above 80%. At the intensity of 2020mW/cm^2, the success rate sinks dramatically to less than 40% and even zero at the intensity of 1010mW/cm^2. On the right side of **Figure 10**, we see the results of the electric resistance measuring. Because of the

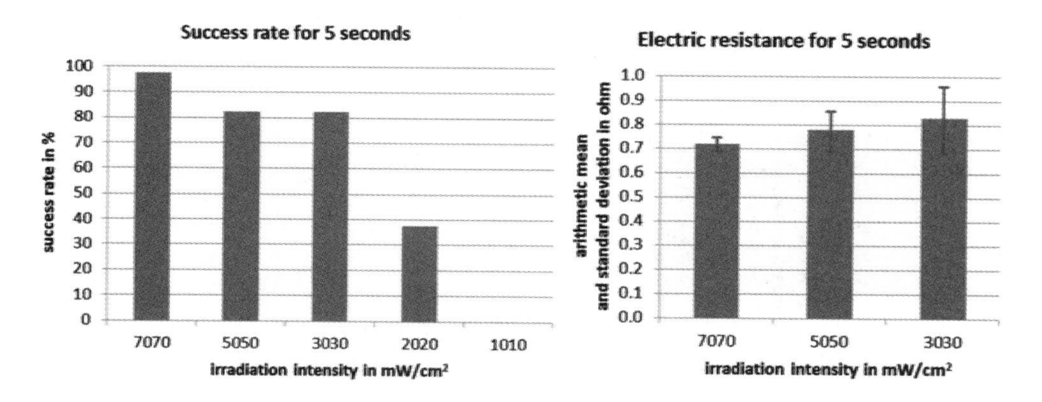

Figure 10. Success rate (left) of optodic bonded chips and their electric resistance (right) in dependence on irradiation intensity.

low success rate, it makes only sense to analyze the specimens which were bonded with the intensity from 7070mW/cm^2 to 3030mW/cm^2. It is evident that the electric conductivity is getting worse with reduced intensity, not only from the aspect of the arithmetic mean value but also as far as the standard deviation is concerned.

To investigate the influence of the irradiation intensity on mechanical strength, shear tests were carried out. Due to the low success rate, the specimens produced from the intensity of 2020mW/cm^2 and 1010mW/cm^2 were not tested. **Figure 11** shows that the shear force decreases appreciably along with the reduced intensity, *i.e.* the mechanical strength is proportional to the applied intensity.

Summarizing from the resistance measuring and shear tests, the irradiation time and irradiation intensity have significant effects on the performance of bonded chips, which are integrated by means of optodic bonding. Long irradiation time ensures good conductivity and high mechanical strength; nevertheless, it must not last longer than necessary. No improvements are achieved by a longer lasting irradiation, if the demanded time has been reached. So even with short irradiation time, production quality can be guaranteed. In this way, higher production efficiency can be achieved. Regarding the influences of irradiation intensity, higher intensity provides better electric conductivity and stronger mechanical connection.

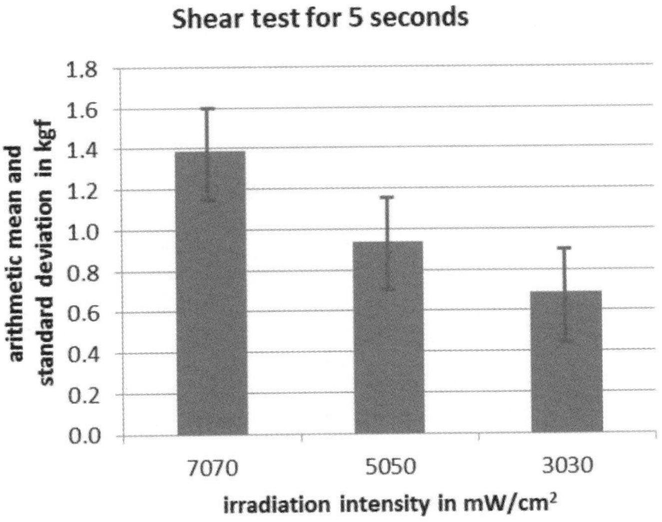

Figure 11. Shear force of optodic bonded chips in dependence on irradiation intensity.

4. Conclusion and Outlook

In this work, we introduced a new flip-chip bonding process, the so-called optodic bonding, to integrate optoelectronic bare chips in polymer optronic systems. To avoid damaging of thermal sensitive polymers, we applied UV curing adhesives as bonding material instead of commonly used heat curing adhesives. The use of an optode was highlighted functioning as a sideway UV source to achieve a higher performance of the curing process. The feasibility of the concept was confirmed through simulations and subsequently realized. In selecting UV lamp we investigated two types and applied one of them to minimize the thermal loading possibly induced by UV irradiation. A high success rate of the integrated chips from manual operation under appropriate settings of process parameters verified the optodic bonding process. Two relevant process factors, irradiation time and irradiation intensity, were investigated by evaluating the electrical and mechanical reliability. The results imply that the optodic bonding process can be an appropriately qualified integration technique for establishing polymer-based flexible and highly functional sensor networks. It is emphasized that the optodic bonding technique allows short curing time, *i.e.* short process time, which has an important impact on mass production.

Future research work will focus on the investigation of further process parameters, such as applied bonding force and positioning accuracy. An optimization of process parameters must be implemented. In addition, besides the sideway optode, a new configuration of optodes, e.g. with UV irradiation from the bottom side, will be attempted. The ultimate aim is to realize a stable, reproducible integration process of optoelectronic chips in polymer optronic systems.

Acknowledgements

We gratefully acknowledge financial support from the Deutsche Forschungsgemeinschaft (DFG) within the framework of the Collaborative Research Center "Transregio 123—Planar Optronic Systems" (PlanOS). The authors would like to thank Anan Dai for his participation in realizing the optode.

Source: Wang Y, Overmeyer L. Low Temperature Optodic Bonding for Integration of Micro Optoelectronic Components in Polymer Optronic Systems [J]. Procedia Technology, 2014, 15:531–540.

References

[1] Heiserich G, Franke S, Fahlbusch T, Overmeyer L, Altmann D. Flip Chip Assembly of UHF-ID Electronics on Flexible Substrates. Proc. 1st Annual RFID Eurasia Conference, Istanbul, 2007, pp. 44–88.

[2] Overmeyer L, Wolfer T, Wang Y, Schwenke A, Sajti L, Roth B, Dikty S. Polymer Based Planar Optronic Systems. The 6th International Congress of Laser Advanced Materials Processing (LAMP2013). Niigata, Japan: Japan Laser Processing Society, 2013.

[3] Harman G. Wire bonding in Microelectronics. 3/E, McGraw Hill Professional, 2000.

[4] Lau JH, Flip Chip Technologies. New York: McGraw-Hill; 1996.

[5] Subramanian KN. Lead-free Electronic Solders. A Special Issue of the Journal of Materials Science: Materials in Electronics. Springer Science+Business Media, LLC, 2007.

[6] Heiserich G, Franke S, Fahlbusch T, Overmeyer L. Optode Based Flip Chip Bonding on Transparent Flexible Substrates. Proc. 2nd CIRP Conference on Assembly Technologies and Systems, Toronto, 2008, pp. 469–477.

[7] Aschenbrenner R, Miessner R, Reichl H. Adhesive flip chip bonding on flexible substrate, 1997, pp. 86-94.

[8] Lai Z, Liu J. Anisotropically Conductive Adhesive Flip-Chip Bonding on Rigid and Flexible Printed Circuit Substrates. IEEE Transactions on components, packaging, and manufacturing technology—part b 1996, 19/3.

Chapter 2

Novel Optoelectronic Devices Based on Single Semiconductor Nanowires (Nanobelts)

Yu Ye[1], Lun Dai[1], Lin Gan[2], Hu Meng[1], Yu Dai[1], Xuefeng Guo[2], Guogang Qin[1]

[1]State Key Lab for Mesoscopic Physics and School of Physics, Peking University, Beijing, China
[2]College of Chemistry and Molecular Engineering, Peking University, Beijing, China

Abstract: Semiconductor nanowires (NWs) or nanobelts (NBs) have attracted more and more attention due to their potential application in novel optoelectronic devices. In this review, we present our recent work on novel NB photodetectors, where a three-terminal metal-semiconductor field-effect transistor (MESFET) device structure was exploited. In contrast to the common two-terminal NB (NW) photodetectors, the MESFET-based photodetector can make a balance among overall performance parameters, which is desired for practical device applications. We also present our recent work on graphene nanoribbon/semiconductor NW (SNW) heterojunction light-emitting diodes (LEDs). Herein, by taking advantage of both graphene and SNWs, we have fabricated, for the first time, the graphene-based nano-LEDs. This achievement opens a new avenue for developing graphene-based nano-electroluminescence devices. Moreover, the novel graphene/SNW hybrid devices can also find use in other applications, such as high-sensitivity sensor and transparent flexible devices in the future.

Keywords: Schottky Junction, Graphene, Nanowires, Nanobelts, Optoelectronics

1. Introduction

Semiconductor single-crystalline nanowires (NWs) or nanobelts (NBs) can be grown on lattice mismatched substrates and constructed into devices with the bottom-up method on basically any substrates[1]. Hence, compared to the conventional ones, semiconductor NW- or NB-based devices have the advantage of versatility in both the material and the device structure. So far, various semicon- ductor NW- or NB-based nanodevices have been emerging continuously[2]-[4]. Developing novel high- performance nano-optoelectronic devices is not only im- portant in diverse device applications, but also has significant meaning in exploring and realizing optoelectronic integration.

In this review, we present our research work on two types of novel optoelectronic devices based on semiconductor NWs (NBs). One is semiconductor NB metal-semiconductor field-effect transistor (MESFET)-based photodetectors[5]. In contrast to the common two-terminal single semiconductor NB (NW) photodetectors, the three-terminal NB MESFET-based photodetector can make a balance among overall performance parameters, which is desired for practical device applications. The other is novel multicolor light-emitting diodes (LEDs) based on graphene nanoribbon (GNR)/semiconductor nanowire (SNW) heterojunctions[6]. Herein, ZnO, CdS, and CdSe NWs were employed for demonstration. At forward biases, the GNR/ SNW heterojunction LEDs emitted light from ultraviolet (380nm) to red (705nm), which were determined by the bandgaps of the involved SNWs. This work opens a new avenue for developing diverse graphene-based optoelectronic devices[7]. These two works may help to promote nano-optoelec- tronic integration in the future.

2. Single CdS NB MESFET Photodetector

Photodetectors, which convert light to electric signals, are essential elements in high-resolution imaging techniques and light-wave communication, as well as in future memory storage[8]. Single NB (NW) photodetectors may find applications as binary switches, light-wave communications, and optoelectronic circuits. So far, most of the reported single NB (NW) photodetectors are two-terminal devices[3][8]-[17]. We summarize the key parameters of the three-terminal MESFET CdS NB photodetector and the reported two-terminal CdS NB (NW) photodetectors in **Table 1**. In general, for the two-terminal NB (NW) photodetectors, there exists a trade-off among the performance parameters, such as current responsivity (R_λ), photoresponse ratio (I_{light}/I_{dark}), and photoresponse time (rise and fall times). For example, Golberg *et al.* reported ohmic contact-based single CdS NB photodetectors with ultrahigh R_λ (approximately 7.3×10^4A/W) and fast response time (approximately 20μs of both rise and fall times); however, the I_{light}/I_{dark} was quite low (approximately 6)[15]. Wang *et al.* reported Schottky contact-based NW photodetectors with a higher I_{light}/I_{dark} (approximately183); however, the response time was not satisfying (approximately 320 ms of fall time)[17]. Compared to the reported two-terminal NB (NW) photodetectors, the MESFET-based photodetector can make a balance among these key parameters and have an overall improvement in the device performance.

A field-emission scanning electron microscopy (FESEM) image of the as-fabricated photodetector was shown in the inset of **Figure 1(a)**. Two In/Au (10:100nm) ohmic contact electrodes were defined on one single CdS NB, while one Au (100nm) Schottky contact electrode was defined in between the ohmic electrodes across the CdS NB. We can see that the NB has a uniform width (500nm) along the entire length (13μm) between the two ohmic contacts.

Typical electrical transport properties of the CdS NB MESFETs are shown in **Figure 1**. During the electrical transport measurements, the source electrodes were grounded. **Figure 1(a)** shows the I-V curve measured between the source and drain electrodes. It shows a linear behavior, confirming the ohmic contacts between the In/Au electrodes and the CdS NB. **Figure 1(b)** shows the *I-V* curve

Table 1. Comparison of the key parameters for the CdS NB (NW) photodetectors with different structures.

Materials	Structure	Rise; fall time	Responsivity (A/W)	Ilight/Idark	Reference
CdS NB	Ohmic contact	746; 794μs	approximately38	6.0×10^3	[9]
CdS NB	Ohmic contact	approximately 20; approximately 20μs	7.3×10^4	6.0	[15]
CdS NW	Schottky contact	-; 320ms	-	approximately 183.0	[17]
CdS NB	MESFET	137; 379μs	Approximately 2.0×10^2	approximately 2.7×10^6	This work

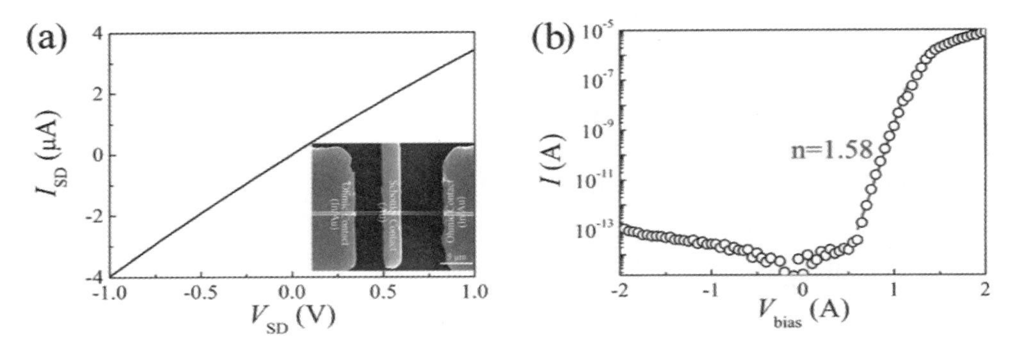

Figure 1. Typical electrical transport properties of the single CdS NB MESFETs. (a) The current-voltage (*I-V*) curve measured between the source and drain electrodes. Inset: a typical FESEM image of a single CdS NB MESFET-based photodetector. (b) The red straight line shows the fitting result with the equation $\ln(I)=\frac{qV}{nkT}+\ln(I_0)$ where I_0 is the reverse saturation current, q is the electronic charge, V is the applied bias, n is the diode ideality factor, k is the Bolzmann constant, and T is the absolute temperature.

measured between the source and gate electrodes on an exponential scale. We can see a good Schottky contact rectification behavior between the Au electrode and the CdS NB. A rectification ratio of approximately 108 is obtained when the voltage changes from +2V to −2V. The turn-on voltage is around 1.25V. The typical transfer characteristic curves of the CdS NB MESFETs with and without light illumination are depicted in **Figure 2(a)**. We can see that the Vth shifts from −2.9V to −3.8V when the light is switched from off-state to on-state. This phenomenon can be understood as follows: there are two processes involved when the as-fabricated device is upon above-bandgap illumination. One is that the channel conductance increases due to the photon-generated electrons and holes; the other is that photon-generated electrons and holes at the Schottky junction are separated by the strong local electric field[[16][18], which may reduce the electron-hole recombination rates and lower the barrier height[16]-[19]. Both processes will make the CdS NB channel more difficult to be depleted, and hence, the V_{th} shifts to a more negative value. The photocurrent response of a control device, a two-terminal CdS NB photodetector, is shown in **Figure 2(b)**. It has a small photoresponse ratio (I_{light}/I_{dark} approximately 2.78) and a long decay tail (tens of seconds). The photocurrent response of the CdS NB MESFET measured at V_G = 0V is shown in **Figure 2(c)**. We can see that the average dark current (light-off) and photocurrent (light-on) are about 1.77μA and 1.86μA, respectively, resulting in an I_{light}/I_{dark} of approximately 1.05. Again, a long decay tail of tens of seconds can be observed.

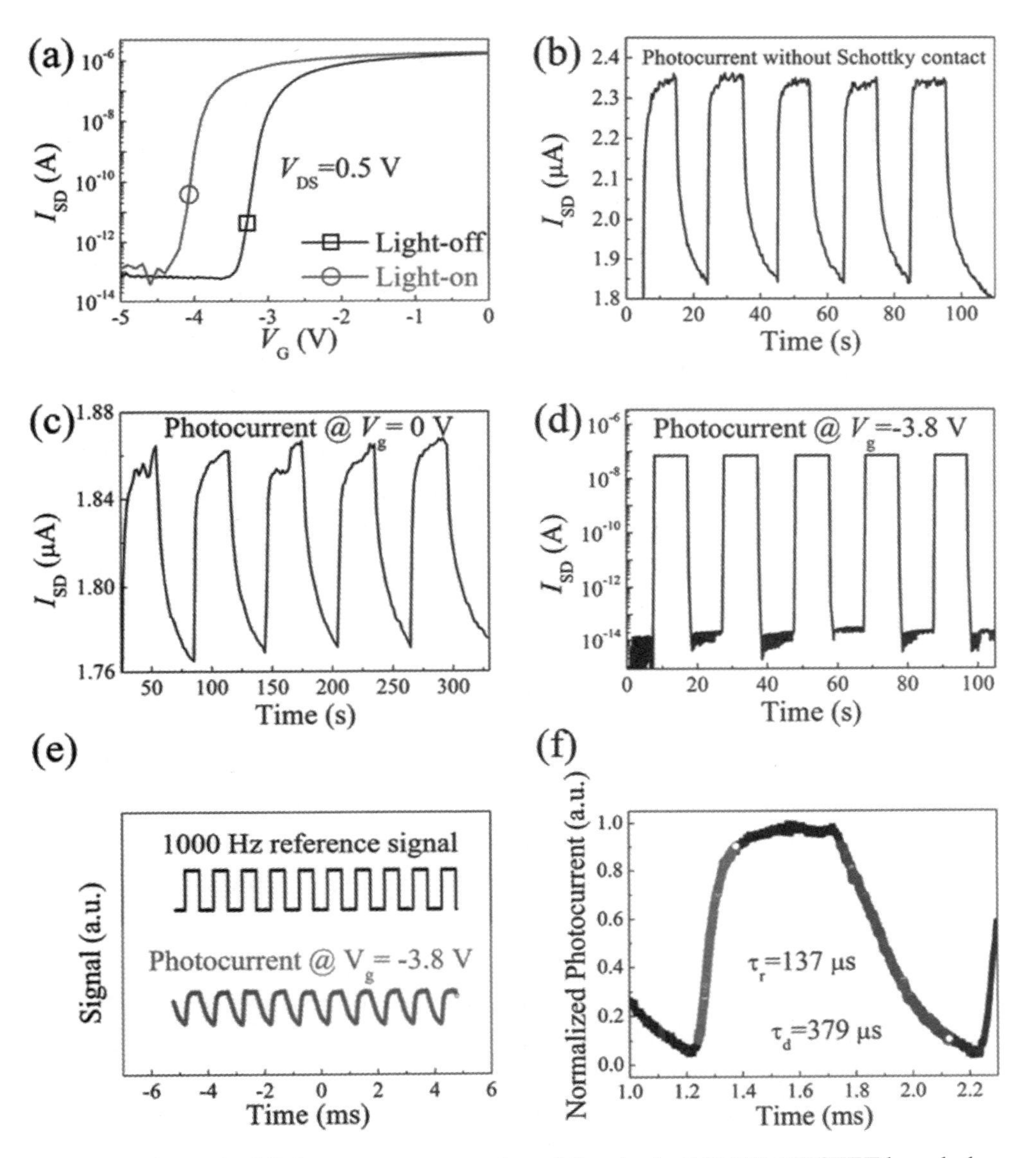

Figure 2. The typical light response properties of the single CdS NB MESFET-based photodetectors. (a) The transfer characteristics of a CdS NB MESFET-based photodetector measured in the dark (black line) and under illumination (red line). (b) On/off photocurrent response of the CdS NB without Schottky contact as a function of time. (c) On/off photocurrent response of the CdS NB MESFET-based photodetector with V_G = 0V as a function of time on a linear scale. (d) On/off photocurrent response of the CdS NB MESFET-based photodetector with V_G = −3.8V as a function of time on an exponential scale. (e) A transient response of the CdS NB MESFET-based photodetector (V_G = −3.8V, V_{DS} = 0.5V) along with a reference signal of the chopped light with a frequency of 1000Hz. (f) A close-up of the result shown in (e).

Figure 2(d) shows the on/off photocurrent response of the CdS NB MESFET-based photodetector measured at $V_G = -3.8V$, which is the threshold voltage of the MESFET under light illumination. We can see that the average dark current and photocurrent are about 26fA and 70nA, respectively, resulting in a I_{light}/I_{dark} as high as approximately 2.7×10^6. To the best of our knowledge, this is so far among the highest reported values for single NB (NW) photodetectors[3][8]-[17]. In addition, the photoresponse processes (both rise and decay processes) are quite fast, which have exceeded the detection limit (0.3s) of the measurement apparatus (Keithley 4200, Cleveland, OH, USA).

The R_λ, defined as the photocurrent generated per unit power of incident light on the effective illuminated area of a photoconductor, and the external quantum efficiency (EQE), defined as the number of electrons detected per incident photon, are two critical parameters for photodetectors. The R_λ and EQE can be calculated with equations $R_\lambda = \frac{\Delta I}{P_\lambda S}$ and $EQE = \frac{hcR_\lambda}{e\lambda}$ [11], respectively. Here, ΔI is the difference between the photocurrent and the dark current, P_λ is the light power density, S is the effective illuminated area, h is Planck's constant, c is the velocity of light, e is the electronic charge, and λ is the light wavelength. Using $\Delta I = 7.0 \times 10^{-8}$A [measured from **Figure 2(d)**], $P_\lambda = 5.3$mW/cm^2, $S = 500$nm $\times$ 13μm [measured from the inset of **Figure 1(a)**], $\lambda = 488$nm, the R_λ and EQE of the CdS NB MESFET photodetector can be estimated to be approximately 2.0×10^2A/W and 5.2×10^2, respectively.

In order to further investigate the detailed photo- response times of the single CdS NB MESFET photodetector, we employed a 200-MHz digital oscilloscope (Tektronix DPO2024, Beaverton, OR, USA) with a 10-MΩ impedance and an optical chopper working at a frequency of 1000 Hz, as shown in **Figure 2(e)**. From the close-up of the measured result shown in **Figure 2(f)**, the rise timer, defined as the time needed for the photocurrent to increase from 10% i_{peak} to 90% i_{peak}, is 137μs and the decay timed, defined analogously, is 379μs.

We attribute the overall high performance of our CdS NB MESFET-based photodetectors to the unique advantage of the MESFET structure. Compared to two-terminal photodetectors, there are two main advantages of the MESFET-based photodetectors. First, it has a much lower dark current because the applied negative gate voltage (in our case, the threshold voltage under illumination) helps to deplete the channel carriers. Second, this gate depletion effect will also cause a fast current recovery when the light is turned off. Consequently, the decay tail, which is normally observed in a two-terminal photodetectors, is suppressed in the MESFET-based photodetectors.

3. Multicolor GNR/SNW Heterojunction LEDs

By taking advantage of both graphene and SNWs, we have fabricated, for the first time, the graphene-based nano-LED[6]. This achievement opens a new avenue for developing graphene-based nano-electroluminescence devices. Moreover, the novel graphene/SNW hybrid devices can also find use in other applications, such as high-sensitivity sensor and transparent flexible devices in the future.

Both the n-type NWs[20]–[22] and the graphene[23] used in this work were synthesized via the CVD method. Before device fabrication, the graphenes were transferred by the stamp method with the help of polymethyl methacrylate[24] to Si/300-nm SiO_2 substrates for Raman and electrical property characterizations, to quartz substrates for transparency characterization, and transferred to carbon-coated grids for high-resolution transmission electron microscopy (HRTEM) characterization (Tecnai F30, FEI, Eindhoven, The Netherlands). Their electrical properties were measured by a Hall effect measurement system (Accent HL5500, York, England).

The HRTEM, Raman, and transparency characterization results for the as-synthesized graphenes (**Figure 3**) demonstrate that the graphenes have high quality, monolayer, and high transparency. The typical sheet resistance, hole concentration, and hole mobility of the graphenes are about 345Ω/sq, $1.84 \times 10^{14}cm^{-2}$, and $98.6cm^2/V.s$, respectively.

The fabrication processes of a GNR/SNW heterojunction LED are shown in **Figure 4**. **Figure 5(a)** shows an FESEM image of an as-fabricated GNR/CdS NW hetero-junction LED. The *I-V* curve [**Figure 5(b)**] of the LED shows an excellent rectification characteristic. An on/off current ratio of approximately 3.4×10^7 can be obtained when the voltage changes from +1.5V to −1.5V. The turn-on voltage

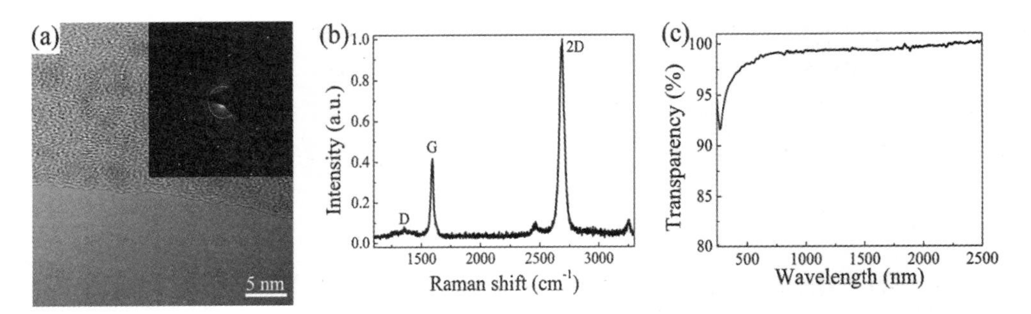

Figure 3. Properties of as-synthesized graphene. (a) Typical HRTEM image of an as-synthesized graphene, indicating the formation of monolayer graphene. Inset: selected area electron diffraction pattern of the graphene. (b) Raman spectrum of an as-synthesized graphene on a Si/300-nm SiO_2 substrate. (c) The transparency spectrum of the graphene on a quartz substrate.

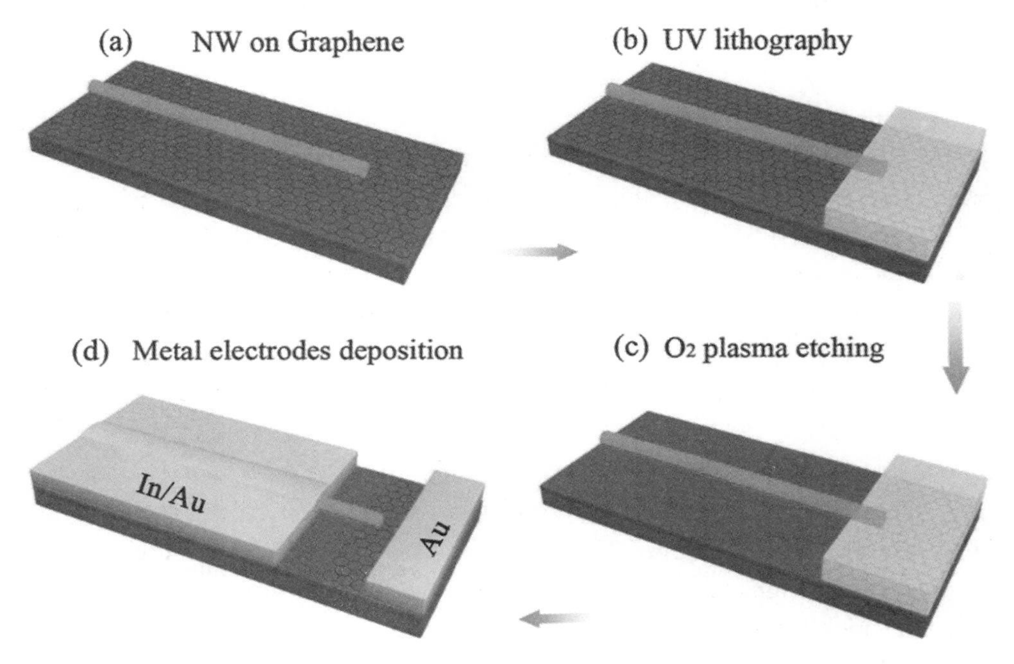

Figure 4. Schematic illustration of the fabrication processes of a GNR/SNW heterojunction LED. (a) The as-synthesized large-scale graphene was transferred to a Si/SiO_2 substrate. After that, SNW suspension was dropped on the graphene. (b) A photoresist pad was patterned to cover one end of a SNW by UV lithography and development processes. (c) Oxygen plasma etching was used to remove the exposed graphene. After that, the GNR formed under the SNW. (d) After removing the photoresist, In/Au and Au ohmic contact electrodes to SNW and graphene pad were defined, respectively. It is worth noting that because an undercut was formed during the oxygen plasma etching process [**Figure 4(c)**][6][27], the In/Au electrode on the SNW will not contact with the GNR beneath.

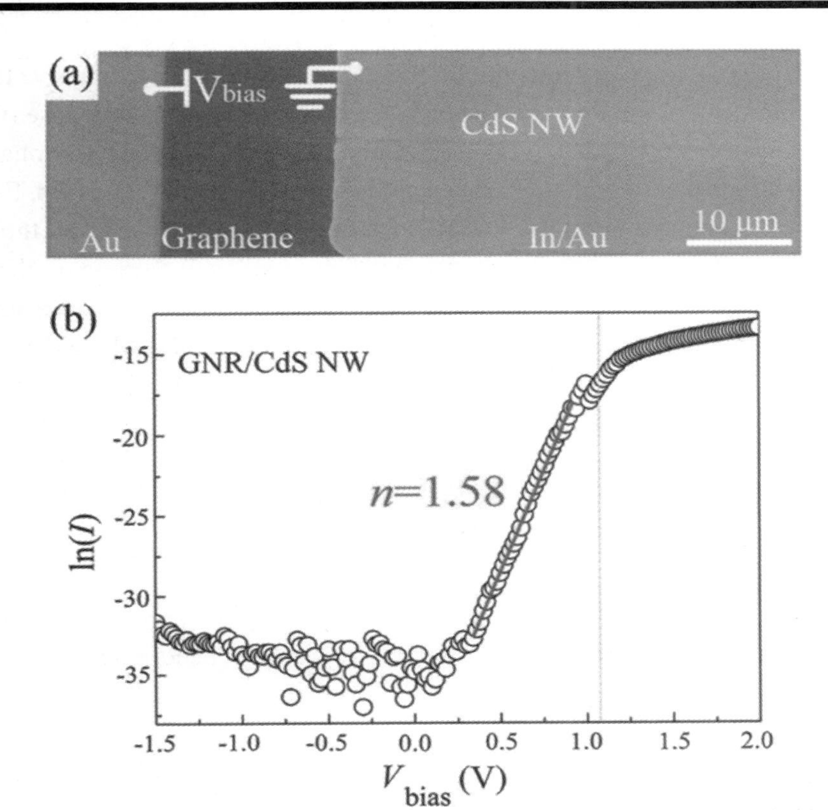

Figure 5. FESEM image and room-temperature I-V characteristic of as-fabricated GNR/CdS NW heterojunction LED. (a) FESEM image of an as-fabricated GNR/CdS NW heterojunction LED. (b) Room-temperature I-V characteristic of the LED in (a) on a semilog scale. The red straight line shows the fitting result of the *I-V* curve by the equation $\ln(I)=\frac{qV}{nkT}+\ln(I_0)$.

is around 1.1V. In view of the high conductivity and near-zero bandgap characteristics of the GNR[25], the heterojunction structure of the GNR/CdS NW can be considered approximately as a metal-semiconductor contact of the Schottky model[26]. We can deduce that the diode ideality factor n = 1.58. Note that the GNR/ ZnO NW and GNR/CdSe NW heterojunctions show similar rectification characteristics as described above, with the turn-on voltages to be about 0.7V and 1.2V, respectively.

Figures 6(a)-6(c) shows the electroluminescence (EL) images (Olympus

BX51M, Shinjuku-ku, Japan) of the GNR/SNW (ZnO, CdS, CdSe, respectively) heterojunction LEDs at a forward bias of 5 V. Except for the ZnO NW case (where the emitting light is invisible ultraviolet light) in **Figure 6(a)**, strong emitting light spots can be seen clearly with naked eyes at the exposed ends of the NWs. For the CdS NW case [**Figure 6(b)**], we can see another glaring light spot on the NW. This may be due to the scattering from the defect or adhered particle on the CdS NW[28]. **Figures 6(d)-6(f)** shows the room-temperature EL spectra measured at various forward biases for the GNR/SNW heterojunction LEDs, where the SNWs are ZnO, CdS, and CdSe NWs, respectively. For all the LEDs, EL intensities increase with the forward biases. The peak wavelength of each EL spectrum [380nm, 513nm, and 705nm, respectively, from (d) to (f)] coincides with the band-edge emission of the SNW involved. This indicates that the radiative recombination of electrons and holes occurs in the SNWs.

We can qualitatively understand the mechanism of the light emitting for the GNR/SNW heterojunction LEDs by studying the energy band diagrams. **Figure 7(a)** shows the thermal equilibrium energy band diagram of a graphene/n-type

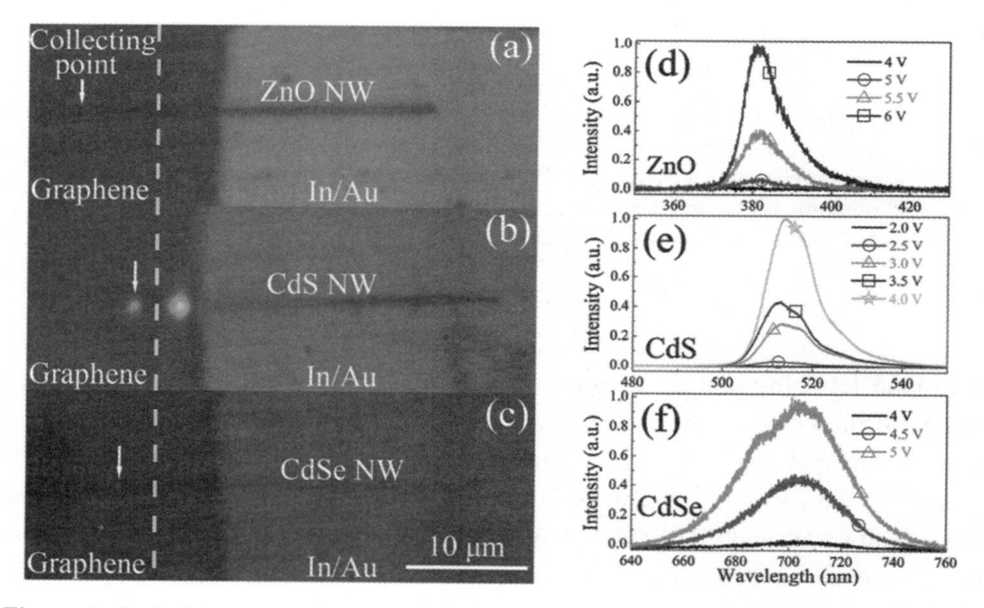

Figure 6. Optical images and room-temperature EL spectra of the GNR/SNW heterojunction LEDs. (a, b, c) The optical images of the GNR/SNW (ZnO, CdS, CdSe, respectively) heterojunction LEDs at a forward bias of 5V. Dashed lines were used to demarcate the graphenes from the substrates. White arrows: the light collecting points during the EL

measurements. (d, e, f) Room-temperature EL spectra for GNR/SNW (ZnO, CdS, CdSe, respectively) heterojunction LEDs at various forward biases.

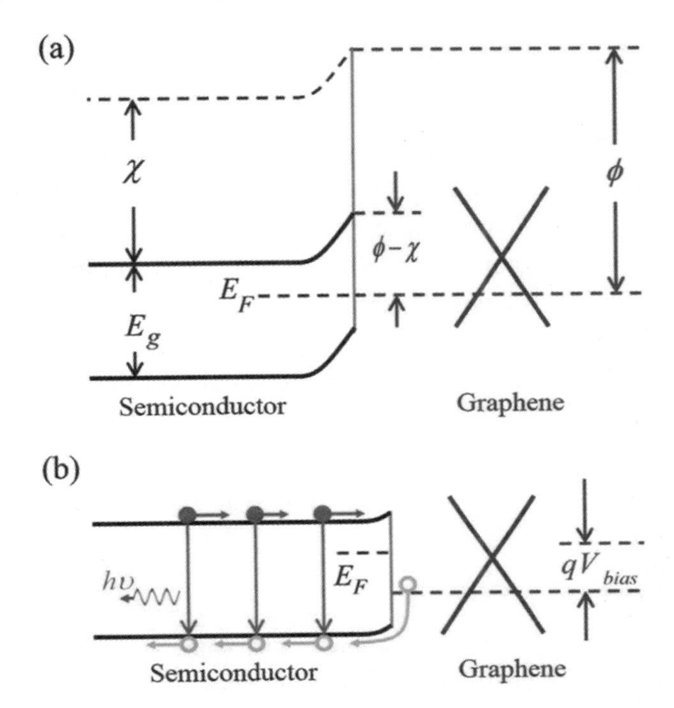

Figure 7. Schematic illustration of the energy band diagrams of a graphene/semiconductor heterojunction. (a) The thermal equilibrium energy band diagram. (b) The energy band diagram of the heterojunction under a forward bias. Φ: the work function of graphene; χ : the electron affinity of the semiconductor.

semiconductor structure, where the work function of graphene is Φ, and the electron affinity of the semiconductor is χ. E_g and E_F correspond, respectively, to the bandgap and the Fermi level of the semiconductor. It is worth noting that because the graphene used in this work has a very high conductivity and can be taken as a metal, the graphene/SNW heterostructure herein can be taken as a kind of Schottky junction. At the thermal equilibrium contacting state, the energy band of the semiconductor will bend upward at the graphene/semiconductor interface due to the difference between their work functions, and the Fermi levels at the two sides are brought into coincidence. Under a forward bias (*i.e.*, a positive bias on graphene), the built-in potential is lowered. Therefore, more electrons will flow from n-type semiconductor to graphene, and simultaneously, more holes will flow from graphene to n-type semiconductor. Herein, the injected holes have a higher radiative recombination with the electrons in the SNW (the direct bandgap semiconductor). Accordingly, the EL spectra are determined mainly by the band-edge emission of the SNWs.

It is worth noting that the GNR/NW structure has clear advantage over the conventional Schottky structure. For comparison, we have fabricated various metal/SNW Schottky structures, where the NWs used are identical to those reported in this work. Unfortunately, no EL can be observed in these structures. We attribute this to the well-known luminescence quenching effect caused by the involved metal[29]. Moreover, in our face-to-face contact LED, the active region, where the radiative recombination occurs, is larger and the series resistance is smaller, compared to the crossed NWs or NW/Si pad heterojunction structures[20][30][31]. These merits may benefit high-efficiency EL and even electrically driven laser in the future.

4. Conclusion

We review two types of novel nano-optoelectronic devices developed in our group recently. One is the photodetector, which converts light to electric signals. Our MESFET-based photodetectors have ultrahigh Ilight/Idark (approximately 2.7×10^6) and fast response (rise time, approximately 137μs; decay time, approximately 379μs) simultaneously. The other is LED, which converts electric power to light. At forward biases, our novel GNR/SNW heterojunction LEDs emitted light with wavelengths varying from ultraviolet (380nm) to red (705nm), which were determined by the bandgaps of the involved SNWs. These two types of nano- optoelectronic devices may find diverse applications in future nano-optoelectronic integration.

Competing Interests

The authors declare that they have no competing interests.

Authors' Contributions

YY carried out the device fabrications, participated in the statistical measurements, and drafted the manuscript. LD and GQ participated in the instruction, discussion, and manuscript revision. LG and XG synthesized the graphene. HM participated in the device design. YD synthesized the CdSe NWs. All authors have read and approved the final manuscript.

Acknowledgements

This work was supported by the National Natural Science Foundation of China (nos. 61125402, 51172004, 11074006, 10874011, 50732001), the National Basic Research Program of China (nos. 2012CB932703, 2007CB613402), and the Fundamental Research Funds for the Central Universities.

Source: Ye Y, Dai L, Lin G, *et al.* Novel optoelectronic devices based on single semiconductor nanowires (nanobelts) [J]. Nanoscale Research Letters, 2012, 7(1): 1–7.

References

[1] Yang PD, Yan RX, Fardy M: Semiconductor nanowire: what's next? Nano Lett 2010, 10:1529.

[2] Duan XF, Huang Y, Agarwal R, Lieber CM: Single-nanowire electrically driven lasers. Nature 2003, 421:241.

[3] Shen GZ, Chen D: One-dimensional nanostructures for photodetectors. Recent Pat Nanotechnol 2010, 4:20.

[4] Zhai TY, Fang XS, Li L, Bando Y, Golberg D: One-dimensional CdS nanostructures: synthesis, properties, and applications. Nanoscale 2010, 2:168.

[5] Ye Y, Dai L, Wen XN, Wu PC, Pen RM, Qin GG: High-performance single CdS nanobelt metal-semiconductor field-effect transistor-based photodetectors. ACS Appl Mater Interfaces 2010, 2:2724.

[6] Ye Y, Gan L, Dai L, Meng H, Wei F, Dai Y, Shi ZJ, Yu B, Guo XF, Qin GG: Multicolor graphene nanoribbon/semiconductor nanowire heterojunction light-emitting diodes. J Mater Chem 2011, 21:11760.

[7] Bonaccorso F, Sun Z, Hasan T, Ferrari AC: Graphene photonics and optoelectronics. Nat Photon 2010, 4:611.

[8] Kind H, Yan HQ, Messer B, Law M, Yang PD: Nanowire ultraviolet photodetectors and optical switches. Adv Mater 2002, 14:158.

[9] Jie JS, Zhang WJ, Jiang Y, Meng XM, Li YQ, Lee ST: Photoconductive characteristics of single-crystal CdS nanoribbons. Nano Lett 1887, 2006:6.

[10] Gao T, Li QH, Wang TH: CdS nanobelts as photoconductors. Appl Phys Lett 2005, 86:173105.

[11] Zhai TY, Fang XS, Liao MY, Xu XJ, Li L, Liu BD, Koide Y, Ma Y, Yao JN, Bando Y, Golberg D: Fabrication of high-quality In2Se3 nanowire arrays toward high-performance visible-light photodetectors. ACS Nano 2010, 4:1596.

[12] Fang XS, Xiong SL, Zhai TY, Bando Y, Liao MY, Gautam UK, Koide Y, Zhang XG, Qian YT, Golberg D: High-performance blue/ultraviolet-light-sensitive ZnSe-nanobelt photodetectors. Adv Mater 2009, 21:5016.

[13] Fang XS, Bando Y, Liao MY, Gautam UK, Zhi CY, Dierre B, Liu BD, Zhai TY, Sekiguchi T, Koide Y, Golberg D: Single-crystalline ZnS nanobelts as ultraviolet-light sensors. Adv Mater 2034, 2009:21.

[14] Zhai TY, Liu HM, Li HQ, Fang XS, Liao MY, Li L, Zhou HS, Koide Y, Bando Y, Golberg D: Centimeter-long V2O5 nanowires: from synthesis to field- emission, electrochemical, electrical transport, and photoconductive properties. Adv Mater 2010, 22:2547.

[15] Li L, Wu PC, Fang XS, Zhai TY, Dai L, Liao MY, Koide Y, Wang HQ, Bando Y, Golberg D: Single-crystalline CdS nanobelts for excellent field-emitters and ultrahigh quantum-efficiency photodetectors. Adv Mater 2010, 22:3161.

[16] Zhou J, Gu YD, Hu YF, Mai WJ, Yeh PH, Bao G, Sood AK, Polla DL, Wang ZL: Gigantic enhancement in response and reset time of ZnO UV nanosensor by utilizing Schottky contact and surface functionalization. Appl Phys Lett 2009, 94:191103.

[17] Wei TY, Huang CT, Hansen BJ, Lin YF, Chen LJ, Lu SY, Wang ZL: Large enhancement in photon detection sensitivity via Schottky-gated CdS nanowire nanosensors. Appl Phys Lett 2010, 96:013508.

[18] Ye Y, Dai L, Wu PC, Liu C, Sun T, Ma RM, Qin GG: Schottky junction photovoltaic devices based on CdS single nanobelts. Nanotechonology 2009, 20:375202.

[19] Jin YZ, Wang JP, Sun BQ, Blakesley JC, Greenham NC: Solution-processed ultraviolet photodetectors based on colloidal ZnO nanoparticles. Nano Lett 2008, 8:1649.

[20] Yang WQ, Huo HB, Dai L, Ma RM, Liu SF, Ran GZ, Shen B, Lin CL, Qin GG: Electrical transport and electroluminescence properties of n-ZnO single nanowires. Nanotechnology 2006, 17:4868.

[21] Ye Y, Dai Y, Dai L, Shi ZJ, Liu N, Wang F, Fu L, Peng RM, Wen XN, Chen ZJ, Liu ZF, Qin GG: High-performance single CdS nanowire (nanobelt) Schottky junction solar cells with Au/graphene Schottky electrodes. ACS Appl Mater Interfaces 2010, 2:3406.

[22] Ye Y, Ma YG, Yue S, Dai L, Meng H, Li Z, Tong LM, Qin GG: Lasing of CdSe/SiO2 nanocables synthesized by the facile chemical vapor deposition method. Nanoscale 2011, 3:3072.

[23] Gan L, Liu S, Li DN, Gu H, Cao Y, Shen Q, Wang ZX, Wang Q, Guo XF: Facile fabrication of the crossed nanotube-graphene junctions. Acta Phys - Chim Sin 2010, 26:1151.

[24] Reina A, Jia XT, Ho J, Nezich D, Son H, Bulovic V, Dresselhaus MS, Kong J: Large area, few-layer graphene films on arbitrary substrates by chemical vapor deposition. Nano Lett 2010, 4:2689.

[25] Castro Neto AH, Guinea F, Peres NMR, Novoselov KS, Geim AK: The electronic properties of graphene. Rev Mod Phys 2009, 81:109.

[26] Thomas D, Boettcher J, Burghard M, Kern K: Photocurrent distribution in graphene-CdS nanowire devices. Small 1868, 2010:6.

[27] Liu C, Dai L, Ye Y, Sun T, Peng RM, Wen XN, Wu PC, Qin GG: High-efficiency color tunable n-CdSxSe1-x/p+-Si parallel-nanobelts heterojunction light-emitting diodes. J Mater Chem 2010, 20:5011.

[28] Ma RM, Wei XL, Dai L, Liu SF, Chen T, Yue S, Li Z, Chen Q, Qin GG: Light coupling and modulation in coupled nanowire ring-Fabry-Pérot cavity. Nano Lett 2009, 9:2679.

[29] Flynn RA, Kim CS, Vurgaftman I, Kim M, Meyer JR, Mäkinen AJ, Bussmann K, Cheng L, Choa FS, Long JP: A room-temperature semiconductor spaser operating near 1.5μm. Opt Express 2011, 19:8954.

[30] Zhong ZH, Qian F, Wang DL, Lieber CM: Synthesis of p-type gallium nitride nanowires for electronic and photonic nanodevices. Nano Lett 2003, 3:343.

[31] Gudiksen MS, Lauhon LJ, Wang JF, Smith DC, Lieber CM: Growth of nanowire superlattice structures for nanoscale photonics and electronics. Nature 2002, 415: 617.

Chapter 3

Symmetric Periodic Solutions of Delay-Coupled Optoelectronic Oscillators

Chunrui Zhang[1], **Hongpeng Li**[2]

[1]Department of Mathematics, Northeast Forestry University, Harbin, P.R. China
[2]College of Mechanical and Electronic Engineering, Northeast Forestry University, Harbin, P.R. China

Abstract: Delay-coupled optoelectronic oscillators are considered. These structures are based on mutually coupled oscillators which oscillate at the same frequency. By taking the time delay as a bifurcation parameter, the stability of the zero equilibrium and the existence of Hopf bifurcations induced by delay are investigated, and then stability switches for the trivial solution are found. Conditions ensuring the stability and direction of the Hopf bifurcation are determined by applying the normal form theory and the center manifold theorem. Using the symmetric functional differential equation theories combined with the representation theory of Lie groups, the multiple Hopf bifurcations of the equilibrium are demonstrated. In particular, we find that the spatio-temporal patterns of bifurcating periodic oscillations will alternate according to the change of the propagation time delay in the coupling. The existence of multiple branches of bifurcating periodic solutions and their spatio-temporal patterns are obtained. Some numerical simulations are used to illustrate the effectiveness of the obtained results.

Keywords: Optoelectronic Oscillators, Symmetric Bifurcation, Stability, Spatio-

Temporal Patterns, Delay

1. Introduction

A system of coupled nonlinear oscillators is capable of displaying a rich dynamics and has applications in various areas of science and technology, such as physical, chemical, biological, and other systems[1]-[3]. When oscillators are coupled, non-negligible coupling delays naturally arise because signals in real systems inevitably propagate from one oscillator to the next with a finite propagation speed. These time delays can induce many new phenomena and complex dynamics. Examples include neuronal networks, biological oscillators, and physical models[4]-[6].

The coupled optoelectronic oscillator is a novel and unique device which simultaneously produces spectrally pure microwave reference signals as in a microwave oscillator[7], and short optical pulses, in a mode locked laser[8]-[10]. In[8], the authors response of two delay-coupled optoelectronic oscillators. Each oscillator operates under its own delayed feedback. They show that the system can display square-wave periodic solutions that can be synchronized in phase or out of phase depending on the ratio between self- and cross-delay times. Furthermore, they show that multiple periodic synchronized solutions can coexist for the same values of the fixed parameters. As a consequence, it is possible to generate square-wave oscillations with different periods by just changing the initial conditions. Furthermore, in[9], the positive delayed feedback is considered, and they show that the scenario arising for positive feedback is much richer than with negative feedback. In[10], the authors model two non-identical delay-line optoelectronic oscillators mutually coupled through delayed cross-feedback. The system can generate multi-stable nanosecond periodic square-wave solutions which arise through a Hopf instability. They show that for suitable ratios between self- and cross-delay times, the two oscillators generate square waves with different amplitude but synchronized in phase, out of phase or with a dephasing of a quarter of the period.

In[11], two laser-pumped fiber-coupled Mach-Zehnder modulators (MZM) with a coupling scheme were experimentally set up, where the output intensity of one MZM is used to modulate the radio frequency (rf) input of the other MZM. In this paper we consider a system of two laser-pumped fiber-couple Mach-Zehnder modulators (MZM) as in[10], where the output intensity of one MZM is used to modulate the radio frequency (rf) input of the other MZM, as shown in **Figure 1**.

This system can be described by two coupled described by two coupled differential equations with nonlinearities $f(x) = \cos^2(m + d\tanh(x)) - \cos^2(m)$:

$$\begin{cases} \dot{x}_1 = -x_1(t) - y_1(t) + \gamma_{12} f(x_2(t-\tau)), \\ \dot{y}_1 = \varepsilon x_1(t), \\ \dot{x}_2 = -x_2(t) - y_2(t) + \gamma_{21} f(x_1(t-\tau)), \\ \dot{y}_2 = \varepsilon x_2(t), \end{cases} \tag{1.1}$$

where x_i $(i = 1, 2)$ account for the dimensionless and scaled variable corresponding to the measured output voltage, respectively; $\varepsilon > 0$ denotes the linear coefficient, and $\gamma_{ij}(i, j = 1, 2)$ the nonlinear coupled coefficient. The effective coupling strengths γ_{12} and γ_{21} take into account all losses and gains in the system and are directly proportional to α_{12} and α_{21} In[11] it was shown in experiment and numerical simulations that this system has oscillatory solutions when the product of the two coupling strengths exceeds a critical value. Beyond the oscillation threshold, the oscillation amplitudes grow smoothly and the dependence of the amplitude on the coupling strengths can be described by a scaling law[11]. Beyond the oscillation threshold, the oscillation amplitudes grew smoothly and they found a scaling law that describes the dependence of the amplitude on the coupling strengths.

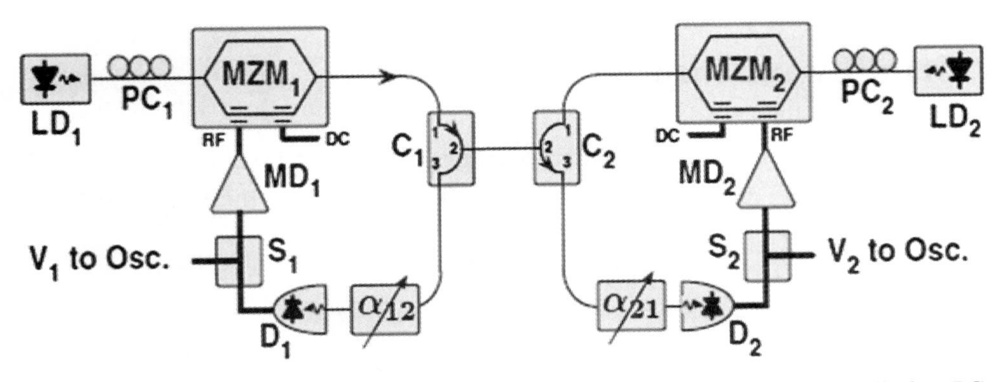

Figure 1. Schematic of the cross-coupled optoelectronic oscillators: LD, laser diodes; PC, polarization controllers; MZM, Mach-Zehndermodulators; C, optic circulators; α, adjustable optic attenuators; D, photodetectors; S, electronic splitters; MD, modulator drivers.

In this paper, considering the special case with $\gamma=\gamma_{12}=\gamma_{21}$, we attempt to analytically investigate how the time delay can affect its stability, the bifurcations of new solutions when stability is lost, and also spatio-temporal patterns of the bifurcating periodic oscillations. Spatio-temporal patterns reflect the relationship of the evolution of the dynamic behaviors of two sub-networks such as in-phase or anti-phase oscillation. The results show that the emerging oscillations can exhibit different spatio-temporal patterns sensitive to the delay. It has been shown that even small, compared to the oscillation period, delays may have a large impact on the dynamics of delay-coupled optoelectronic oscillators.

The plan for the article is as follows. In Section 2, we consider the linear stability of Equation (1.1) and present some theorems about the region of stability of the trivial solution as a function of the physical parameters in the model. We find some new phenomena such as the stability switch for Equation (1.1) which is not mentioned in[8]-[11]. Coupling can lead synchronization, phase trapping, phase locking, amplitude death, chaos, bifurcation of oscillators, and so on[12][13]. Since two identical oscillators are coupled symmetrically, the most typical patterns of behavior are perfect synchrony or perfect antisynchrony (in which the oscillators are half a period out of phase with each other); see[14][15]. In Section 3, we give the Z_2-equivariant property of Equation (1.1) and the existence of multiple periodic solutions (synchronous (respectively, anti-phased). Then we investigate the direction and stability of the Hopf bifurcating periodic solutions in Section 4. In the final section, we present some numerical simulation to support our analytical results. The last section concludes the paper.

2. Stability and Bifurcation Analysis

It is clear that (0, 0, 0, 0) is an equilibrium point of Equation (1.1). Linearizing system (1.1) at the trivial solution (0, 0, 0, 0) leads to the following linear system:

$$\begin{aligned}
\dot{x}_1 &= -x_1(t)-y_1(t)+\gamma f'(0)x_2(t-\tau),\\
\dot{y}_1 &= \varepsilon x_1(t),\\
\dot{x}_2 &= -x_2(t)-y_2(t)+\gamma_{21} f'(0)x_1(t-\tau),\\
\dot{y}_2 &= \varepsilon x_2(t).
\end{aligned}\tag{2.1}$$

The characteristic equation associated with the linearization of (2.1) is

$$\{\Delta = \left[\lambda^2 + \lambda + \varepsilon + a\lambda e^{-\lambda\tau}\right]\left[\lambda^2 + \lambda + \varepsilon - a\lambda e^{-\lambda\tau}\right] = \Delta_1\Delta_2 = 0, \tag{2.2}$$

where

$$\Delta_1 = \lambda^2 + \lambda + \varepsilon + a\lambda e^{-\lambda\tau},$$
$$\Delta_2 = \lambda^2 + \lambda + \varepsilon - a\lambda e^{-\lambda\tau},$$

and $a = \gamma f'(0)$. The distribution of zeros of Equation (2.2) determines the dynamic properties of (1.1). In the following, the analysis on the distribution of the roots to Equation (2.2) is based on the conclusion: the sum of the order of the zeros of Equation (2.2) in an open right-plane can change only if a zero appears or crosses the imaginary axis as parameter is varied.

In the following, we assume that:

(H_1) either $1 - a^2 > 2\varepsilon$ or $0 < 1 - a^2 < 4\varepsilon$;

(H_2) $a > 1$

Consider equation $\Delta_1 = 0$. Let $i\omega(\omega > 0)$ be a root of $\Delta_1 = 0$, then plugging $i\omega$ into $\Delta_1 = 0$ to get $-\omega^2 + i\omega + \varepsilon + ia\omega e^{-i\omega\tau} = 0$. Solving ω we have: if (H_1) holds, then $\Delta_1 = 0$ has no purely imaginary roots; if (H_2) holds, then $\Delta_1 = 0$ has a pair of purely imaginary roots $\pm i\omega^+$ ($\pm i\omega^-$, respectively) at $\tau = \tau_j^{1+}$ $\left(\tau = \tau_j^{1-}, \text{respectively}\right)$, where

$$\begin{cases} \tau_j^{1+} = \dfrac{1}{\omega^+}\left(2j\pi + \pi - \arccos\dfrac{1}{a}\right), \\ \tau_j^{1-} = \dfrac{1}{\omega^-}\left(2j\pi + \pi + \arccos\dfrac{1}{a}\right), \end{cases} j = 0, 1, 2, \cdots, \tag{2.3}$$

$$\omega^{\pm} = \sqrt{\frac{-1 + a^2 + 2\varepsilon \pm \sqrt{\left(-1 + a^2 + 2\varepsilon\right)^2 - 4\varepsilon^2}}{2}}. \tag{2.4}$$

Similarly, under the condition (H_2), we have: $\Delta_2 = 0$ has a pair of purely imaginary roots $\pm i\omega^+$ ($\pm i\omega^-$, respectively) at $\tau = \tau_2^{2+}\left(\tau = \tau_j^{2-}, \text{respectively}\right)$, where

$$\begin{cases} \tau_j^{2+} = \dfrac{1}{\omega^+}\left(2k\pi + 2\pi - \arccos\dfrac{1}{a}\right), \\ \tau_j^{2-} = \dfrac{1}{\omega^-}\left(2j\pi + \arccos\dfrac{1}{a}\right), \end{cases} \quad j = 0, 1, 2, \cdots, \tag{2.5}$$

and $\omega^{\pm}$ meets Equation (2.4).

Rewrite Equation (2.3) and Equation (2.5):

$$\begin{cases} \tau_j^{+} = \dfrac{1}{\omega^+}\left((j+1)\pi - \arccos\dfrac{1}{a}\right), \\ \tau_j^{-} = \dfrac{1}{\omega^-}\left(j\pi + \arccos\dfrac{1}{a}\right), \end{cases} \quad j = 0, 1, 2, \cdots, \tag{2.6}$$

Then we have the following lemma.

Lemma 2.1

1) If (H_1) holds, then Equation (2.2) has no purely imaginary roots.

2) If (H_2) holds, then Equation (2.2) has a pair of purely imaginary roots $\pm i\omega^+$ ($\pm i\omega^-$, respectively) at $\tau = \tau_j^{+}\left(\tau = \tau_j^{-}, \text{respectively}\right)$. Furthermore, the transversality condition is satisfied at $\tau_j^{\pm}\left(j = 0, 1, 2, \cdots\right)$:

$$\begin{cases} \left.\dfrac{d\,\mathrm{Re}\left(\lambda(\tau)\right)}{d\tau}\right|_{\tau = \tau_j^{+}, \omega = \omega^{+}} > 0, \\ \left.\dfrac{d\,\mathrm{Re}\left(\lambda(\tau)\right)}{d\tau}\right|_{\tau = \tau_j^{-}, \omega = \omega^{-}} < 0. \end{cases}$$

when $\tau = 0$, Equation (2.2) has at least one root with positive real part. Since we are concerned about the stability of Equation (1.1), we prefer to investigate the relation of τ_0^+ and τ_0^-. From Equation (2.6), we have $\tau_0^+ > \tau_0^-$ if and only if $\frac{\omega-\pi}{\omega^+ + \omega^-} > \arccos\frac{1}{a}$.

Applying the above conclusions and bifurcation theorems for functional differential equations, we have the following results presenting the stability and existence bifurcations to the symmetric system (1.1).

Theorem 2.1

1) If either $1 - a^2 > 2\varepsilon$ or $0 < 1 - a^2 < 4\varepsilon$, then the zero solution of Equation (1.1) is unstable for all $\tau \geq 0$.

2) If $a > 1$ and $\frac{\omega - \pi}{\omega^+ + \omega^-} > \arccos\frac{1}{a}$, then $\tau_0^- < \tau_0^+$, and there exists a positive integer j such that $\tau_{j-1}^- < \tau_{j-1}^+ < \tau_j^+ < \tau_j^-$. Then the zero solution of Equation (1.1) is asymptotically stable for $\tau \in \left(\tau_0^-, \tau_0^+\right) \cup \left(\tau_1^-, \tau_1^+\right) \cup \cdots \cup \left(\tau_{j-1}^-, \tau_{j-1}^+\right)$ and unstable for $\tau \in \left[0, \tau_0^-\right) \cup \left(\tau_{j-1}^-, +\infty\right)$.

3) If $a > 1$, then for any $k, l \in \{0, 1, 2, \cdots\}$, the Equation (1.1) undergoes a Hopf bifurcation at τ_k^- or τ_l^+ for $\tau_k^- \neq \tau_l^+$.

Remark 2.1 Theorems 2.1 shows that under the conditions $a > 1$, and $\frac{\omega-\pi}{\omega^+ + \omega^-} \rangle \arccos\frac{1}{a}$, there are j switches from stability to instability to stability.

3. Existence of Multiple Periodic Solutions

In the following, we consider the symmetric properties of Equation (1.1). Using the theories of functional differential equations, (1.1) can be written as

$$\dot{x}(t) = Lx_t + Fx_t,$$
$$L\varnothing = \begin{pmatrix} A_1 & O \\ O & A_1 \end{pmatrix} \varnothing(0) + \begin{pmatrix} O & A_2 \\ A_2 & O \end{pmatrix} \varnothing(-\tau) \tag{3.1}$$

where $x_t = x(t+\theta)$ for $-\tau \le \theta \le 0$

$$A_1 = \begin{pmatrix} -1 & -1 \\ \varepsilon & 0 \end{pmatrix}, \quad A_2 = \begin{pmatrix} \gamma f'(0) & 0 \\ 0 & 0 \end{pmatrix},$$

and

$$F\varnothing = \begin{pmatrix} \frac{rf''(0)}{2}\varnothing_3^2(-\tau) + \frac{rf'''(0)}{6}\varnothing_3^3(-\tau) \\ 0 \\ \frac{rf''(0)}{2}\varnothing_1^2(-\tau) + \frac{rf'''(0)}{6}\varnothing_1^3(-\tau) \\ 0 \end{pmatrix}$$

for $\varnothing = (\varnothing_1, \varnothing_2, \varnothing_3, \varnothing_4) \in C([-\tau, 0], R^4)$.

It is clear that the system (3.1) is Z_2-equivariant with

$$(\rho U)_r = U_{r+1}(\text{mod}\,2).$$

for any U_r in R^2. It is very interesting to consider the spatio-temporal patterns of bifurcating periodic solutions. For this purpose, we give the concepts of some spatio-temporal symmetric types of symmetry: spatial and temporal.

The oscillators $(u_1(t), v_1(t))$ and $(u_2(t), v_2(t))$ are synchronized if the state takes the form

$$(u(t), v(t), u(t), v(t))$$

for all times t. On the other hand, the oscillator $(u_1(t), v_1(t))$ is half a period out of phase with the (anti-synchronous) oscillator $(u_2(t), v_2(t))$, meaning the state takes the form

$$\left(u(t),\, v(t),\, u\left(t+\frac{T}{2}\right),\, v\left(t+\frac{T}{2}\right)\right).$$

Now, we explore the possible (spatial) symmetry of the system (3.1). Consider the action of $Z_2 \times S^1$ on $\left(\left[-\tau,0\right],R^4\right)$ with

$$(r,\theta)x(t) = rx(t+\theta),(r,\theta)\in Z_2\times S^1,$$

where S^1 is the temporal. Let $T=\dfrac{2\pi}{\omega^+}$ or $T=\dfrac{2\pi}{\omega^-}$, and denote by P_T the Banach space of all continuous T-periodic functions $x(t)$. Denoting by SP_T the subspace of P_T consisting of all T-periodic solution of system (3.1) with $\tau=\tau_k^{\pm}$, then for each subgroup $\sum\subset Z_2\times S^1$,

$$Fix(\Sigma,SP_T)=\{X\in SP_T,(r,\theta)x=x, for\ all(r,\theta)\in\Sigma\}$$

is a subspace.

Theorem 3.1 The trivial solution of system (3.1) undergoes a Hopf bifurcation giving rise to one branch of synchronous (respectively, anti-phased) periodic solutions.

Proof Let $\pm i\omega^+$ (respectively, $\pm i\omega^-$) satisfy Equation (2.4). The corresponding eigenvectors of $\Delta(\lambda)$ can be chosen as

$$q_1(\theta)=\left(\upsilon_1(\theta)^T,\upsilon_1(\theta)^T\right)^T,$$

where $\upsilon_1(\theta)$ satisfies $\left(A_1+e^{-i\omega^{\pm}\tau_j^{2\pm}}A_2\right)v_1(\theta)=i\omega^{\pm}v_1(\theta);$

$$q_2(\theta)=\left(v_2(\theta)^T,-v_2(\theta)^T\right)^T,$$

and $\upsilon_2(\theta)$ satisfies $\left(A_1-e^{-i\omega^{\pm}\tau_j^{1\pm}}A_2\right)v_2(\theta)=i\omega^{\pm}v_2(\theta)$

The isotropic subgroup of $Z_2\times S^1$ is $z_2(p)$, the center space associated to eigenvalues $\pm i\omega\pm$, which implies that it is spanned by $q_1(\theta)$ and $\overline{q}_1(\theta)$, and the bifurcated periodic solutions are synchronous, taking the form

$$\left(u(t),v(t),u(t),v(t)\right)$$

Similarly, $Z_2 \times S^1$ has another isotropic subgroup $z_2(\rho,\pi)$, the center space associated to eigenvalues $\pm i\omega^{\pm}$ is spanned by $q_2(\theta),\overline{q}_2(\theta)$, which implies that the bifurcated periodic solutions are anti-phased, *i.e.*, taking the form

$$\left(u(t),v(t),u\left(t+\frac{T}{2}\right),v\left(t+\frac{T}{2}\right)\right)$$

where T is a period.

4. Direction and Stability of the Hopf Bifurcation

In this section, we shall study the direction, stability, and the period of the bifurcating periodic solutions. We first focus on the case $\Delta_1 = 0$, because the other case can be dealt with analogously. We re-scale the time by $t \mapsto t/\tau$, to normalize the delay of the system (3.1) and introduce the new parameter $v = \tau - \tau_k^{1\pm}$ such that $v = 0$ is a Hopf bifurcation value. Then Equation (3.1) can be written as

$$\begin{aligned} &\dot{u}(t) = L_v u_t + F u_t, \\ &L_v \varnothing = \tau\left[\begin{pmatrix} A_1 & O \\ O & A_1 \end{pmatrix}\varnothing(0) + \begin{pmatrix} O & A_2 \\ A_2 & O \end{pmatrix}\varnothing(-1)\right], \end{aligned} \tag{4.1}$$

where $u_t = x(t+\theta)$ for $-1 \le \theta \le 0$,

$$A_1 = \begin{pmatrix} -1 & -1 \\ -\varepsilon & 0 \end{pmatrix}, A_2 = \begin{pmatrix} \gamma f'(0) & 0 \\ 0 & 0 \end{pmatrix},$$

and

$$F\varnothing=-\tau\begin{pmatrix}\frac{\gamma f''(0)}{2}\varnothing_3^2(-1)+\frac{\gamma f'''(0)}{6}\varnothing_3^3(-1)\\0\\\frac{\gamma f''(0)}{2}\varnothing_1^2(-1)+\frac{\gamma f'''(0)}{6}\varnothing_1^3(-1)\\0\end{pmatrix}$$

For $\varnothing=(\varnothing_1,\varnothing_2,\varnothing_3,\varnothing_4)\in C([-1,0],R^4)$

By the Riesz representation theorem, there exists a function $\eta(\theta,\mu)$ $(0\le\theta\le1)$, whose elements are of bounded variation such that

$$L_v\varnothing=\int_{-1}^{0}\mathrm{d}\eta(\theta,v)\varnothing(\theta),\varnothing\in C$$

In fact, we choose

$$\eta(\theta,v)=(\tau_j^{1\pm}+v)A\delta(\theta)+(\tau_j^{1\pm}+v)B\delta(\theta+1)$$

where δ is determined by

$$\delta(\theta)=\begin{cases}1, & \theta=0,\\0, & \theta\ne0.\end{cases}$$

For $\varnothing\in C^1([-1,0],R^4)$, define

$$A(v)\varnothing=\begin{cases}d\varnothing(\theta)/d\theta, & \theta\in[-1,0),\\\int_{-1}^{0}d\eta(t,v)\varnothing(t), & \theta=0,\end{cases}$$

and

$$R(v)\varphi=\begin{cases}0, & \theta\in[-1,0),\\F(v,\varphi), & \theta=0.\end{cases}$$

Then system (4.1) is equivalent to the following operator equation:

$$\dot{u}_t = A(v)u_t + R(v)u_t, \tag{4.2}$$

where $u_t = u(t+\theta), \theta \in [-1,0]$. For $\psi \in C^1\left([0,1],\left(R^4\right)^*\right)$, define

$$A^*\psi(s) = \begin{cases} -d\psi(s)/ds, & s \in (0,1], \\ \int_{-1}^{0} d\eta^T(s,v)\varnothing(-s), & s = 0, \end{cases}$$

and a bilinear form

$$\langle \psi(s), \varnothing(\theta) \rangle = \bar{\psi}(0)\varnothing(0) - \int_{\theta=1}^{0}\int_{\varepsilon=0}^{\theta} \bar{\psi}(\xi-\theta) d\eta(\theta)\varnothing(\xi) d\xi,$$

where $\eta(\theta) = \eta(\theta,0)$, then $A(0)$ and A^* are adjoint operators. From Section 3, $q(\theta)$ and $q_1^*(s)$ are the eigenvectors of $A(0)$ and A^* corresponding to $i\tau_j^{1\pm}\omega^{\pm}$ and $-i\tau_j^{1\pm}\omega^{\pm}$ where $q_1(\theta) = \left(-\varepsilon, i\omega^{\pm}, -\varepsilon, i\omega^{\pm}\right)^T$, $q^*(s) = \bar{D}\left(i\omega^{\pm}, 1, i\omega^{\pm}, 1\right)$, and $D = \tau_j^{1\pm} e^{i\omega^{\pm}\tau_j^{1\pm}}$. Then $\langle q^*, q \rangle = 1$ and $\langle q^*, \bar{q} \rangle = 0$.

Let u_t be the solution of Equation (4.2) when $v = 0. z(t) = \langle q^*, u_t \rangle$, $W(t,\theta) = u_t(\theta) - 2\mathrm{Re}\{z(t)q(\theta)\}$. On the center manifold C_0 we have $W(t,\theta) = W(z(t), \bar{z}(t), \theta)$, where

$$W(z,\bar{z},\theta) = W_{20}(\theta)\frac{z^2}{2} + W_{11}(\theta)z\bar{z} + W_{02}(\theta)\frac{z^2}{2} + W_{30}(\theta)\frac{z^3}{6} + \cdots$$

z and z^- are local coordinates for the center manifold C_0 in the direction of q^* and $\bar{q}^*$. Note that W is real if ut is real. We only consider real solutions. For the solution $u_t \in C_0$ of Equation (4.2), since $v = 0$, we have

$$z'(t) = i\omega^{\pm} z + \left\langle q^*(\theta), F\left(W + 2\mathrm{Re}\{z(t)q(\theta)\}\right)\right\rangle$$
$$\overset{def}{=} i\omega^{\pm} z + \bar{q}^*(0)F_0(z,\bar{z}).$$

We rewrite this equation as

$$z'(t) = i\omega^{\pm} z(t) + g(z, \bar{z}) \tag{4.3}$$

With

$$\begin{aligned} g(z,\bar{z}) &= \bar{q}^{*}\tilde{F}\left(W(z,\bar{z},0) + 2\operatorname{Re}\{z(t)q(0)\}\right) \\ &= g_{20}\frac{z^2}{2} + g_{11}z\bar{z} + g_{02}\frac{\bar{z}^2}{2} + g_{21}\frac{z^2\bar{z}}{2} + \cdots \end{aligned} \tag{4.4}$$

It follows from (4.3) and (4.4) that

$$W' = u_t' - z'q - \bar{z}'\bar{q} = \begin{cases} AW - 2\operatorname{Re}\{\bar{q}^{*}(0)Fq(\theta)\}, & \theta \in [-1,0), \\ AW - 2\operatorname{Re}\{\bar{q}^{*}(0)Fq(\theta)\} + F, & \theta = 0. \end{cases}$$

Comparing of coefficients we have

$$g_{20} = -\frac{\gamma f''(0)\bar{D}}{2} i\omega^{\pm}\varepsilon^2 e^{-2i\omega^{\pm}\tau_k^{1\pm}},$$

$$g_{11} = -\gamma f''(0)\bar{D} i\omega^{\pm}\varepsilon^2,$$

$$g_{02} = -\frac{\gamma f''(0)\bar{D}}{2} i\omega^{\pm}\varepsilon^2 e^{2i\omega^{\pm}\tau_k^{1\pm}}$$

$$g_{21} = \frac{\gamma f'''(0)\bar{D}}{6}\left[i\omega^{\pm}\varepsilon\left(W_{11}^{(3)}(-1) + W_{11}^{(1)}(-1)\right)e^{-2i\omega^{\pm}\tau_k^{1\pm}} + i\omega^{\pm}\varepsilon^3 e^{-i\omega^{\pm}\tau_k^{1\pm}} \right],$$

where

$$W_{11}(\theta) = \frac{g_{11}}{\omega^{\pm}} q(0)e^{i\omega^{\pm}\theta} - \frac{\bar{g}_{11}}{i\omega^{\pm}}\bar{q}(0)e^{-i\omega^{\pm}\theta} + E_2$$

Moreover, E_2 satisfies the following equations:

$$\begin{pmatrix} -1 & -1 & \gamma f'(0) & 0 \\ \varepsilon & 0 & 0 & 0 \\ \gamma f'(0) & 0 & -1 & -1 \\ 0 & 0 & -\varepsilon & 0 \end{pmatrix} E_2 = \gamma f''(0)\bar{D}\begin{pmatrix} \varepsilon^2 \\ 0 \\ \varepsilon^2 \\ 0 \end{pmatrix}$$

Then we can compute the following quantities:

$$\begin{aligned} c_1(0) &= \frac{i}{2\omega^{\pm}}\left(g_{20}g_{11} - 2|g_{11}|^2 - \frac{1}{3}|g_{02}|^2\right) + \frac{g_{21}}{2}, \\ v_2 &= -\frac{\operatorname{Re} c_1(0)}{\operatorname{Re} \lambda'\left(\tau_k^{1\pm}\right)}, \\ \beta_2 &= 2\operatorname{Re} c_1(0). \end{aligned} \tag{4.5}$$

Theorem 4.1 *Assume that* Δ_1 *crosses critical values* 0. *In* (4.5), *the sign of* v_2 *determines the direction of the Hopf bifurcation*: *if* $v_2 > 0(v_2 < 0)$, *then the Hopf bifurcation is supercritical* (*subcritical*) *and the bifurcating periodic solution exists for* $\tau > \tau_j^{1\pm}\left(\tau < \tau_j^{1\pm}\right)$. β_2 *determines the stability of the bifurcating periodic solution*: *the bifurcating periodic solution is stable* (*unstable*) *if* $\beta_2 < 0(\beta_2 > 0)$.

We also can focus on the case $\Delta_2 = 0$. Since the conclusions are the same, we shall omit it.

5. Numerical Simulations

In this section, we use some numerical simulations to illustrate the analytical results we obtained in the previous sections.

Let $m = \frac{\pi}{4}, d = 1$. Then $f(x) = \cos^2\left(\frac{\pi}{4} + \tanh(x)\right) - \cos^2(m)$.

For parameters $\gamma = 1.155, \varepsilon = 1,$ using the algorithm in Section 3, we obtain $\omega^+ = 1.33, \omega^- = 0.75$. Then $\tau_0^- = 0.696 < \tau_0^+ = 1.97 < \tau_1^+ = 4.33 < \tau_1^- = 4.87$. From Theorem 4.1, the zero solution is asymptotically stable with $\tau_0^- < \tau < \tau_0^+$. When τ

meets $\tau_j^{1\pm}$ or $\tau_j^{2\pm}$, the anti-phased or synchronous periodic solutions will appear.

Figure 2, **Figure 3** show that a branch of anti-phased periodic solutions is bifurcated from the trivial solution with $\tau \le \tau_0^- = 0.696$.

Figure 4, **Figure 5** show that the zero solution is asymptotically stable when $\tau_0^- = 0.696 < \tau < \tau_0^+ = 1.97$.

Figure 6, **Figure 7** show a branch of synchronous periodic solutions is bifurcated from the trivial solution when $\tau_0^+ = 1.97 \le \tau \le \tau_1^+ = 4.33$. **Figure 8** shows the anti-phased periodic solutions appear again from the synchronous periodic solutions.

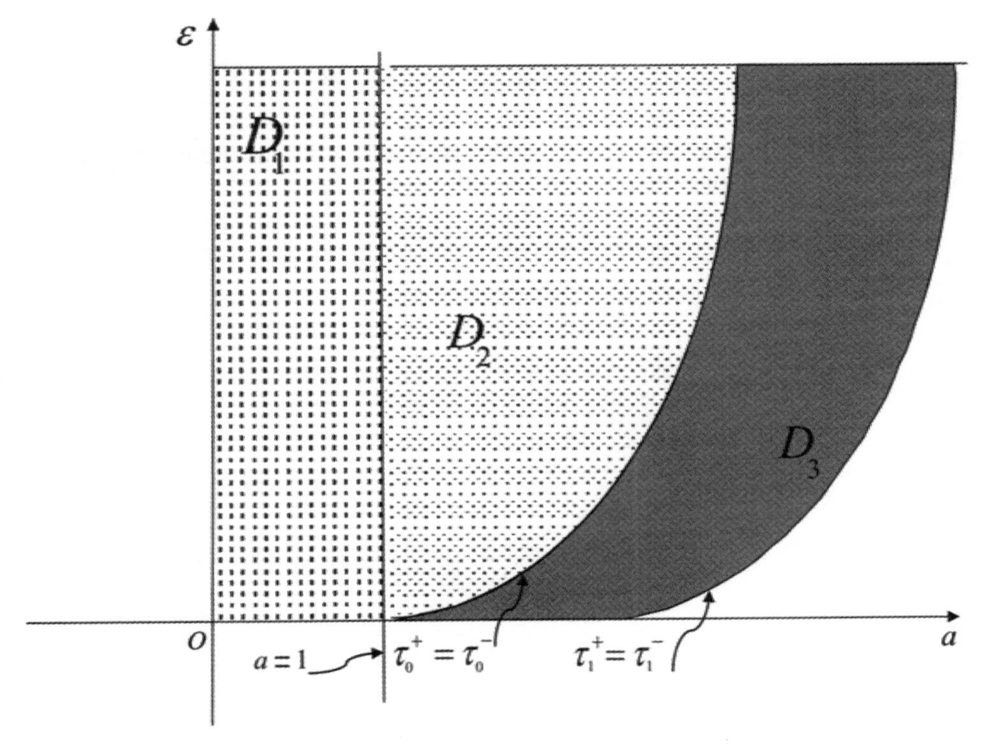

Figure 2. The bifurcation diagram for Equation (2.1). The zero solution is unstable for all $\tau \ge 0$ when $(a, \varepsilon) \in D_1$. When $(a, \varepsilon) \in D_2$, the zero solution is unstable for all $\tau \ge 0$ and system undergoes a Hopf bifurcation at $\tau = \tau_j^{\pm} \, (j = 0, 1, 2, \cdots)$. When $(a, \varepsilon) \in D_3$, $\tau_0^+ < \tau_0^- < \tau_1^+$ and the zero solution is stable.

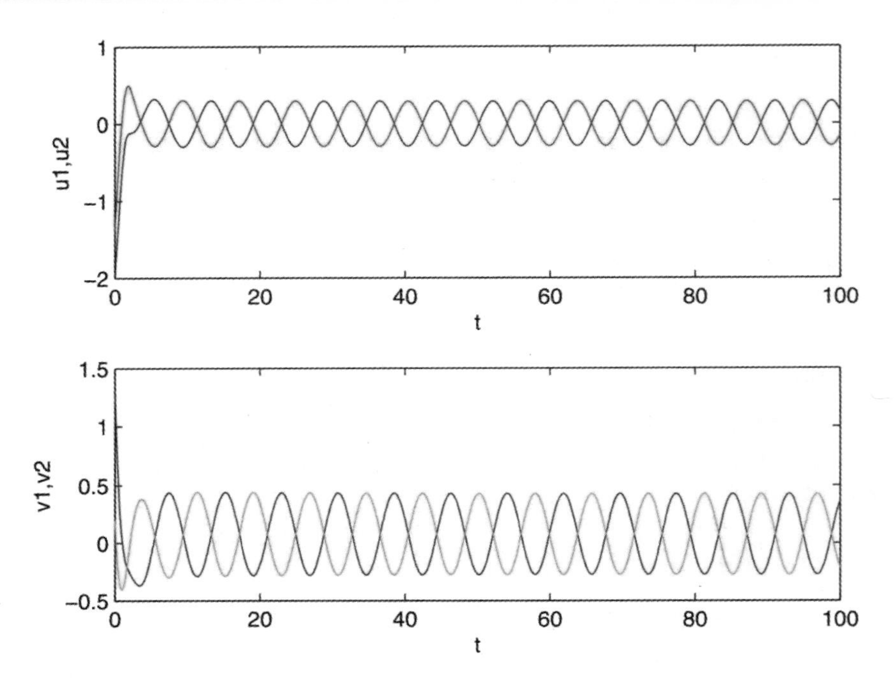

Figure 3. A branch of the anti-phased periodic solution is bifurcated from the trivial solution with $\tau = 0.5$ and initial condition $(-2, 1.3, -1.4, 0.2)$.

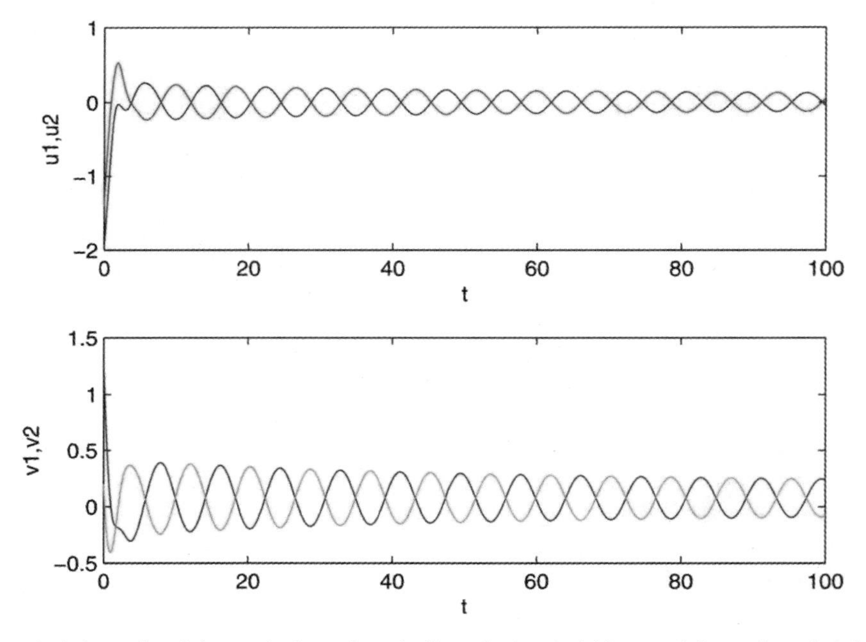

Figure 4. A branch of the anti-phased periodic solution is bifurcated from the trivial solution with $\tau = 0.697$ and initial condition $(-2, 1.3, -1.4, 0.2)$.

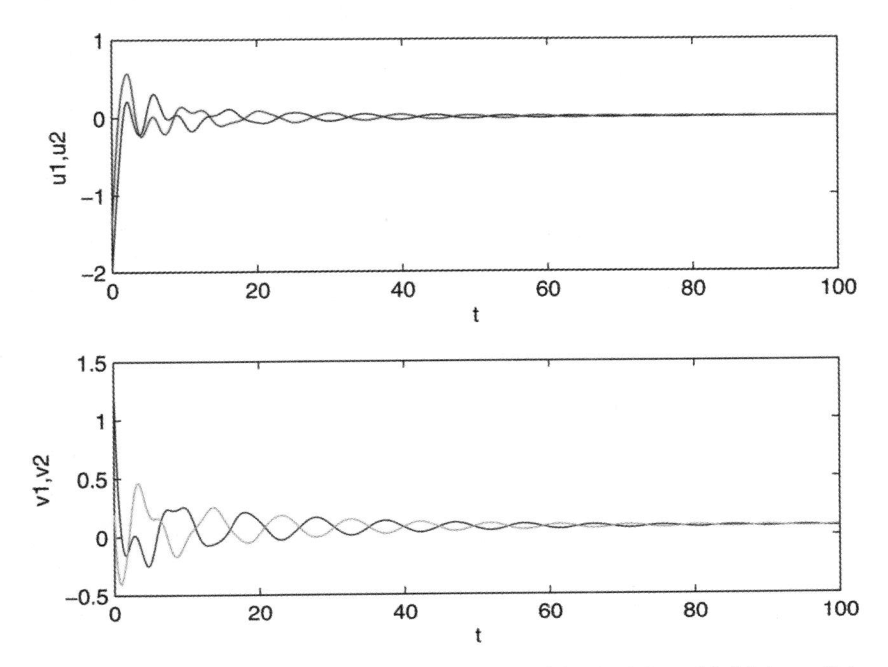

Figure 5. The zero solution is asymptotically stable with $\tau = 1.2$ and initial condition (−2, 1.3, −1.4, 0.2).

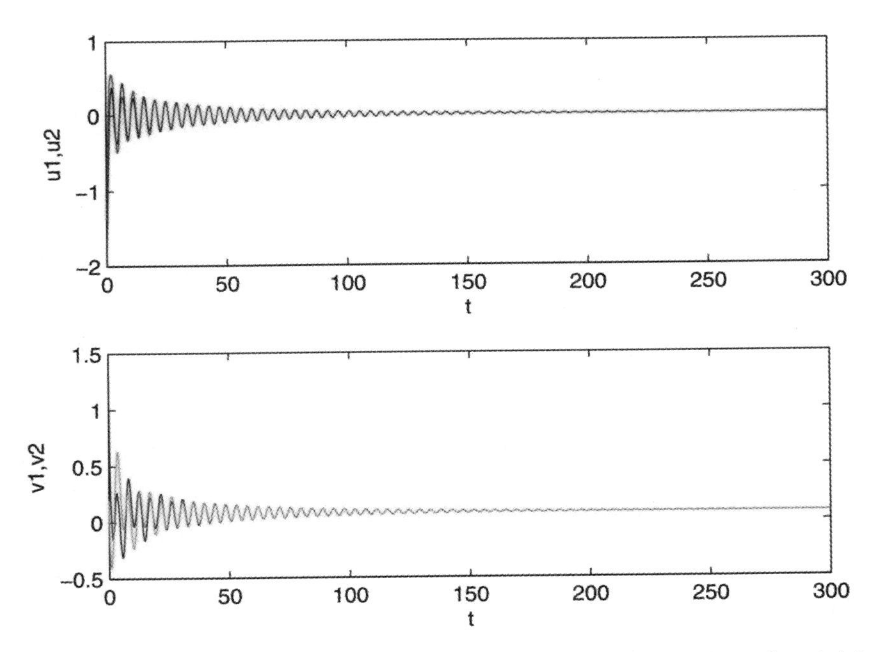

Figure 6. A branch of the synchronous periodic solution is bifurcated from the trivial solution with $\tau = 1.97$ and initial condition (−2, 1.3, −1.4, 0.2).

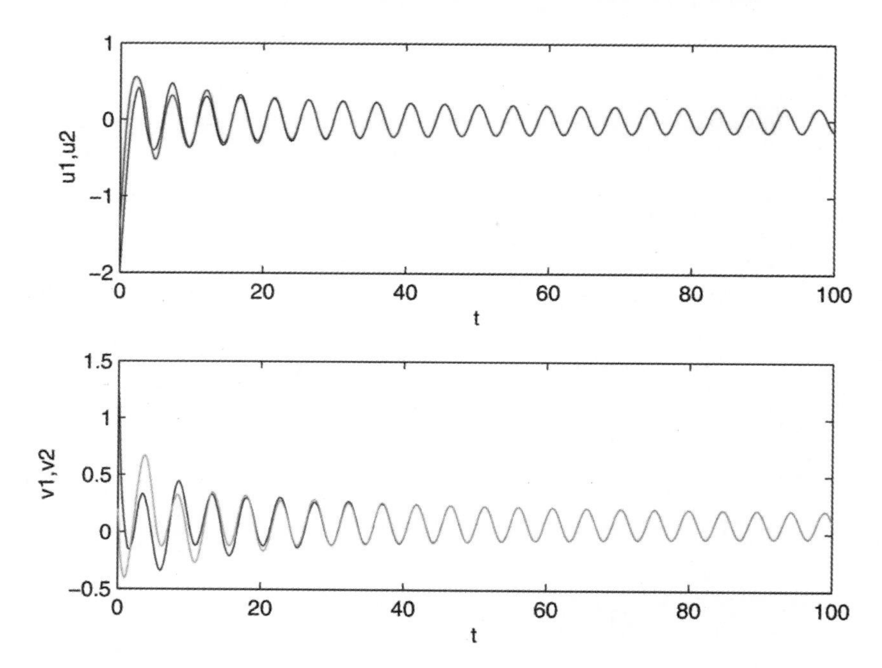

Figure 7. The zero solution is asymptotically stable with $\tau = 1.8$ and initial condition (−2, 1.3, −1.4, 0.2).

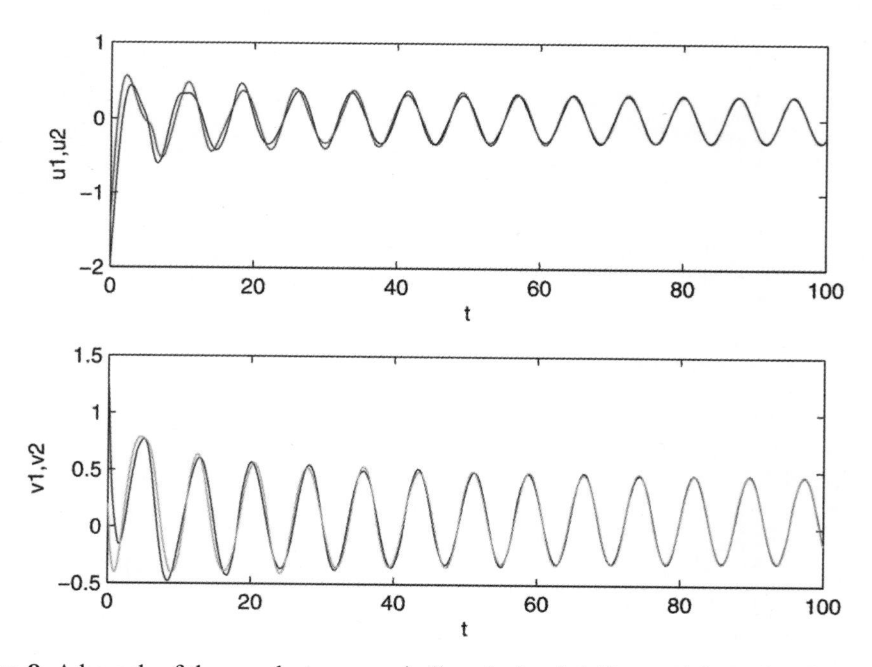

Figure 8. A branch of the synchronous periodic solution is bifurcated from the trivial solution with $\tau = 4.33$ and initial condition (−2, 1.3, −1.4, 0.2).

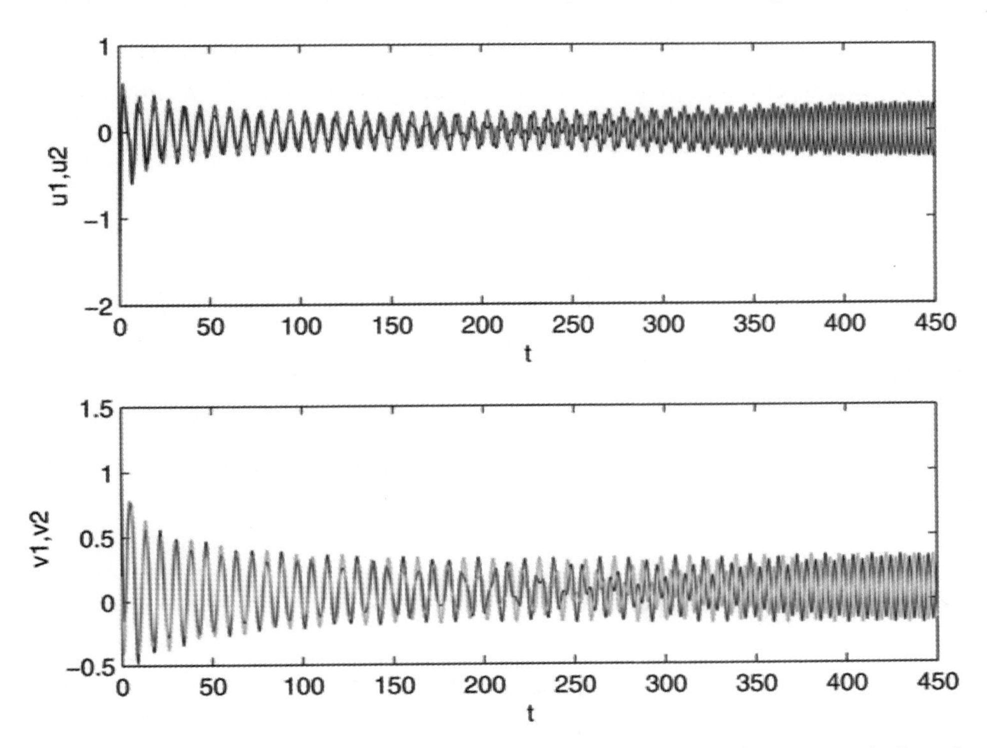

Figure 9. An anti-phased periodic solution appears again from synchronous periodic solutions with $\tau = 4.87$ and initial condition $(-2, 1.3, -1.4, 0.2)$.

6. Conclusion

In this paper, a delay-coupled optoelectronic oscillators described by (1.1) is considered. The effect of the time delay on the linear stability of the system is investigated. Based on the standard Hopf bifurcation theory, we find that as the time delay increases and crosses through the critical values, there exist stability switches in a certain region of the plane of the linear coefficient ε and the coupling strength γ and a branch of periodic solutions bifurcate from the zero equilibrium. By means of the general symmetric local Hopf bifurcation theorem, we not only investigated the effect of a delay of the signal transmission on the pattern formation of model (1.1), but we also obtain some important results about the spontaneous bifurcation of multiple branches of periodic solutions and their spatio-temporal patterns. A remarkable finding is that the spatio-temporal patterns between the two output voltage depend not only on the critical value τ of the coupling time delay, but also on the parameter region where the bifurcation occurs. There are two types of spatio-temporal patterns: one is in phase and the other is in antiphase.

Competing Interests

The authors declare that they have no competing interests.

Authors' Contributions

The authors have made equal contributions to each part of this paper. All the authors read and approved the final manuscript.

Acknowledgements

The authors would like to thank the support of Heilongjiang Province Natural Science Foundations (A2015016).

Source: Zhang C, Li H. Symmetric periodic solutions of delay-coupled optoelectronic oscillators [J]. Advances in Difference Equations, 2016, 2016(1):1–12.

References

[1] Martínez-Llinàs, J, Colet, P, Erneux, T: Tuning the period of square-wave oscillations for delay-coupled optoelectronic systems. Phys. Rev. E 89, 042908 (2014).

[2] Atay, FM: Oscillator death in coupled functional differential equations near Hopf bifurcation. J. Differ. Equ. 221, 190–209 (2006).

[3] Dias, APS, Lamb, JSW: Local bifurcation in symmetric coupled cell networks. Physica D 223, 93–108 (2006).

[4] Marcus, CM, Westervelt, RM: Stability of analog neural network with delay. Phys. Rev. A 39, 347–359 (1989).

[5] Perlikowski, P, Yanchuk, S, Popovych, OV, Tass, PA: Periodic patterns in a ring of delay-coupled oscillators. Phys. Rev. E 82, 036208 (2010).

[6] Bonnin, M: Waves and patterns in ring lattices with delays. Physica D 238, 77–87 (2009).

[7] Son, R, Solodkov, O, Tchijikova, E: Low-frequency model of the microwave frequency (phase) detector with amplitude modulator and shift oscillator. Radioelectron. Commun. Syst. 7, 363–370 (2009).

[8] Martínez-Llinàs, J, Colet, P, Erneux, T: Tuning the period of square-wave oscillations for delay-coupled optoelectronic systems. Phys. Rev. E 89, 042908 (2014).

[9] Martínez-Llinàs, J, Colet, P, Erneux, T: Synchronization of tunable asymmetric square-wave pulses in delay-coupled optoelectronic oscillators. Phys. Rev. E 91, 032911 (2015).

[10] Martínez-Llinàs, J, Colet, P: In-phase, out-of-phase and T/4 synchronization of square waves in delay-coupled non-identical optoelectronic oscillators. Opt. Express 23, 24785–24799 (2015).

[11] Illing, L, Hoth, G, Shareshian, L, May, C: Scaling behavior of oscillations arising in delay-coupled optoelectronic oscillators. Phys. Rev. E 83, 026107 (2011).

[12] Drubi, F, Ibáñez, S, Ángel, J, Íguez, R: Coupling leads to chaos. J. Differ. Equ. 239, 371–385 (2007).

[13] Wu, J: Symmetric functional differential equations and neural networks with memory. Trans. Am. Math. Soc. 350, 4799-4838 (1998).

[14] Golubitsky, M, Stewart, IN, Schaeffer, DG: Singularities and Groups in Bifurcation Theory: Vol. 2. Appl. Math. Sci. 69, Springer, New York (1988).

[15] Song, L: Hopf bifurcation and spatio-temporal patterns in delay-coupled van der Pol oscillators. Nonlinear Dyn. 63, 223–237 (2011).

Chapter 4

An Optoelectronic Monitoring System for Aviation Hydraulic Fluids

Andreas Helwig[1], Konrad Maier[1], Gerhard Müllera Torsten Bley[2], Jörg Steffensky[2], Horst Mannebach[2]

[1]AIRBUS Group Innovations, D-81663 München, Germany
[2]HYDAC ELECTRONIC GmbH, D-66128 Saarbrücken, Germany

Abstract: A non-dispersive infrared (NDIR) sensor system has been designed to monitor the state of contamination and degradation of aviation hydraulic fluids. Core part of this system is a MOEMS subsystem consisting of a micro cuvette, a MEMS thermal IR emitter and a four-channel thermopile array. For ruggedness, this MOEMS subsystem is integrated into a high-pressure resistant metal package. The MOEMS subsystem measures the transparency of the fluid under test at four specifically chosen spectral channels. Three of these channels allow the water content, the total acidity number and the remaining acid scavenger reserve to be monitored. The forth channel serves for calibration and self-test purposes. Fluid monitoring systems of this kind will form key components in an innovative AIR-BUS aircraft maintenance system.

Keywords: Hydraulic Fluid Monitoring, Optical Sensors, Non-Dispersive Infrared, Water Content, Total Acid Number, Scavenger Depletion

1. Introduction

In commercial aircrafts, flats, slats, tail plane fins and landing gears, *i.e.* all kinds of safety-critical mechanical subsystems, are powered by hydraulic actuators. The mechanical power produced by these actuators stems from pressurized hydraulic fluids, which for reasons of passenger safety need to be fire-resistant. Phosphate-esters fluids (PEF), which are widely employed for this reason, are hygroscopic in nature and therefore susceptible to thermal degradation as flight operations are carried out[1][2].

PEF, in the first step, take up water from the environment through air-pressurized reservoirs and through unavoidable small leaks in pipes, pumps and actuators. Absorbed water in connection with Joule heat, in turn, can damage PEFs through hydrolysis. Reaction products which emerge from such degradation reactions are various alcohols, which compromise the force-transmitting potential of the PEF, and—more importantly—a range of phosphorous-containing acidic compounds, which can react with the remaining water contamination to produce corrosive hydronium ions. As such corrosion damage is hard to localize and expensive to repair, an effective quality monitoring of aviation hydraulic fluids forms an important part of the current aircraft maintenance schemes. To date this monitoring is performed discontinuously during C-checks in specialized laboratories, roughly with an annual frequency. A risk of this discontinuous scheme is that airplanes are being "over-serviced" or that they might get grounded to perform urgent and unscheduled maintenance actions in locations badly suited for this purpose.

Our sensor system allows all above-mentioned fluid degradation steps to be monitored in a quasi-continuous manner. With this information being available, water can be withdrawn from the PEF in overnight ground stops before any significant follow-on damage to the fluid has occurred. In case irreversible damage has already been inflicted on the PFE, its amount and its rate of introduction can be assessed. Fluid exchange operations therefore can be strategically planned to occur at times and in locations of choice, which minimize the interruption of the normal flight schedules. Innovative sensor-based aircraft maintenance schemes with a high cost-cutting potential can therefore be implemented.

2. Optical Fluid Degradation Features

In order to arrive at optical signatures of water contamination, several commercial brands of PEF were contaminated with controlled amounts of semiconductor grade deionized water and optical measurements were performed as described below. Additionally, the water content of the contaminated samples was determined using standard Karl-Fischer titrations[3]. A fraction of the water-contaminated samples was sealed in small cuvettes and thermally treated in an oven at temperatures in the range from 120°C to 180°C for different lengths of time. Depending on those treatment parameters a fraction of the added water was consumed and thermal degradation products (acidic partial phosphates and split-off alcohols) were generated from the base fluid. Before optical measurements, the water content and the total acidity number were determined by means of standard titration treatments[1][3].

All of the above samples were put into a thin cuvette with optically transparent CaF_2 windows and Fourier Transform Infra-Red (FTIR) spectra were taken with a Bruker Vertex 80V spectrometer. The measurements were taken with the light entering and leaving perpendicular to the CaF_2 windows. Throughout our measurements the fluid optical path was kept constant at dopt = 100μm. Measurements of this kind yield values of the normalized optical transmission t as a function of the wavelength (λ) or the wavenumber (WN) of the incoming light. Some representative results are shown in **Figure 1(a)**. This first set of data shows that different spectral regions are affected when water (W) is added to a PEF base fluid and when it is further degraded. **Figure 1(b)** shows the spectrum of a heavily degraded fluid sample decomposed into several spectral components, which turn up in the course of water contamination and fluid degradation. Also shown in **Figure 1(b)** are four spectral windows (red circles) which were chosen for the NDIR monitoring of the corresponding spectral components in the NDIR sensor system described below.

3. Sensor System

The layout of the newly developed NDIR sensor system is shown in **Figure 2**. Key concern in its design was providing the stability and ruggedness mandated

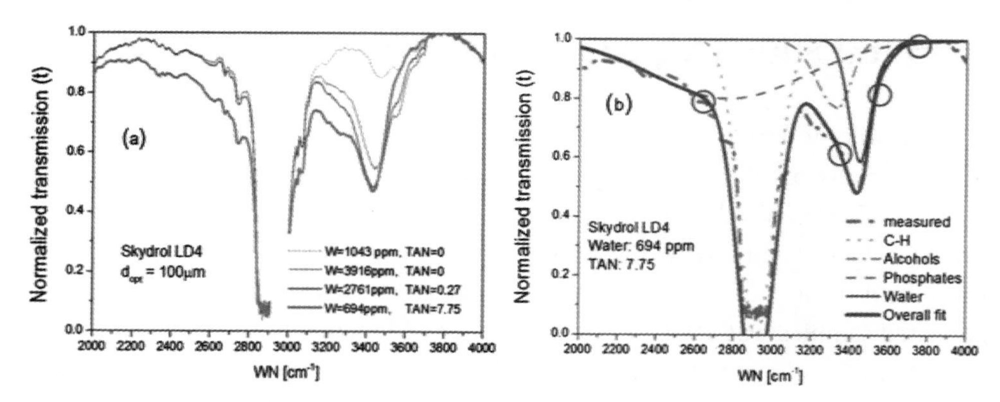

Figure 1. (a) FTIR spectra of Skydrol LD4 samples in different states of water contamination (W) and fluid degradation (TAN). All spectra were normalized at a wavenumber of $3750cm^{-1}$; (b) Deconvolution of measured FTIR spectra into C-H, TAN- and water-related spectral features. The colored circles denote sensor interrogation points used in the NDIR sensor prototype described in chapter 3.

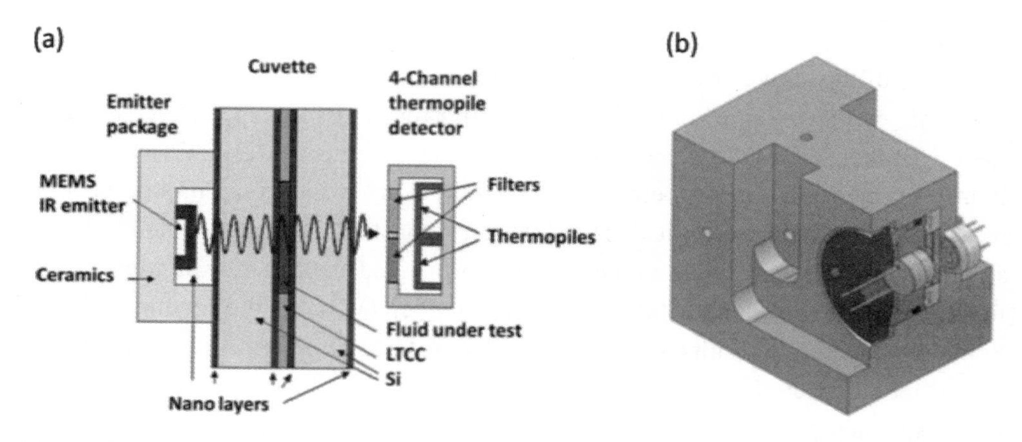

Figure 2. (a) Micro-cuvette containing the fluid under test fitted in between a MEMS IR emitter and a 4-thermopile sensor array; (b) Mechanical set-up containing the MEMS IR emitter, the micro-cuvette (yellow) and the thermopile detector array inside a common metal block (grey). Emitter and Si-cuvette are fixed in position by a pretensioning screw (red) and a specifically formed brass item (magenta).

by the intended application in mobile hydraulic systems. **Figure 2(a)** shows the core components of the NDIR system, consisting of a micro cuvette in the center, a MEMS thermal IR emitter on the left and a four-thermopile sensor array on the right-hand side. In this latter array each thermopile was covered with a specific bandpass filter with its center frequency matching one of those spectral bands indicated in **Figure 1(b)**. The micro cuvette, in turn, consists of a pair of low-doped, thermally oxidized silicon wafers, separated by a LTCC spacer (LTCC: low-temperature co-fired ceramics) to form a fluid layer with a thickness of 150μm. In order to achieve a high level of mechanical strength and ruggedness, the entire NDIR system was inserted into a metal block as shown in **Figure 2(b)**.

4. Sensor Response to Water Contaminated and Thermally Degraded Fluids

With the hardware demonstrator shown above, extensive assessment tests were performed[4]. Some of these results are shown in **Figure 3**. In these experiments, fluids with water contents ranging from 970ppm to 5400ppm and TAN values

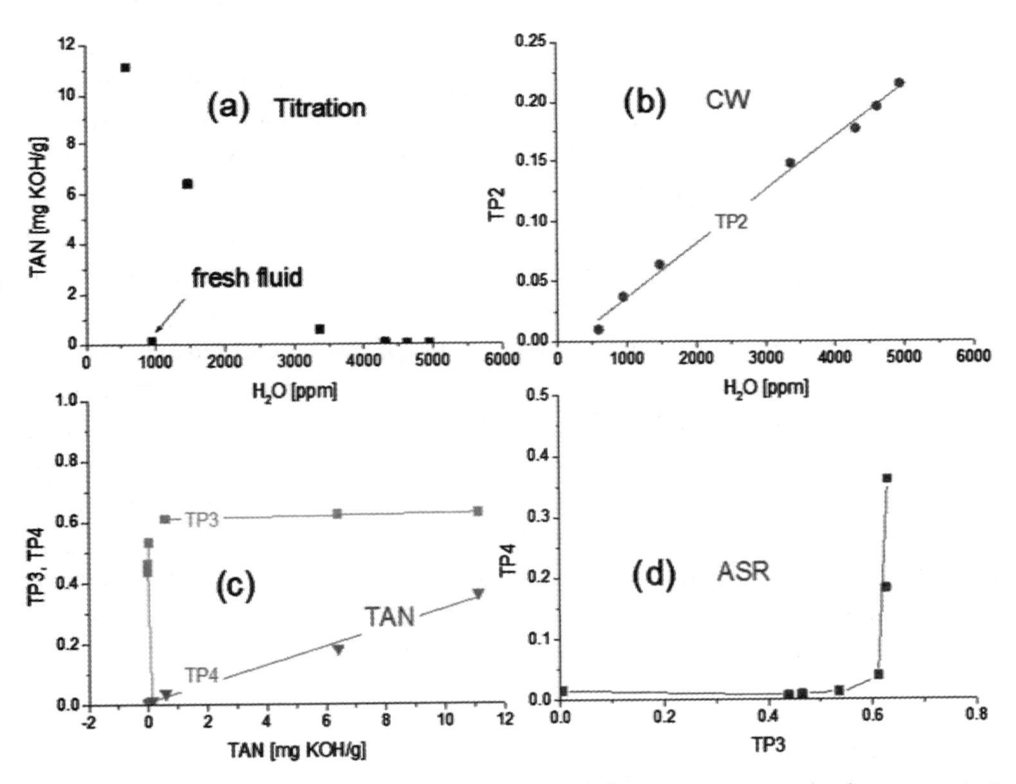

Figure 3. (a) Correlation of TAN values and remaining water content in the sensor test fluids; (b) Output signal of thermopile channel TP2 as a function of the remaining water content in the test fluids; (c) Outputs of thermopile channels TP3 (butanol/phenol) and TP4 (partial phosphates) in response to the TAN value of the test fluids; (d) Correlation of sensor readings TP3 and TP4.

ranging from 0 up to 2.68mg KOH/g were used [**Figure 3(a)**]. During each measurement the fluids were circulated through the sensor system using a membrane pump. After each measurement, the system was cleaned to reset it for a new measurement run. The thermopile voltage readings TP1 to TP4, produced in this way, were mathematically pre-processed to read in units of normalized absorption a, *i.e.* with a ranging from zero to one[4]. **Figure 3(b)** shows that the thermopile centering on the high-energy wing of the water absorption ($3559cm^{-1}$) directly scales with the titrated amount of water in the fluid samples. Similarly, the TP4 absorption, which centers on the broad feature of partial phosphate absorptions ($2632cm^{-1}$), directly scales with the titrated TAN value. These two measurements yield the two parameters of key interest in the currently used maintenance schemes.

Perhaps, the most interesting new result is the variation of the TP3 and TP4 signals with TAN [**Figure 3(c)**]. Keeping in mind that in each thermal degradation event a single partial phosphate is created alongside with a single split-off alcohol, it is surprising at first sight that the corresponding absorptions vary in distinctly different manners. Replotting the TP4 signals as a function of the TP3 ones [**Figure 3(d)**], it is seen that thermal degradation of the PEF causes an immediate rise in the alcohol (TP3) absorption and then in a belated sudden rise in the partial phosphate (TP4) one. This dissimilar behavior is caused by the presence of acid scavengers inside the PEF, which had been added to prevent an early rise of the TAN values into the critical range. The downside of this kind of chemical protection is a fairly sudden and often unpredictable rise in the TAN into the range of critical values after the acid scavengers had become depleted. The key innovative feature of our sensor system is that it allows an ongoing thermal degradation of the PEF to be detected right from its beginning simply by monitoring the rate of increase in the TP3 absorption. With this knowledge, predictions can be made as to when the strong increase in the TP4 phosphate absorption will occur [**Figure 3(d)**]. In other words, our sensor allows the remaining lifetime of the PEF to be estimated and unscheduled and costly maintenance stops to be avoided.

Acknowledgements

The authors gratefully acknowledge financial support under the BMBF project NAMIFLU (16SV5357). The authors also benefitted from discussions and support from colleagues at EADS, in particular from Dr. Volker Baumbach

(AIRBUS Bremen).

Source: Helwig A, Maier K, Müller G, *et al.* An Optoelectronic Monitoring System for Aviation Hydraulic Fluids [J]. Procedia Engineering, 2015, 120(2): 233– 236.

References

[1] G.E. Totten, Handbook of Hydraulic Fluid Technology Mechanical Engineering, Marcel Dekker Inc., 2000.

[2] S. Paul, W. Legner, A. Krenkow, G. Müller, T. Lemettais, F. Pradat, D. Hertens, Chemical contamination sensor for phosphate ester hydraulic fluids, Int. J. Aerosp. Eng. 2010 (2010).

[3] K. Fischer, Neues Verfahren zur maßanalytischen Bestimmung des Wassergehaltes von Flüssigkeiten und festen Körpern, Angew. Chemie. 48 (1935) 394–396.

[4] A full account of our work will be given in a forthcoming publication in Sensors and Actuators B.

Chapter 5

Compositional and Optical Properties of SiO_x Films and (SiO_x/SiO_y) Junctions Deposited by HFCVD

Diana E. Vázquez-Valerdi, Jose A. Luna-López, Jesús Carrillo-López, Godofredo García-Salgado, Alfredo Benítez-Lara, Néstor D. Espinosa-Torres

ICUAP BUAP (BENEMERITA UNIVERSIDAD AUTONOMA DE PUEBLA, in Puebla México)

Abstract: In this work, non-stoichiometric silicon oxide (SiO_x) films and (SiO_x/SiO_y) junctions, as-grown and after further annealing, are characterized by different techniques. The SiO_x films and (SiO_x/SiO_y) junctions are obtained by hot filament chemical vapor deposition technique in the range of temperatures from 900°C to 1,150°C. Transmittance spectra of the SiO_x films showed a wavelength shift of the absorption edge thus indicating an increase in the optical energy band gap, when the growth temperature decreases; a similar behavior is observed in the (SiO_x/SiO_y) structures, which in turn indicates a decrease in the Si excess, as Fourier transform infrared spectroscopy (FTIR) reveals, so that, the film and junction composition changes with the growth temperature. The analysis of the photoluminescence (PL) results using the quantum confinement model suggests the presence of silicon nanocrystal (Si-nc) embedded in a SiO_x matrix. For the case of the as-grown SiO_x films, the absorption and emission properties are correlated with quantum effects in Si-nc and defects. For the case of the as-grown (SiO_x/SiO_y) junctions, only the emission mechanism related to some kinds of defects was considered, but silicon nanocrystal embedded in a SiO_x matrix is present. After thermal annealing, a phase separation into Si and SiO_2 occurs, as the FTIR spectra illustrates, which has repercussions in the absorption and emission properties of the films and junctions, as shown by the change in the A and B band positions on the PL spectra. These results lead to good possibilities for proposed novel applications in optoelectronic devices.

Keywords: HFCVD, PL, FTIR, Transmittance, Thermal Annealing, Si-nc

1. Introduction

Silicon is the semiconductor material predominant in the microelectronics industry. However, it has been long considered unsuitable material for optoelectronic applications[1], due to its indirect band gap, which means it is a poor light emitter. Though, after discovery of visible light emission at room temperature in the porous silicon by Canham[2] in 1990, many investigators have studied emission properties of materials with silicon compounds as the non-stoichiometric silicon oxide (SiO_x), which has gained increasing interest in the research community due to the formation of silicon nanocrystals embedded in the matrix, implying low-dimensional effects and thus determines an efficient emission of visible light, even at room temperature[3][4]. In the SiO_x films, the absorption and emission properties are correlated with quantum effects in silicon nanoparticles, and also associated with defects[5]. From the technological standpoint, the average size of a silicon nanoparticle (Si-np) offers band gap widths, which opens the possibility to tune the emission of light using nanostructured thin films in novel optoelectronic devices. The goal of this work is to study and investigate the compositional and optical properties of SiO_x films and (SiO_x/SiO_y) junctions obtained by hot filament chemical vapor deposition (HFCVD) technique, as-grown and after a further annealing, in order to have a broad and solid view of the behavior of the material by varying the growth temperature and mainly its thickness, which opens the possibility for proposed novel applications in a future work.

2. Methods

SiO_x films and (SiO_x/SiO_y) junctions were obtained by HFCVD technique in the range of temperatures from 900°C to 1,150°C and deposited on quartz and n-type silicon (100) substrates with 1 to 10 Ω cm resistivity, using quartz and porous silicon as the reactive sources. The deposition time was 5min for the SiO_x films and of 10min for the (SiO_x/SiO_y) junctions due to that the time for each film of the junction was of 5min. The chemical reaction for the grown non-stoichiometric silicon oxide in the HFCVD technique is pictured in[6][7]. The relation- ship between the filament temperature (approximately 2,000°C) and the variation of the source-substrate distance (dss) of 4mm, 5mm, and 6mm provides a change in the growth temperature (Tg) of 1,150°C, 1,020°C, and 900°C, respectively, which was measured with a thermocouple in the reaction zone. The SiO_x films were deposited at 1,150°C, 1,020°C, and 900°C. Three samples of each kind have been studied. The (SiO_x/SiO_y) junctions were made with two films deposited at two different temperatures, obtaining six possible combinations, 1,150°C/1,020°C, 1,150°C/900°C, 1,020°C/1,150°C, 1,020°C/900°C, 900°C/1,150°C, and 900°C/1,020°C. Three samples of each kind have been studied. The changes in the dss and Tg, consequently, modify the silicon excess and defects in the non-stoichiometric SiO_x films. The substrates were carefully cleaned with a metal oxide semiconductor standard cleaning process and the native oxide was removed with an HF buffer solution before being introduced into the reactor. The thermal annealing was made using a nitrogen atmosphere at 1,100°C for 1h. Several spectroscopic characterization techniques were used. The film thickness was measured using a Dektak 150 profilometer (Veeco Instruments Inc., Plainview, NY, USA). Room-temperature transmittance of the SiOx films was measured using a UV-Vis-NIR Cary 5,000 system (Agilent Technologies Inc., Santa Clara, CA, USA). The transmittance signal was collected from 200nm to 1,000nm with a resolution of 0.5nm. FTIR spectroscopy measurements were done using a Bruker system model vector 22 (Bruker Instruments, Bellirica, MA, USA). Photoluminescence (PL) response was measured at room temperature using a Horiba Jobin Yvon spectrometer model FluroMax 3 (Edison, NJ, USA) with a pulsed xenon source whose detector has a multiplier tube, which is controlled by a computer. The samples were excited using a 250-nm radiation, and the PL response was recorded between 380nm and 1,000nm with a resolution of 1nm.

3. Results and Discussion

The thicknesses of the SiO_x films and (SiO_x/SiO_y) junctions as a function of the growth temperature (Tg) are shown in **Figure 1** as box statistic charts, where the mean, maximum, and minimum values are shown, the width of the box denotes the standard deviation of the measurements. The Tg and the deposition time affect the thickness of the SiO_x films and (SiO_x/SiO_y) junctions. When Tg was decreased and the deposition time was less, thinner samples and a more uniform deposition were obtained due to lesser amount of volatile precursors likely to deposit. **Figure 1(b)** shows that the junctions have a better uniformity than the films [**Figure 1(a)**] because the standard deviation (width of the box) is less. This could be due to a possible "annealing" of the first SiO_x film deposited. In this case, the first layer has received annealing twice, which would certainly lead to a different formation. Then, these results show that the double annealing procedure and the order of layers have a very strong impact on the thickness. The tendency is that the more annealing, the thicker the layer. For the SiO_x films, average thicknesses were obtained of 500nm to 100nm and for the (SiO_x/SiO_y) junctions of 750nm to 200nm.

Figure 2 shows the UV-Vis transmittance spectra of SiO_x films and (SiO_x/SiO_y) junctions as-grown deposited on quartz. All the samples exhibited a relatively high transmittance (>70%) between 600nm and 1,000nm. The change in the growth temperature produces a change in the stoichiometry of the SiO_x films and thickness, which produces a shift of the absorption edge. In a previous work, where the SiO_x films were characterized by XEDS[7], the information on change

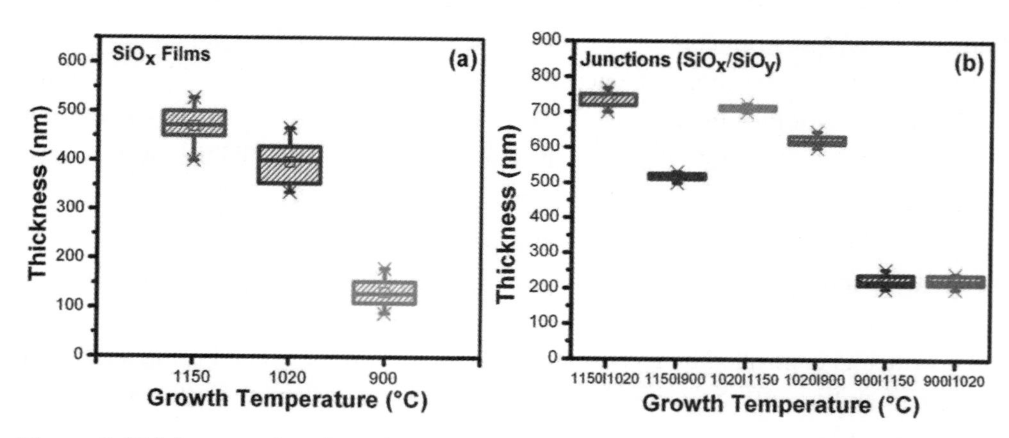

Figure 1. Thickness as function of Tg of the SiO_x films (a) and (SiO_x/SiO_y) junctions (b).

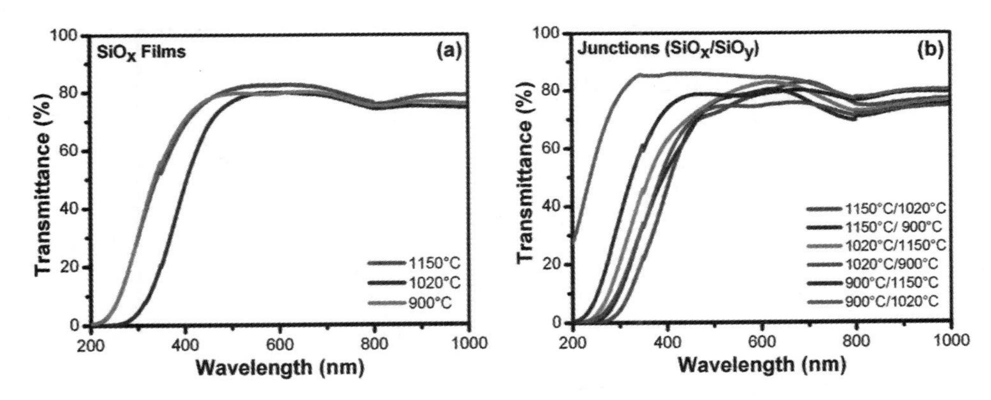

Figure 2. Transmittance spectra as function of Tg of the SiO_x films (a) and (SiO_x/SiO_y) junctions (b).

in stoichiometry of SiO_x were reported; where x for the film deposited at 1,150°C was 1.27, at 1,020°C was 1.81, and at 900°C was 1.66; these values are approximates. **Figure 2(a)** shows a similar absorption edge for 900°C and 1,150°C, while 1,020°C is shifted; therefore, a random result is obtained. **Figure 2(b)** shows that there is an extremely large difference between junctions; this seems to imply that the double annealing procedure and the order of layers have a very strong impact on optical properties. The overall trend seems as the more annealing is received, the less transparent the sample is in visible range—*i.e.*, annealing is facilitating formation of absorbing/emitting states (defects or NCs)[8].

The approximate values of the energy band gap (Eg) were obtained by the relationship known as Tauc plot[9]-[11] as shown in **Figure 3**. The methodology for obtaining these approximate values of Eg has been described in a previous work[7], where the absorption coefficients $\alpha(\lambda)$ were determined from transmission spectra and was obtained in the order of 103cm^{-1} to 104cm^{-1}. The Eg decreases as the growth temperature increases; the optical energy band gap of the SiO_x films is in the range of energies of 2.15eV to 1.8eV and of the (SiO_x/SiO_y) junctions is in the range of energies of 2.25eV to 1.8eV. When x decreases from 2.0 in a-SiO_x, the valence band edge moves up, as the increased Si-Si bond states are gradually overlapped with the oxygen nonbonding states (ONS) and finally spread out into the Si valence band. Simultaneously, the conduction band edge also moves down. The final result is that the optical band gap decreases nonlinearly when Si concentration continually increases[5]. Therefore, we may assume that the Si excess increases as the growth temperature increases.

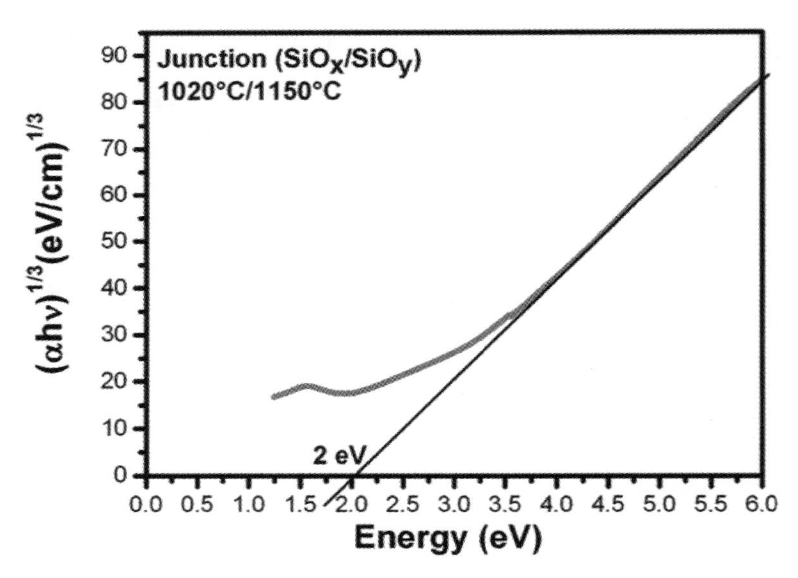

Figure 3. $(\alpha hv)^{1/3}$ versus energy (hv). Example to obtain the approximate value of Eg by the relationship known as Tauc plot.

Figure 4 and **Figure 5** show the FTIR spectra from SiO_x films and (SiO_x/SiO_y) junctions as-grown and after a further annealing, respectively. The IR vibration bands are listed in **Table 1** for the SiO_x films and in **Table 2** for the (SiO_x/SiO_y) junctions. All the spectra show the absorption peaks characteristic of SiO_2, which correspond to the vibration modes of stretching (1,082cm^{-1}), bending (812cm^{-1}), and rocking (458cm^{-1}) of Si-O-Si[12][13]. Before thermal annealing, these peaks show a shift to lower wavenumber with decreasing of the growth temperature (Tg), indicating the change in the silicon excess. After thermal annealing, these shifts disappear and the position of the vibration modes corresponding to SiO_2, thus indicating a phase separation. Before thermal annealing, a vibration band approximately at 885cm^{-1}, assigned to Si-H bending and Si-O of Si-O-H[14]-[16], appeared with a shoulder at 810cm^{-1} related to the Si-O bending vibration band. When the Tg was higher, the intensity of the Si-O bending vibration band increased and shift at the same time that the peak at 870cm^{-1} decreased, and the Si-O bending vibration band became more apparent. These bonds are present in the films due to the incorporation of hydrogen in the experimental process, but after the films were thermally annealed, the band at 885cm^{-1}, disappeared due to the desorption of hydrogen to high temperature and the characteristic absorption peaks of SiO_2 were more noticeable. The Si-O stretching vibration band shifted to higher frequencies after thermal annealing indicating a phase separation

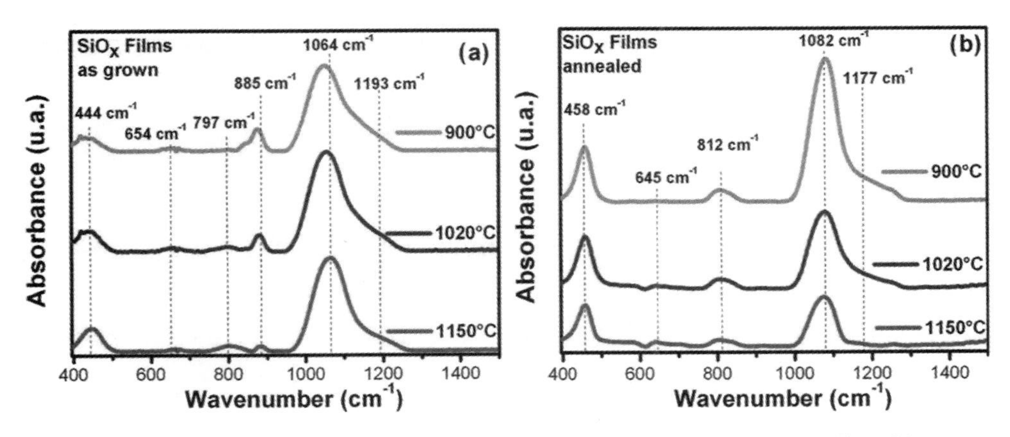

Figure 4. FTIR spectra from SiO_x films as-grown (a) and after further annealing (b).

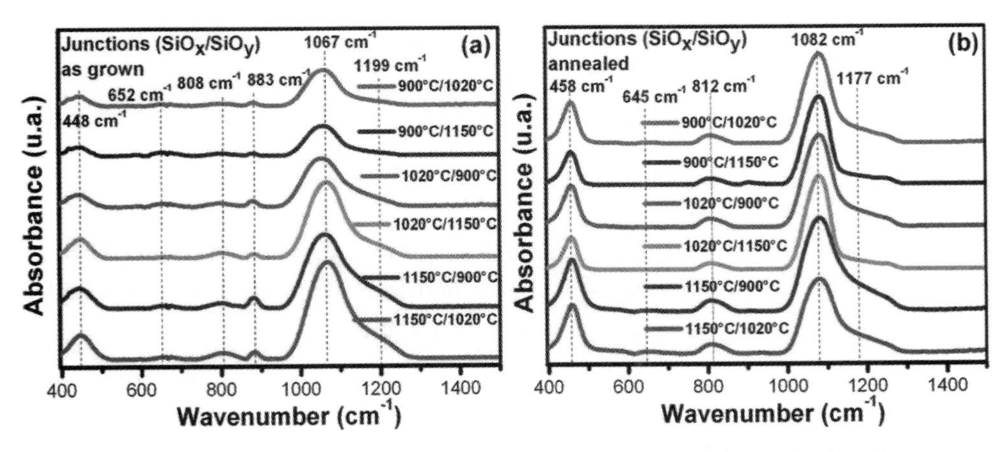

Figure 5. FTIR spectra from (SiO_x/SiO_y) junctions as-grown (a) and after further annealing (b).

Table 1. IR vibration bands[12]-[17] of the SiO_x films before and after thermal annealing.

Vibration mode	Peak position (cm^{-1})					
	As grown			Annealed		
	1,150°C	1,020°C	900°C	1,150°C	1,020°C	900°C
Si-O rocking	444	436	429	458	458	458
Si-O bending	797	800	810	812	812	812
Si-O stretching	1,064	1,055	1,048	1,082	1,082	1,082
Si-H wagging	654	649	645	645	645	-
Si-H bending	885	879	875	-	-	-

Table 2. IR vibration bands[12]-[17] of the (SiO_x/SiO_y) junctions before and after thermal annealing.

	Vibration mode	Peak position (cm^{-1})					
		1,150°C/ 1020°C	1,150°C/ 900°C	1,020°C/ 1,150°C	1,020°C/ 900°C	900°C/ 1,150°C	900°C/ 1,020°C
As grown	Si-O rocking	447	444	442	442	440	440
	Si-O bending	808	803	800	800	797	796
	Si-O stretching	1,067	1,063	1,063	1,055	1,055	1,054
	Si-H wagging	664	659	645	-	652	651
	Si-H bending	883	883	880	879	879	877
Annealed	Si-O rocking	458	458	458	458	458	458
	Si-O bending	812	812	812	812	812	812
	Si-O stretching	1,082	1,082	1,082	1,082	1,082	1,082
	Si-H wagging	645	640	-	637	-	-
	Si-H bending	-	-	-	-	-	-

and its width reduced due to an increment in the oxygen concentration. Also, all samples exhibit a peak around $645cm^{-1}$ to $664cm^{-1}$, which has been associated with Si-H wagging bonds or neutral oxygen vacancies[17]. These bonds are almost imperceptibly present in both cases.

Figure 6 and **Figure 7** show the PL spectra from SiO_x films and (SiO_x/SiO_y) junctions as-grown and after further annealing, respectively.

A wide PL spectrum with a similar shape to a Gaussian curve is observed for all the samples. For this reason, the deconvolution was applied to the PL spectra, with which several peaks were obtained as shown in **Figure 8**; each of the peaks represent the emission individually, which are the components of the PL spectra. Therefore, each peak has different origins.

In general, in the PL spectra of the SiO_x films and (SiO_x/SiO_y) junctions as-grown, there are several emission bands covering a wide spectral range from 380nm to 850nm (violet-near infrared range). After further annealing, the PL spectra there are two main bands, band A of 380nm to 495nm (violet-blue range) with a relatively weak PL intensity and band B of 590nm to 875nm (orange-near infrared

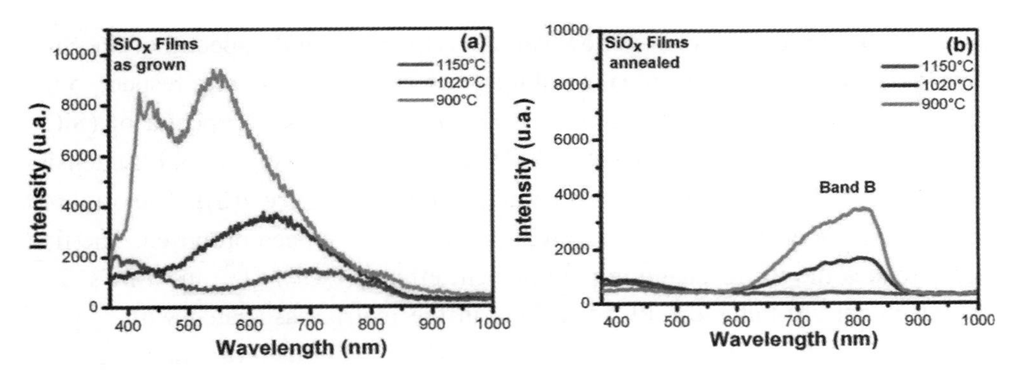

Figure 6. PL spectra from SiO_x films as-grown (a) and after further annealing (b).

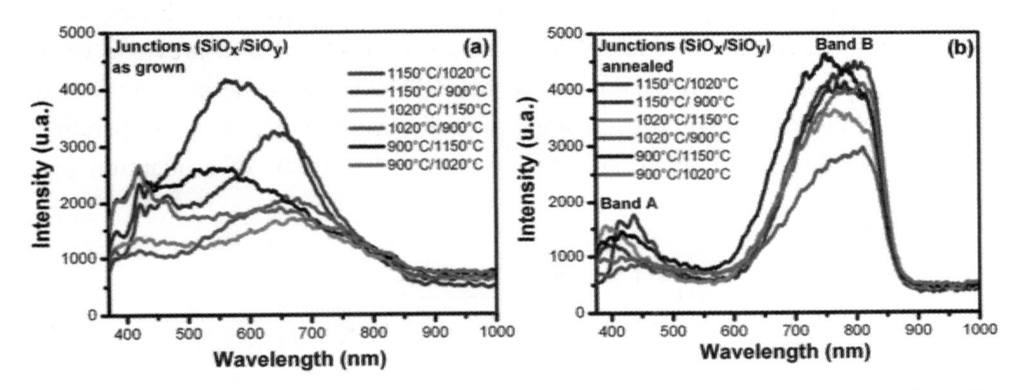

Figure 7. PL spectra from (SiO_x/SiO_y) junctions as-grown (a) and after further annealing (b).

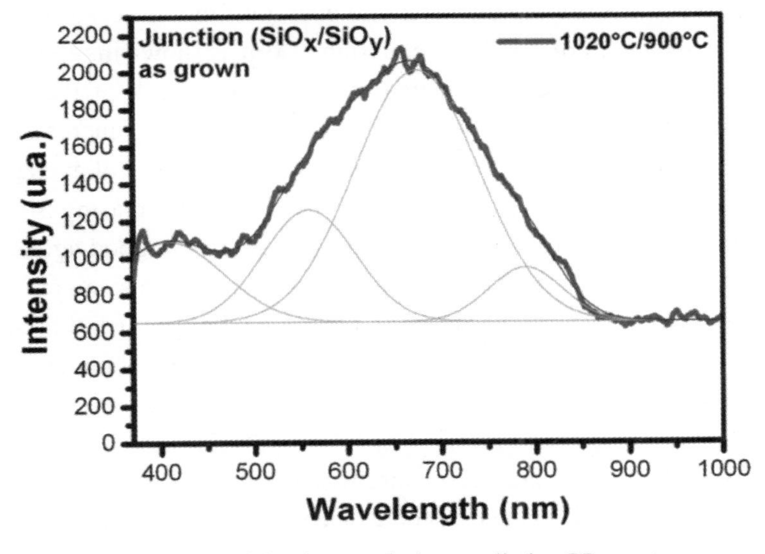

Figure 8. Example of the deconvolution applied to PL spectra.

range). In this case, the green-yellow band disappears. The PL spectra of the SiO_x films, as-grown [see **Figure 6(a)**] exhibit a higher PL intensity with respect to the PL intensity after further annealing [see **Figure 6(b)**]. The PL spectra of (SiO_x/ SiO_y) junctions as-grown [see **Figure 7(a)**] illustrate a relatively weak PL intensity with respect to PL intensity after further annealing [see **Figure 7(b)**]. The origin of the PL emission is still object of debate; two models have been proposed. The first model relates the PL to quantum confinement effects (QCE)[18][19] in Si-nc's. The second model attributes the PL to defects in the matrix associated with oxygen vacancies or defects in the interface SiO_2/Si-nc[16][19]. **Figure 6(a)** shows that the SiO_x films, as-grown, have a maximum emission peak that suffers a blue shift and an increase in the intensity as the growth temperature decreases. This behavior could be related to the presence of Si-nc's as was observed in a previous work, where the SiO_x films were characterized by HRTEM[7], and the average size of Si-nc was obtained, which reduced along with the decreasing growth temperature. Therefore, in this case, the PL spectra are analyzed in terms of a quantum confinement model[18][19]:

$$\lambda(nm) = \frac{1.24\mu m}{E_N(eV)} = \frac{1.24\mu m}{\left[1.12 + \left(3.73/d^{1.39}\right)\right]}$$

It corresponds to the radiative recombination of electron-hole pairs in the Si-nc, where d and EN are the diameter and energy of the Si-nc, respectively, and λ(nm) is the wavelength of the Si-nc emission. **Table 3** shows the theoretical values of average sizes of Si-nc calculated from the main peak of the PL spectra, where the size of the Si-nc reduces and the energy band gap increases with decreasing the growth temperature. **Table 3** also shows the values of average sizes of Si-nc obtained by HRTEM[7]. So the blue shift of PL with decreasing Si concentration and growth temperature, *i.e.*, decreasing Si-nc size, is in good agreement with the QCE. Besides, in these SiO_x films, the absorption and emission are correlated with quantum effects in Si-nc because the theoretical values of EN are similar to the approximate values of the optical band gap (Eg) that were obtained by the relationship known as Tauc plot (see **Figure 3**).

However, PL spectra are very wide, which indicates that two possible mechanisms are involved. Then, the second mechanism of light emission that was considered in the SiO_x films is related to some kinds of defects produced during the growth process, as shown in the FTIR spectra, such as, weak oxygen bonds (WOB), neutral oxygen vacancy (NOV), non-bridging oxygen hole center (NBOHC), positively charged oxygen vacancy (E' center), interstitial oxygen molecules, and peroxide radicals[19]-[24]. Some of these defects, such as NOV and NBOHC are the principal radiative recombination centers or the luminescence centers. So, the different peaks that were defined by the deconvolution of the PL spectra are related to different kinds of defects, as listed in **Table 4**.

Figure 7(a) shows that the PL spectra of the (SiO_x/SiO_y) junctions, as-grown, emit in a wide spectral range. This is due to the junction of two films with different properties and stoichiometries, so these spectra have a behavior mixed with respect to the SiO_x films [see **Figure 6(a)**], where again the first film deposited dominates the behavior of the junction. For this case, the PL spectra are not analyzed in terms of a quantum confinement model, because the PL spectra have not a maximum emission peak that suffers a blue shift. So only the second mechanism of light emission was considered in the (SiO_x/SiO_y) junctions, which is related to some kinds of defects produced during the growth process, as shown in the FTIR spectra and listed in **Table 5**.

Table 3. Theoretical and experimental values.

Tg (˚C)	Gap E_N of Si-nc (eV)	Diameter of Si-nc (nm)	Diameter of Si-nc (nm) by HRTEM[7]
1,150	1.67	3.96	5.5
1,020	1.89	3.11	4
900	2.28	2.31	2.5

Values of the E_N and the diameter of Si-nc calculated from the PL spectra of the SiO_x films as-grown as a function of the Tg.

Table 4. Defect types[19]-[24] linked with the peak position.

Defect type	Peak position (nm)		
	As grown		
	1,150˚C	1,020˚C	900˚C
WOB	400	410	
NOV defect			436
E'δ defect	-	540	543
NBOHC	-	-	658
None identified	739	758	-

Position obtained by deconvolution from PL spectra of the SiO_x films as function of the Tg.

Table 5. Defect types linked[19]-[24] with the peak position.

Defect type	Peak position (nm)					
	As grown					
	1,150°C/ 1,020°C	1,150°C/ 900°C	1,020°C/ 1,150°C	1,020°C/ 900°C	900°C/ 1,150°C	900°C/ 1,020°C
WOB	-	-	-	408	406	414
NOV defect	449	438	432	-	-	-
E'δ defect	532	534	-	559	542	514
NBOHC	648	622	667	674	672	644
None identified	805	782	795	789	770	780

Peak position obtained by deconvolution from PL spectra of the (SiO_x/SiO_y) junctions as function of the Tg.

Figure 6(b) and **Figure 7(b)** show the PL spectra of the SiO_x films and (SiO_x/SiO_y) junctions with thermal annealing, respectively. It is observed that when the thermal an-nealing is applied to the samples, a restructuring occurs (as shown in the FTIR spectra) because the temperature causes a phase separation and growth of Si-nc's. For the case of the SiO_x films, the defect-related PL is almost eliminated and the near-infrared PL intensity (band B) is a little enhanced after annealing, *i.e.*, the PL is dominated by the Si-nc's embedded in the SiO_x matrix. For the case of the (SiO_x/SiO_y) junctions, what is most interesting is that they emit not only in the orange-near in-frared range (band B) but also in the violet-blue range (band A), and the green-yellow band disappears. In this case, the PL intensity of band A slightly declines with respect to the PL intensity of the (SiO_x/SiO_y) junctions as-grown. So the dominant radiative defects have changed from E'δ or NBOHC defects to Si-O species (*i.e.*, WOB and NOV defects). With respect to the PL intensity of band B, this is greatly enhanced after thermal annealing. It is reasonable to consider that for the (SiO_x/SiO_y) junctions, the two mechanisms are not contradictory but run parallel in the samples. In summary, two PL bands, band A and band B, have been observed after thermal annealing. According to the PL spectra, the former band A is ascribed to Si-O-related species[19][20][24], and band B is ascribed to the Si-nc's embedded in the SiO_x matrix[9]. Annealing seems to advance the formation of crystalline silicon (c-Si) as well as the formation of (defect) structures responsible for PL emission. The difference in behavior of the PL intensity before and after annealing between the SiO_x films and junctions is interesting; this difference could be due to the fact that the single layers have a combination of defects and formed Si-nc's which generates intense PL. When annealing SiO_x

films, the defects are desorbed and generated, leading to the de- crease in PL intensity. In double layers, there are several factors, one is the double annealing procedure and the other is the order of layers, which have a very strong impact on optical properties, where the defects of the first layer are preserved due to the growth of the second layer. These layers have a higher concentration of defects and excess silicon; when annealing (SiO_x/SiO_y) junctions, some defects are desorbed or modified, mainly on the second layer, where the PL is improved in the red. According to the analysis of the compositional and optical properties of the SiO_x films and (SiO_x/SiO_y) junctions, we can correlate the evolution of the stoichiometry (obtained by FTIR and XEDS) with the reduction of the Eg (obtained by transmittance), where the oxygen content decreased along with the increase in growth temperature and in turn with the reduction of the Eg. So, we may assume that the Si excess increases as the growth temperature increases.

4. Conclusions

In the present work, SiO_x films and (SiO_x/SiO_y) junctions were obtained by HFCVD technique with different growth temperatures; their compositional and optical properties were studied as-grown and after further annealing. For the case of the SiO_x films, as-grown, the absorption and emission are correlated with quantum effects in silicon nanoparticles and also associated with defects. For the case of the (SiO_x/SiO_y) junctions, as-grown, only the second mechanism of light emission was considered, which is related to some kinds of defects produced during the growth process. After thermal annealing, the PL spectra of the SiO_x films only show band B (740nm), and the PL spectra of the (SiO_x/SiO_y) junctions show bands A (375nm to 450nm) and B (740nm), where the PL band A is ascribed to Si-O-related species, and the PL band B to the quantum size effect of the Si-nc's embedded in the SiO_x matrix. Vibration bands related to Si-H was observed in SiO_x films and (SiO_x/SiO_y) junctions before annealing (as shown in the FTIR spectra), which disappeared after thermal annealing. The behavior of the change in intensity and shift of the FTIR and PL spectra is ascribed to changes of phase in this material, associated with several defects and Si-np's. The (SiO_x/SiO_y) junctions showed an interesting behavior due to the double annealing procedure and 'order' of layers, which have a very strong impact on optical and compositional properties. This opens the possibility of tuning the spectra range for a specific application.

Competing Interests

The authors declare that they have no competing interests.

Authors' Contributions

DEVV conducted the SiO_x growth. DEVV and JALL carried out the FTIR and PL measurements and drafted the manuscript. JCL coordinated the FTIR study. GGS coordinated and participated in the growth of the SiO_x films. ABL conducted the thermal annealing. NDET conducted the UV measurements. JALL provided the idea and supervised the study. All authors read and approved the final manuscript.

Authors' Information

DEVV is currently a PhD student in the Science Institute-Center of Investigation in Semiconductors Devices (IC-CIDS) from Autonomous University of Puebla, Mexico. She started to work on the growth and characterization of non-stoichiometric silicon oxide obtained by HFCVD. Her research interests include experiments and structural, optical and electrical characterization of SiO_x films and MOS structures. JALL is currently a researcher and professor in the Science Institute-Center of Investigation in Semiconductors Devices (IC-CIDS) from Autonomous University of Puebla, Mexico. He started to work on structural, electrical and optical characterization of materials and MOS structures. His research interest is the physics and technology of materials nanostructures and silicon devices. Moreover, his research interests are, too, the nanotechnology, material characterization, and optoelectronic devices such as sensors, LEDs, and solar cells. GGS received his PhD in the Electronic and Solid State Department from the Center of Research and Advanced Studies, National Polytechnic Institute, Mexico City in 2003. He started to work on the growth and characterization of non-stoi-hiometric silicon oxide. His current research interests include metallic oxides obtained by the HFCVD technique, GaN obtained by the metal organic CVD technique and porous silicon gas sensor devices.

Acknowledgements

This work has been partially supported by CONACyT-154725, PROMEP, and VIEP-BUAP-2013. The authors acknowledge INAOE and IFUAP laboratories for their help in the sample measurements.

Source: Vázquezvalerdi D E, Lunalópez J A, Carrillolópez J, *et al.* Compositional and optical properties of SiO_x films and (SiO_x/SiO_y) junctions deposited by HFCVD [J]. Nanoscale Research Letters, 2014, 9(1):1–8.

References

[1] Pai PG, Chao SS, Takagi Y, Lucovsky G: Infrared spectroscopic study of silicon oxide (SiO_x) films produced by plasma enhanced chemical vapor deposition. J Vac Sci Technol A 1986, 4(3):689–694.

[2] Canham LT: Silicon quantum wire array fabrication by electrochemical and chemical dissolution of wafers. Appl Phys Lett 1990, 57(10):1046–1048.

[3] Dong D, Irene EA, Young DR: Preparation and some properties of chemically vapor deposited Si-Rich SiO_2 and Si_3N_4films. J Electrochem Soc 1978, 125(5):819–823.

[4] Fang YC, Li WQ, Qi LJ, Li LY, Zhao YY, Zhang ZJ, Lu M: Photoluminescence from SiO_x thin films: effects of film thickness and annealing temperature. Nanotechnology 2004, 15:495–500.

[5] Chen XY, Lu Y, Tang LJ, Wu YH, Cho BJ, Dong JR, Song WD: Annealing and oxidation of silicon oxide films prepared by plasma-enhanced chemical vapor deposition. J Appl Phys 2005, 97:014913.

[6] Luna-López JA, García-Salgado G, Díaz-Becerril T, Carrillo López J, Vázquez-Valerdi DE, Juárez-Santiesteban H, Rosendo-Andrés E, Coyopol A: FTIR, AFM and PL properties of thin SiOx films deposited by HFCVD. Mater Sci Eng 2010, 174:88–92.

[7] Luna López JA, Carrillo López J, Vázquez Valerdi DE, García Salgado G, Díaz Becerril T, Ponce Pedraza A, Flores Gracia FJ: Morphological, compositional, structural, and optical properties of Si-nc embedded in SiOx films. Nanoscale Res Lett 2012, 7:604.

[8] Matsumoto Y, Godavarthi S, Ortega M, Sánchez V, Velumani S, Mallick PS: Size

modulation of nanocrystalline silicon embedded in amorphous silicon oxide by Cat-CVD. Thin Solid Films 2011, 519:4498–4501.

[9] Wang L, Han K, Tao M: Effect of substrate etching on electrical properties of electrochemical deposited CuO. J Electrochem Soc 2007, 154:D91.

[10] Pankove JI: Optical Process in Semiconductors. Prentice Hall: Englewood; 1971.

[11] Gordillo Delgado F, Mendoza Álvarez JG, Zelaya Ángel O: Actividad fotocatalítica con luz visible de películas de TiO2 crecidas por RF sputtering reactivo. Revista Colombiana de Física 2006, 38:129.

[12] Ay F, Aydinly A: Comparative investigation of hydrogen bonding in silicon based PECVD grown dielectrics for optical waveguides. Opt Mater 2004, 26:33–46.

[13] Jutarosaga T, Jeoung JS, Seraphin S: Infrared spectroscopy of Si-O bonding in low-dose low-energy separation by implanted oxygen materials. Thin Solid Films 2005, 476:303–311.

[14] Benmessaoud A: Caracterización De Subóxidos De Silicio Obtenidos Por Las Técnicas De PECVD, PhD Thesis. Departamento de Física: Universidad Autónoma de Barcelona; 2001.

[15] McLean FB: A framework for understanding radiation-induced interface states in SiO_2 MOS structures. IEEE Trans Nucl Sci 2001, 27(6):1651–1657.

[16] Shimizu-Iwayama T, Hole DE, Boyd IW: Mechanism of photoluminescence of Si nanocrystals in SiO_2 fabricated by ion implantation: The role of interactions of nanocrystals and oxygen. J Phys Condens Matter 1999, 11(34):6595–6604.

[17] Morioka T, Kimura S, Tsuda N, Kaito C, Saito Y, Koike C: Study of the structure of silica film by infrared spectroscopy and electron diffraction analyses. Mon Not R Astron Soc 1998, 299:78–82.

[18] Delerue C, Allan G, Lannoo M: Theoretical aspects of the luminescence of porous silicon. Phys Rev B 1993, 48:11024.

[19] Gong-Ru L, Chung-Jung L, Chi-Kuan L, Li-Jen C, Yu-Lun C: Oxygen defect and Si nanocrystal dependent white-light and near-infrared electroluminescence of Si-implanted and plasma-enhanced chemical-vapor deposition-grown Si-rich SiO_2. J Appl Phys 2005, 97:094306-1–094306-8.

[20] Tong S, Liu X-N, Gao T, Bao X-M: Intense violet-blue photoluminescence in as-deposited amorphous Si:H:O films. Appl Phys Lett 1997, 71:698.

[21] Luna-López JA, Aceves-Mijares M, Malik O, Yu Z, Morales A, Dominguez C, Rickards J: Compositional and structural characterization of silicon nanoparticles embedded in silicon rich oxide. Rev Mex Fis 2007, S53(7):293.

[22] Luna López JA, Morales-Sanchez A, Aceves Mijares M, Yu Z, Dominguez C: Analysis of surface roughness and its relationship with photoluminescence properties of silicon-rich oxide films. J Vac Sci Technol A 2009, 27(1):57.

[23] Kenjon AJ, Trwoga PF, Pitt CW: The origin of photoluminescence from thin films of silicon-rich silica. J Appl Phys 1996, 79(12):9291.

[24] Wehrspohn RB, y Godet C: Visible photoluminescence and its mechanisms from a-SiOx: H films with different stoichiometry. J Lumin 1999, 80:449–453.

Chapter 6

Conformity Analysis in the Measurement of Machined Metal Surfaces with Optoelectronic Profilometer

E. Cuesta[1], B. J. Álvarez[1], M. García-Diéguez[1], D. González-Madruga[2], J. A. Rodríguez-Cortés[3]

[1]Dept. of Manufacturing Engineering. University of Oviedo, Asturias, Spain
[2]Dept. of Manufacturing Engineering. University of Leon, Leon, Spain
[3]PRODINTEC Foundation, Asturias, Spain

Abstract: The objective of this work is to analyse the conformity of the roughness parameters measured with an optoelectronic profilometer in order to make equivalent the measurement results with those obtained with traditional contact devices. The working parameters of the optoelectronic profilometer are based on computational filters which are controlled by software working with a 3D stratified colour map (chromatic fragmentation of the white light). However, these parameters substantially differ from the usual contact profilometers that work with 2D roughness profiles (cut-off, evaluation length, contact stylus radii). This work pursues to find the optical profilometer parameters, and its values, that ensure the best quality measurement for a wide range of machining process.

Keywords: Optoelectronic Profilometer, Surface Quality, Roughness Measurement

1. Introduction

Optoelectronic profilometers employing white light have considerable potential and measuring capabilities, allowing to create real 3D surface maps of a relatively large areas. Depending on the lens used on the profilometer, an area up to a few squared centimetres can be processed with a sub-micrometric scale. Thanks to these maps, surface quality parameters such us amplitude, spacing and hybrid parameters on the roughness and waviness profiles can be measured without any contact with the part. In addition this technology allows for an elevated data acquiring speed, around thousand points per second, what enable a significant reduction in the operation time and consequently in the cost of the inspection task.

In spite of the above advantages, there are still some drawbacks that need to be overcome in order to face up to the conventional methods for roughness measurement. Apart from differences in price and other commercial trades, comparison between technologies was previously studied for specific metallic materials [Durakbasa *et al.* (2011), Viotti *et al.* (2008), and Yiin-Kuen *et al.* (2012)] and non-metallic surfaces, Vorburger *et al.* (2012), but a deeper comparison for covering the most common machining processes is needed. [Demircioglu *et al.* (2011), and Kumar *et al.* (2005)], especially when a wide range of machining processes are considered. There are many parameters that have influence on the measurement; apart from the parameters related with the sensor, other optical parameters must be considered such as the light wavelength, interference phenomena, diffraction, reflection, surface optic characteristics, orientation and machining patterns, colour or brightness, similar to what happens with other optic inspection processes like laser triangulation sensors, Cuesta *et al.* (2009).

The objective of this work is to analyse the conformity of the roughness parameters measured with an optoelectronic profilometer in order to make equivalent the measurement results with those obtained with traditional contact devices. The working parameters of the optoelectronic profilometer are controlled by software and based on filters which are modified depending on the measurement basis, chromatic fragmentation of the white light in this case. However, these parameters substantially differ from the usual contact profilometers (cut-off, evaluation length, contact stylus radii) that work with 2D roughness profiles. Therefore, the equivalence between both optic and contact methods needs to be analysed by measuring a wide range of test parts with different surface finishes and from different machining processes. This work pursues to find the parameters, and its values, that ensure the most accurate measurement for each machining process.

Nomenclature

Ra: arithmetic average roughness parameter, in μm

LCT: cut-off length, in mm

NCT: number of cut-offs

D: scanning density, in μm

EDM: Electro-Discharge Machining

2. Methodology and Experimental Procedure

The available equipment has a singular relevance in this study because it somehow restricts or defines the methodology and the experimental design scope. The profilometer under study is a Solarius Viking that uses chromatic probes based on white light confocal technology (**Figure 1**). The use of chromatic "splitting"

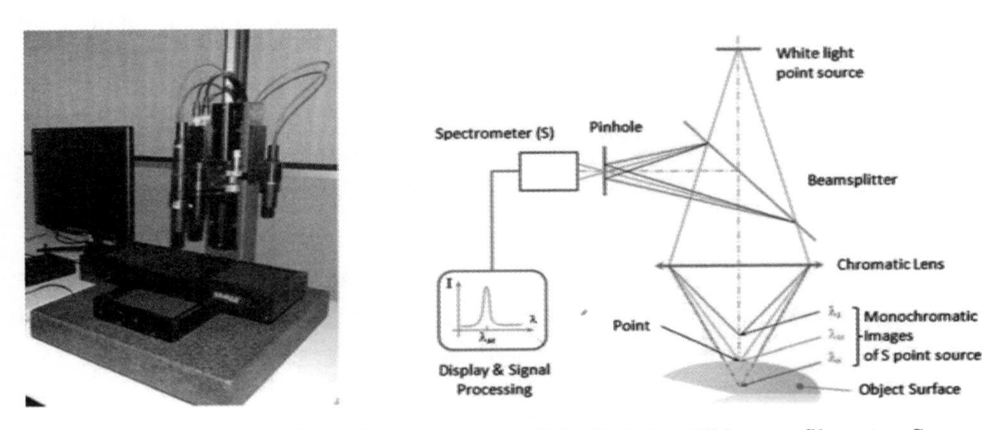

Figure 1. Image and principle of measurement of the Solarius Viking profilometer. Source: Stil S.A., Aix-en-Provence, France.

allows for sub-division of the white light spectrum to correspond to unique Z-axis displacement levels. Each wavelength will be perfectly focused at a different Z height using the confocal principle, all out of focus light is rejected resulting in only the in-focus point been registered.

This type of profilometers has several colour sensors with capture ranges between 110μm and 20mm, resolution between 0.005μm and 0.6μm respectively. In this case only two white light chromatic probes were available, although sufficient for the current study. These models are WLC4 and WLC2, whose main characteristics are shown in **Table 1**. The probe signal is initially processed by the chromatic confocal sensor CHR150 and then, the software (CHR setup utility®, STIL S.A.) calibrates and performs a previous analysis of the signal, thus obtaining a 3D chromatic map. Thereafter, the signal is filtered and finally analysed be the 3D SolarMap software.

Before comparing the measurement results of optoelectronic profilometers and contact roughness testers, a criterion for determining what is the best quality or best fit measurement must be defined. The purpose is to establish the manufacturing processes and the range of surface roughness where the best fit measurement is reached (conformance zone). In order to analyse the influence of the parameters of roughness measurement on an optoelectronic profilometer, the following methodology was carried out:

- Previous study of the optoelectronic profilometer parameters, proper optical probe selection, design of experiments for pre-classifying the parameters under study.

Table 1. Chromatic confocal parameters of the two sensors.

Features	WLC2	WLC4
Range	300μm	2.5mm
Working distance	11mm	16.4mm
Resolution	0.012μm	0.075μm
Accuracy	0.06μm	0.4μm
Angle	±28°	±22°
Spot	2.6μm	8μm

- Selection of roughness gauges or standards that represent significantly the majority of surface finishes that may be found in machined parts (manufacturing processes based on chip removal).

- Development of a two-step procedure: 1st) preliminary probing of the specimens in order to approximate to the optimal solution and, 2nd) in-depth search within the conformance zone of the parameters.

- Analysis of the results, establishing a complete map of the parameters and their conformance zones.

- Generation of working guidelines for measuring surface roughness with high accuracy and estimation of the measurement uncertainty associated with every surface finish, comparing it with the uncertainty of the profilometer assured by the manufacturer.

In the first and second stages optimum filter values were established for the profilometer parameters by taking into account guidelines from the equipment manufacturer as well as standards about surface quality and later by performing tests with different filter types. Thereafter, in order to extend the study to a wider range of machining processes and to a wider range of surface quality, roughness gauges of Rugotest type (TESA©) were tested, classified into roughness grades from N3 to N9 (ISO/R468 and ISO2632-1.2) and related to different machining processes, mainly metal removal processes. The different processes and range of surface finishes analysed in this work are shown in **Table 2**.

All these 54 gauges were measured in a controlled room, in temperature (20°C ± 0.5°C) and relative humidity, using an optimum sampling frequency of 300Hz with a discrete variation of the scanning density (D) between 0.1 and 30 or 45 micrometers (depending on the process). The density is a crucial parameter directly related with the scanning speed (mm/s) and the frequency. With a few exceptions, the more accurate sensor, WLC-2, was adequate for most of the gauges of ISO roughness lower than N8 grade ($Ra < 1.6\mu m$) whereas the gauges of ISO grades N8 and N9 required the use of WLC-4 sensor. With regard to the third stage and in order to find the parameters values that ensure the best quality measurement for all the machining process available, a definition about the best "quality of measurement" is needed. Obviously, the best quality measurements will be those that best fit (statistically) to the measurement results obtained with a contact profilometer.

On the other hand, roughness gauges are classified according ISO grades that do not specify a single value for the arithmetic mean roughness (Ra) but an interval of values. For example, ISO N4 corresponds to an interval of [0.1μm–0.2μm], ISO N5 corresponds to [0.2μm–0.4μm], etc. This fact implies a double check; in first place, measurement results may vary quite within the interval of the ISO grade of the roughness gauge and in second place, the deviation between a series of measurements (even sampling on adjacent regions) must be low enough to consider the measurement as "best quality". The parameter values that lead to a minimum deviation will be taken as the optimum parameter values for the process. This idea that constitutes the third phase of the experimentation is subdivided in two stages (**Figure 2**).

2.1. Test Design—Part 1: Profilometer Parameters and Conformance Zone Establishment

This first step starts from the previous study, so that the optimal working

Table 2. Processes and roughness gauges available.

Processes	"ISO N" grade available	*Ra* (μm)
Electro-Discharge Machining (EDM)	N5-N6-N7-N8-N9	0.4–6.3
Shot-blasting (spherical grain)	N6-N7-N8-N9	0.8–6.3
Shot-blasting (sharp grain)	N6-N7-N8-N9	0.8–6.3
Free hand Grinding (flat specimen)	N6-N7-N8-N9	0.8–6.3
Shaping hand filling (straight & crossed pattern)	N6-N7-N8	0.8–3.2
Vertical face milling (frontal)	N5-N6-N7-N8-N9	0.4–6.3
Turning	N5-N6-N7-N8-N9	0.4–6.3
Horizontal face milling	N7-N8	1.6–3.2
Planing	N6-N7-N8-N9	0.8–6.3
Flat Grinding	N3-N4-N5-N6-N7-N8	0.1–3.2
Cylindrical grinding	N3-N4-N5-N6-N7-N8	0.1–3.2
Lapping	N3-N4-N5	0.1–0.4
Superfinishing (cylindrical surface)	N3-N4-N5	0.1–0.4

Figure 2. Roughness gauge on 1st step measurement (left) and 2nd iterations sampling (right).

distances, lighting conditions, filter type depending on the process and optical sensor have been already set. In this step (**Figure 2** left), all the roughness gauges have been scanned varying the sampling density from the minimum recommended value (0.1μm) up to a value close to the lowest possible density (45μm). However, as only few roughness gauges allow for testing sampling densities larger than 30μm, a final upper limit for sampling density of 30μm was selected.

Figure 3 shows an example of measurement of a roughness gauge that correspond to a turning process and an ISO grade N4. Measurement was carried out employing the WLC-2 sensor, applying a Gaussian filter over a sampling length of 12.5mm, with a cut-off length (LCT) of 2.5mm, so the number of cut-offs (NCT) were 5. At the left of **Figure 3** the unfiltered profile can be seen whereas the figure at the right shows the profile after leveling and filtering simulating the probe ball radius. Finally, the resulting value for Ra was 0.175μm.

2.2. Test Design—Part 2: Measurement Iterations and Refining Conformance Zone

After performing the first group of tests, it is possible to set a conformance zone, that is, to establish those values of sampling density (sampling density range) that make coincide the Ra measured value with the defined interval by the ISO grade of each roughness gauge. A second iteration (**Figure 2** right), that consisted on performing more tests for each value of sampling density within the conformance zone, allowed for determining the sampling quality, *i.e.*, the dispersion between the measured values for an specific sampling density.

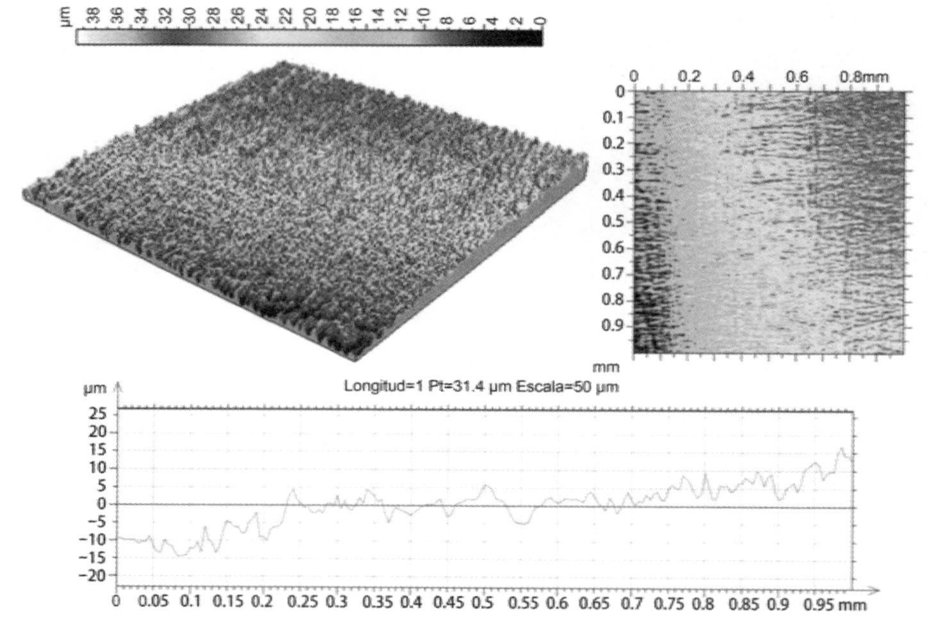

Figure 3. Chromatic 3D map and extracted 2D profile obtained from "turning rugotest" gauge.

It must be noted that it is impossible to scan exactly the same region with both methods, contact scanning and white light scanning. Therefore a statistical sampling is needed. In this case 10 scans were carried out for each roughness gauge, taking care that each scan were performed over an adjacent region different than the previous scanned region (**Figure 2** left).

Table 3 shows a summary of the test stages performed over a Lapping ISO N5 gauge. As can be seen in the table, first group of tests include all the sampling densities possible, from 0.1µm up to 30µm, whereas for the second group of tests, only those densities that provided Ra values within the "conformance zone" (*i.e.*, in concordance with the ISO grade of roughness gauge) were analysed. For this range of selected sampling densities, 10 new scans were carried out in order to obtain the standard deviation, range and mean value of Ra over the whole surface of the roughness gauge (that is, scanning different zones). Finally, it can be appreciated that the optimal scanning parameters for this process and this surface finish are sampling densities of 2µm, 3µm or 5µm. Among these values, the densities of 2µm and 3µm are preferred because the standard deviation of their measurement is lower (0.026µm).

Table 3. Results for 1st and 2nd test groups on ISO N5 lapping gauge (NCT = 5mm, LCT = 0.8mm).

1st test			2nd test												
Scanning density *D*[μm]	*Ra*[μm] primary profile	*Ra*[μm] level & filtered	1	2	3	4	5	6	7	8	9	10	Ra Mean	Ra Std.Dev.	Ra Range
0.1	1.650	0.525													
0.2	1.490	0.514													
0.3	1.510	0.514													
0.4	1.510	0.516													
0.5	1.520	0.505													
0.7	1.330	0.478													
1	1.320	0.472	0.582	0.495	0.553	0.492	0.559	0.470	0.497	0.540	0.528	0.491	0.521	0.03681	0.112
2	**1.360**	**0.364**	**0.413**	**0.414**	**0.424**	**0.381**	**0.418**	**0.427**	**0.446**	**0.377**	**0.399**	**0.358**	**0.406**	**0.02675**	**0.088**
3	**0.913**	**0.295**	**0.380**	**0.312**	**0.357**	**0.346**	**0.351**	**0.341**	**0.385**	**0.358**	**0.331**	**0.288**	**0.345**	**0.02930**	**0.097**
5	**0.252**	**0.236**	**0.284**	**0.324**	**0.335**	**0.237**	**0.310**	**0.249**	**0.302**	**0.254**	**0.273**	**0.247**	**0.282**	**0.03490**	**0.098**
10	0.310	0.196	0.202	0.212	0.238	0.212	0.210	0.189	0.265	0.207	0.188	0.218	0.214	0.03288	0.077
15	0.170	0.154													
20	0.221	0.135													
25	0.182	0.179													
30	0.216	0.161													

3. Results and Conclusions

The methodology explained above was applied to the measurement of the 54 roughness gauges shown in **Table 2** that correspond to 13 different machining processes (most of them are chip removal processes) and 7 different surface finish grades (from ISO N3 to ISO N9). The test results were represented graphically (**Figure 4** and **Figure 5**) for each process, for all of the sampling densities and for every surface finish available. In this type of graphs every curve represents, for the same roughness gauge, the measured Ra value obtained for the different sampling densities.

A first conclusion that can be deduced from these graphs is that a plain curve (more or less horizontal) indicates that the profilometer is able to measure the Ra value independently of the sampling density. Therefore, faster scans may be carried out with low sampling densities for the same level of accuracy. Even more,

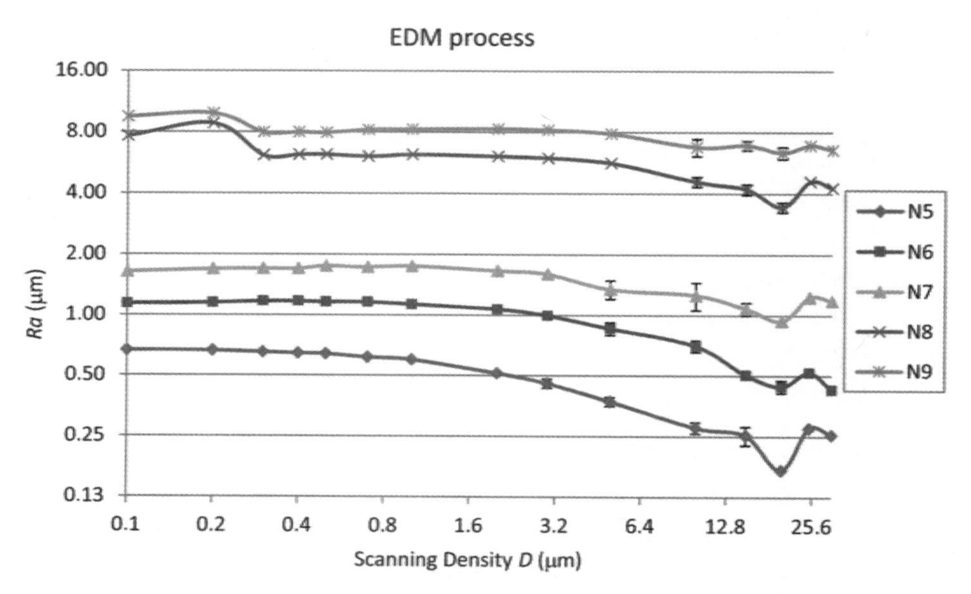

Figure 4. Measured values of Ra for different scanning density measurements over the EDM roughness gauge.

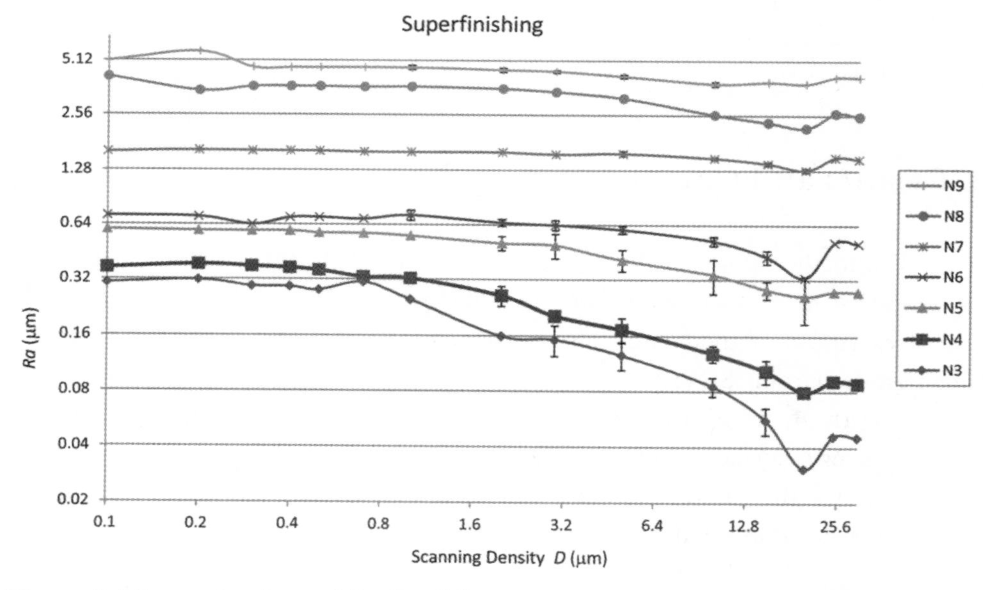

Figure 5. Measured values of Ra for different scanning density measurements over the Superfinishing (N9 to N6) and Honing (N5 to N3) roughness gauges.

for those processes where the optimal scanning density is low (D > 15µm, as it occurs for the ISO N8 and N9 grades of EDM gauges of **Figure 4**) there is no need to test the lower values of sampling density, avoiding the typical trial & error tests that imply elevated measurement times and costs.

Finally mention that in the conformance zone, where the Ra obtained with the profilometer clearly lies within ISO grade of roughness, **Figure 4** and **Figure 5** show the standard deviation as vertical bars error. This standard deviation is calculated from the 10 iterations of the second step, performed on 10 different regions of the gauge. Again, these error bars also provide information on the scan quality; if the error bar is very small or negligible for a specific surface finish, it is likely to be measured with an optimum density and with very high accuracy (that is, with less uncertainty in the determination of Ra). On the other hand, larger values of the vertical error bar within the conformance zone indicate that the profilometer is not capable of measuring accurately the gauge roughness. In **Figure 4** and **Figure 5**, note that logarithmic scales both in the X axis (density) and Y axis (Ra) have been used in order to facilitate the representation. Therefore, a large error bar or a large slope indicates that the profilometer may make an unacceptable mistake, even overlap another ISO grade, as it happens with N3, N4 and N5 gauges from Superfinishing by Honing process (**Figure 5**).

The aim of these tests is to propose working guidelines for measurement of roughness with "high accuracy". With these guidelines, metrologists can take advantage of the high potential of optical profilometer as an alternative to the contact one, and with high reliability. In order to achieve this objective, a complete map with the optimum densities for each machining process and for all tested gauges was developed. **Table 4** summarizes this map, which shows the density values needed for each process and surface finish, *i.e.*, where the measured values of Ra match on both profilometers. Numbers in bold display the value of preferred scan density, meaning that they had lower density deviation. Processes that correspond to smaller ranges of optimum density, even a single value, are those with fairly flat performance curves where the standard deviations of the Ra (obtained in the second group of tests) are relatively large (Std Dev = 0.1μm–0.8μm). Such is the case of specimens EDM processes, Vertical Face Milling, Planing, Shaping hand filing and, to a lesser extent, the Shot-Blasting process. On the other hand, processes with a wide range of optimum density that generate large conformance zones, correspond to decreasing curves and with lower standard deviations (Std Dev = 0.01μm–0.05μm). This is the case of the finishing processes like Lapping, Grinding and Superfinishing (Honing).

Table 4. Recommended interval values for optimum scanning density.

Processes	"ISO N" grade						
	3	4	5	6	7	8	9
EDM			5-**10**	**15**	**15**	**20**	15-**20**
Shot-blasting (spherical grain)				**5**-10-15	**20**	15-**20**	**15**-20
Shot-blasting (sharp grain)				**10**-15	**25**	10-**15**-20	**15**-20
Shaping hand filing (str.&cross.)				**10**	**15**	**20**	
Vertical Face milling			**3**-5-10	3-**5**-10	5-**10**	10-**15**	10-**15**
Turning			1-**2**-3	5-**10**	5-**10**	1-2-3-**5**	1-2-3-**5**
Horizontal Face milling					5-10-**15**	3-5-10-**15**	
Planing				15-**20**	1-2-3-5-**10**	5-**10**-15	**5**-10
Free hand Grinding				1-2-3-**5**-10	5-**10**-15-20	**5**-10-15	5-**10**-15
Flat Grinding (CNC)	**3**-5-10-15	2-3-5-**10**-15	10-**15**-20	3-5-**10**-15-20	**5**-10-15	5-10-**15**-20	
Cylindrical Grinding	15-20-25-**30**	2-3-5-10-**15**	2-3-**5**-10	3-5-10-**15**	5-**10**-15-20	10-15-**20**	
Lapping	3-5-**10-15**	3-5-**10-15**	1-**2-3-5**-10				
Superfinishing	3-5-**10-15**	**3-5**-10-15	**5-10-15**-20	**1-2-3-5**-10-15	5-**10**-15-20	5-**10**-15-20	1-2-**3-5**-10

4. Summary and Future Works

Extensive tests were carried out over parts with a wide range of surface finish (roughness). An in-depth study of the test results allowed for finding the main parameters that relates the contact profilometer measurement and the optoelectronic profilometer measurement. Operation time savings, avoiding "trial & error" tests for finding the optimum setup of a particular measurement, justifies the importance of this study.

The following survey will be to compare the values of the standard deviations between processes and between surface finishes, so the study can be extend to answer the question about what is the best process, and its best finish, that allows for the best fit with minimum error.

The ultimate objective of this knowledge is to identify the best process and surface finish to manufacture certain geometries that can be used as standard gauges for optical measurement devices like those based on laser triangulation, structured white light, or interferometers. The geometry of these standard gauges must cover not only spheres, but also (and especially) cylinders, cones and planes in order to allow the use of these geometries for calibrating multisensor Coordinate Measuring Machines, both contact and contactless

Acknowledgements

Authors gratefully acknowledge the support provided by the Prodintec Foundation and also to the Spanish Ministry of Economy and Competitiveness to grant the project "Quality assurance and knowledge Modeling applied to portable coordinate measuring systems" (ref. DPI2012-36642-C02-01), due to part of the results will have influence on the cited research.

Source: Cuesta E, Álvarez B J, García-Diéguez M, *et al.* Conformity Analysis in the Measurement of Machined Metal Surfaces with Optoelectronic Profilometer [J]. Procedia Engineering, 2013, 63:463–471.

References

[1] Durakbasa, M.N., Osanna, P.H., Demircioglu P., 2011. The factors affecting surface roughness measurements of the machined flat and spherical surface structures—The geometry and the precision of the surface. Measurement 44, pp. 1986–1999.

[2] Viotti, M.R., Albertazzi, A., Fantin, A.V. Dal Pont. A., 2008. Comparison between a white-light interferometer and a tactile form tester for the measurement of long inner cylindrical surfaces, Optics and Lasers Engineering 46, pp. 396–403.

[3] Yiin-Kuen Fuh, KuoChanHsu, JiaRenFan, 2012. Roughness measurement of metals using a modified binary speckle image and adaptive optics. Optics and Lasers in Engineering 50, pp. 312–316.

[4] Vorburger, T.V., Rhee, H.G., Renegar, T.B., Song, J.F., Zheng. A., 2007. Comparison of optical and stylus methods for measurement of surface texture, International Jour-

nal of Advanced Manufacturing Technology 33, pp. 110–118.

[5] Demircioglu, P., Durakbasa. M.N., 2011. Investigations on machined metal surfaces through the stylus type and optical 3D instruments and their mathematical modelling with the help of statistical techniques, Measurement 44 (4), pp. 611–619.

[6] Kumar R., Kulashekar, P. Dhanasekar, B., Ramamoorthy. B., 2005. Application of digital image magnification for surface roughness evaluation using machine vision, Machine Tools & Manufacture 45, pp. 228–234.

[7] Cuesta, E., Rico, J.C., Fernández, P., Blanco, D., Valiño. G., 2009. Influence of roughness on surface scanning by means of a laser stripe system, International Journal of Advanced Manufacturing Technology 43, pp. 1157–1166.

Chapter 7

Modeling of Optoelectronic Devices with One-Band Effective Mass Equation: Nonequilibrium Green's Function Approach

Andrzej Kolek

Department of Electronics Fundamentals, Rzeszow University of Technology, Al. Powstancow Warszawy 12, 35-959 Rzeszow, Poland

Abstract: One-band one-dimensional effective mass model useful in the simulations of layered n-type devices is proposed. The model preserves nonparabolicity both in transport and in-plane directions and enables calculations of intersubband absorption. It is integrated into nonequilibrium Green's function method which is used to simulate quantum cascade laser.

Keywords: One-Band Effective Mass Approximation, In-Plane Dispersion Nonequilibrium Green's Function, Quantum Cascade Laser, Intersubband Absorption

1. Introduction

When modeling unipolar devices, a one-band effective mass equation (EME) is the first choice. The main benefit of this approach is the saving of computer resources which can be then used for more complex description that takes into account quantum coherence and scattering. For optoelectronic devices utilizing intersubband optical transition in lm range, the energy levels are raised well above the band bottom, where they are strongly influenced by the remote bands. One-band description can account for this effect, e.g., by the use of energy-dependent effective mass. For layered n-type devices, an efficient approximation is one dimensional (1D) energy-dependent EME

$$\frac{-\hbar^2}{2}\frac{d}{dz}\frac{1}{m(E,z)}\frac{df}{dz}+\left(E_c(z)+\frac{\hbar^2k^2}{m(E,z)}\right)f=Ef, \tag{1}$$

which uses bulk effective mass (m) at the total energy (E) for both in-plane and longitudinal (z) kinetic energy terms. The efficiency of Equation (1) stems from the fact that it preserves in-plane nonparabolicity that matches well the results of an 8-band kp model (Faist 2013). However, the solutions of Equation (1) are not orthonormal and so annot be used for the evaluation of momentum matrix elements which requires orthonormal wavefunctions. Usually, this difficulty is overcome by solving the two-band (TB) Hamiltonian which gives orthogonal, two-component wavefunctions (f, g) composed of the function $f(z)$ completed with the fictitious valence band component $g(z)$ (Leavitt 1991; Sirtori *et al.* 1994). For the Hamiltonian described by Equation (1), its TB counterpart

$$\begin{bmatrix} E_c(z)+\frac{\hbar^2k^2}{2m(E)} & \frac{\hbar^2}{2m^*\sqrt{\gamma}}\frac{d}{dz} \\ -\frac{\hbar^2}{2m^*\sqrt{\gamma}}\frac{d}{dz} & E_v(z) \end{bmatrix}\begin{bmatrix} f \\ g \end{bmatrix}=E\begin{bmatrix} f \\ g \end{bmatrix} \tag{2}$$

is still energy-dependent, and so the solutions (f, g) are still not orthogonal and useless in the calculations of dipole matrix elements.

In this paper, another effective mass approximation is proposed which, on the one hand, matches well the in-plane dispersion predicted by the 8-band kp model and, on the other hand, reduces to the longitudinal Hamiltonian which has an energy-independent TB counterpart. Then, intersubband absorption can be rigorously treated because the two-component eigenfunctions (f, g) of such Hamiltonian are orthogonal.

2. Anisotropic Nonparabolicity

Instead of Equation (1), one may consider EME

$$\frac{-\hbar^2}{2}\frac{d}{dz}\frac{1}{m(E_z,z)}\frac{df}{dz}+\left(E_c(z)+\frac{\hbar^2k^2}{m(E,z_{aw})}\right)f=Ef, \quad (3a)$$

$$E_z=E-\frac{\hbar^2k^2}{2m(E,z_{aw})}, \quad (3b)$$

where different effective masses are used in transport and in-plane directions. The in-plane dispersion

$$E=\frac{E_g}{2}\left\{-1+\frac{E_i}{E_g}+\sqrt{\left(1+\frac{E_i}{E_g}\right)^2+\frac{2\hbar^2k^2}{m^*E_g}}\right\}, \quad (4)$$

for a subband originated at energy level $E_z = E_i$ can be obtained inserting $m(E, z_{aw}) = m^*[1 + (E - E_c)/E_g]$ into Equation (3b). The plots $E(k)$ predicted by Equation (4) for an exemplary quantum well with two confined states at $E_1 \cong 0.15$ eV and $E_2 \cong 0.413$ eV are shown in **Figure 1**. The deviation from the dispersion relation predicted by the 8-band kp model for the lower state is even smaller than that for Equation (1). Moreover, there is no evidence which approximation is better because the 8-band kp model is also an approximation which introduces inaccuracy at high in-plane momentum. This is evident when compared to more exact results predicted by a 14-band kp model (Ekenberg 1989) (see **Figure 1**).

The subscript '*aw*' in Equation (3) abbreviates 'active wells', so z_{aw} points at the wells where a major optical transition takes place. Fixing z at z_{aw} in Equation (3b) makes longitudinal energy loose its spatial dependence. Then E_z becomes a real number and Equation (3a) can be rewritten in the form

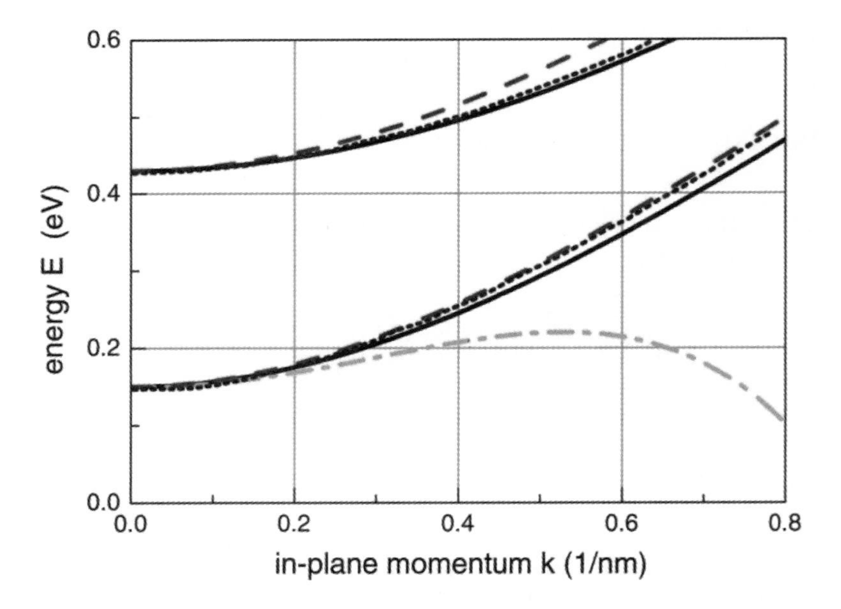

Figure 1. In-plane dispersion relation calculated for Equation (1) (solid black lines), or Equation (3) (dashed red lines) for 4.8 nmwide InGaAs/AlInAs quantum well (lattice matched to InP) compared to the 8-band kp model calculations of Faist (2013) (dotted blue lines) and the 14-band kp model (dash-dotted green lines) of Ekenberg (1989). (Color figure online).

$$\frac{-\hbar^2}{2}\frac{d}{dz}\frac{1}{m(E_z,z)}\frac{df}{dz}+E_c(z)f=E_zf, \tag{5}$$

which has the energy-independent TB counterpart

$$\begin{bmatrix} E_c(z) & \frac{\hbar^2}{2m^*\sqrt{\gamma}}\frac{d}{dz} \\ -\frac{\hbar^2}{2m^*\sqrt{\gamma}}\frac{d}{dz} & E_v(z) \end{bmatrix}\begin{bmatrix} f \\ g \end{bmatrix}=E_z\begin{bmatrix} f \\ g \end{bmatrix}. \tag{6}$$

Equation (6) provides the orthonormal solutions (f, g) which can be used in the calculations of the momentum matrix elements (Leavitt 1991; Sirtori *et al.* 1994).

3. Nonequilibrium Green's Function (NEGF) Implementation

Equation (3) defines a k-dependent Hamiltonian which can be used in Non-equilibrium Green's Function (NEGF) method. In the real space implementation, the conduction band Green's functions (GFs) G^R, $G^<$ are the four-parameter functions of positions z, z', energy E, and in-plane momentum modulus k. For discretized Hamiltonians, useful in numerical simulations and offering a compact form of integral equations, the retarded *GF*, G^R (E, k), is the solution of the Dyson equation (Kubis *et al.* 2009)

$$\left(E\mathrm{I}-H-\Sigma^{R}\right)G^{R}\left(E,k\right)=\mathrm{I} \tag{7}$$

where the self-energy term Σ^R comprises scatterings and couplings to the leads. The valence component of the retarded GF is given by Kolek (2015)

$$\begin{aligned} G_{vc}^{R}\left(E,k\right)&=\left(E_{z}-E_{v}\right)^{-1}H_{cv}^{t}G^{R}\left(E,k\right),\\ G_{vc}^{R}\left(E,k\right)&=\left(E_{z}-E_{v}\right)^{-1}\left[\mathrm{I}+H_{cv}^{t}G_{cv}^{R}\left(E,k\right)\right], \end{aligned} \tag{8}$$

where $H_{c\upsilon}$ is the discretized differential operator

$$\frac{\hbar^{2}}{2m^{*}\sqrt{\gamma}}\frac{d}{dz}\rightarrow \mathrm{H}_{cv} \text{ and } E_{c}\left(z\right)-E_{g}\left(z\right)=E_{v}\left(Z\right)\rightarrow E_{v}$$

Lesser Green's functions can be obtained from the Keldysh equation

$$\begin{bmatrix} G^{<} & G_{cv}^{<} \\ G_{vc}^{<} & G_{v}^{<} \end{bmatrix}=\begin{bmatrix} G^{R}\Sigma^{<}G^{R^{*}} & G^{R}\Sigma^{<}G_{vc}^{R^{*}} \\ G_{vc}^{R}\Sigma^{<}G^{R^{*}} & G_{vc}^{R}\Sigma^{<}G_{vc}^{R^{*}} \end{bmatrix}, \tag{9}$$

where $\Sigma^<$ is the conduction band lesser self-energy. Equations (7)–(9) together with the Poisson equation should be iterated. In this iteration, the GFs valence components G_{υ}^{R}, $G_{\upsilon}^{<}$ should be included as they contribute to the density of states (DOS) and the density of electrons (DOE)

$$N(E,k,z) = -\frac{1}{\pi a}\mathrm{Im}\left[G^R(E,k,z,z) + G_v^R(E,k,z,z)\right],$$
$$n(E,k,z) = -\frac{i}{2\pi a}\left[G^<(E,k,z,z) + G_v^<(E,k,z,z)\right],$$

and thus influence the GFs through the Poisson equation. Once the self-consistency is achieved, the optical absorption can be evaluated making use of the approach developed by Wacker (2002) applied to the TB Hamiltonian (Kolek 2015).

4. Quantum Cascade Laser

The model described in Sect. 2 is especially useful for a quantum cascade laser (QCL)—a device consisting of tens of layers with carefully tuned width what calls for a very dense discretization mesh in the simulations. This generates huge matrices and makes multiband modeling hardly possible. On the contrary, the one-band approach keeps the matrices in reasonable size and enables the use of the NEGF method, which treats quantum tunneling and scattering on equal footing (Kubis *et al.* 2009) what is necessary for this type of devices.

The model of Sect. 2 and its implementation of Sect. 3 were used to calculate electronic transport and optical gain in QCL emitting at $\cong 5\,\mu\mathrm{m}$ $(h\nu \cong 0.25\,\mathrm{eV})$, designed by Evans *et al.* (2007). The equations of the NEGF formalism were solved for one QCL module with the contact self-energies which mimic periodic boundary conditions (Haldas *et al.* 2011; Kolek *et al.* 2012). Other self-energies included into formalism represent electron-phonon (LO, LA), interface roughness, alloy disorder and ionized impurity scatterings. Carrier-carrier interaction was included as the mean-field through the Poisson equation. The discretization mesh spacing was 0.6 nm. In **Figure 2**, which shows exemplary results, the focus is paid on the k-resolved quantities which demonstrate in-plane non-parabolicity inherently contained in the model. Preserving this feature is crucial for realistic modeling of intersubband gain. As discussed by Faist *et al.* (1996), in QCLs the optical gain can emerge due to the local (in k-space) population inversion which is not destroyed by the absorption at higher k-values due to the in-plane nonparabolicity. Indeed, our simulations show that substantial gain emerges (see **Figure 2** inset) (due to $4 \to 3$ transitions) at the energy $h\nu \cong 0.26\,\mathrm{eV}$ that agrees very well with the experimental lasing wavelength. This is in sharp contrast with purely one-band

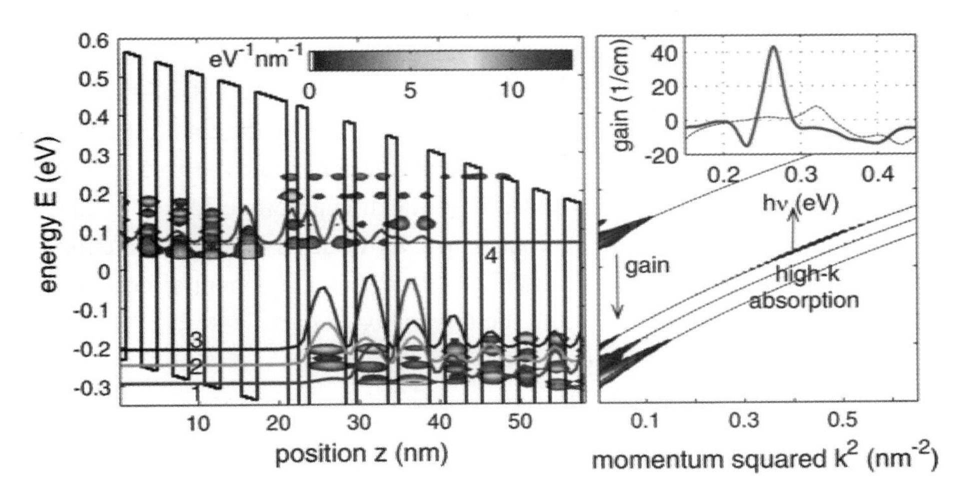

Figure 2. (left) Spectral function at $k = 0$, and (right) the subband occupation in active wells of QCL design of Evans *et al.* (2007). Population inversion is observed only at low-k momenta, where the occupation of the upper laser subband 4 exceeds the occupation of the lower laser subband 3. $4 \rightarrow 3$ transitions at highk values, where the occupation is normal, do not destroy gain because they absorb photons with lower energy: $h\nu_{high-k} < h\nu_{low-k}$. These transitions burn a hole in front of the gain peak in the gain spectrum as shown in the inset. Blue (dashed) line in the inset shows gain spectrum (multiplied $\times 10$) calculated for the one-band model which uses constant effective masses in well and barrier materials (calculations were made for the same value of electric field). (Color figure online).

model which predicts both wrong lasing energy and almost no gain: for this case, the gain peak of merely 0.8cm^{-1} occurs at $h\nu \cong 0.26\,\text{eV}$ Results of one-band simulations are also shown in **Figure 2** for the comparison.

When gain emerges due to local population inversion (which is our case) the most important is an accurate modeling of the in-plane dispersion at low and medium k-values, where intersubband transitions occur at nearly the same values of the photon energy and thus contribute to the gain spectrum. The photons emitted at remote k's have the energies well apart from the energy emitted in the transitions at $k \cong 0$ and so hardly influence the value of the gain peak. Then, going beyond the accuracy of 8-band description, which one may concern looking at the deviations in **Figure 1**, seems to be both hard and impractical, at least in the case of QCLs. The model defined by Equation (3) preserves the accuracy of 8-band model at low k-values which, as shown, is sufficient enough to provide reliable results.

The gain in **Figure 2** inset was calculated as the negative absorption coefficient averaged over one laser period. Both the conduction and the valence components of GFs were involved in these calculations. The amount of the valence component can be estimated, e.g., from the contribution to the total electron density. The plots in **Figure 3** allow to estimate that it contributes to $\cong 5$ percent of the total charge. This number illustrates how much charge one looses if does not care about the valence component what is usually done in one-band calculations. Worth to note is that through the Poisson equation this component affects not only DOE but all other quantities including the intersubband gain.

In summary, a one-band effective mass model was proposed which allows for *k*-resolved calculations which preserve correct nonparabolicity both in transport and in-plane directions and allows for rigorous calculations of intersubband absorption which takes these (nonparabolicities) into-account.

Acknowledgements

The work was supported by the Polish research supporting agencies: the National Science Center (NCN), under the project SONATA 2013/09/D/ST7/03966

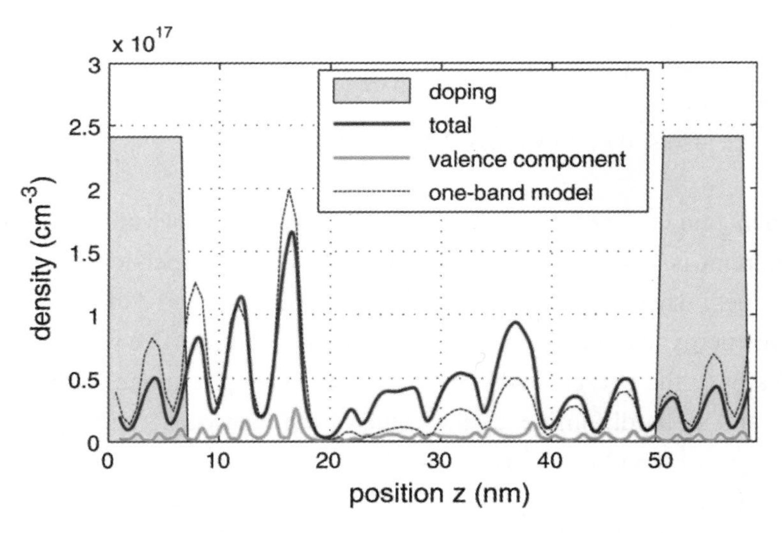

Figure 3. Total density of electrons (red line) in one module of QCL. The valence contribution is marked with green line. Shaded area shows doping. Dashed (blue) line shows electron density calculated for purely oneband model (for the same electric field). (Color figure online).

and the National Center for Research and Development (NCBR) under the grant PBS 1/B3/2/2012 (EDEN).

Source: Kolek A. Modeling of optoelectronic devices with one-band effective mass equation: nonequilibrium Green's function approach [J]. Optical & Quantum Electronics, 2016, 48(2):1–6.

References

[1] Ekenberg, U.: Nonparabolicity effects in a quantum well: Sublevel shift, parallel mass, and Landau levels. Phys. Rev. B 40, 7714–7726 (1989).

[2] Evans, A., Darvish, S.R., Slivken, S., Nguyen, J., Bai, Y., Razeghi, M.: Buried heterostructure quantum cascade lasers with high continuous-wave wall plug efficiency. Appl. Phys. Lett. 91, 071101 (2007).

[3] Faist, J., Capasso, F., Sirtori, C., Sivco, D.L., Hutchinson, A.L., Hybertsen, M.S., Cho, A.Y.: Quantum cascade lasers without intersubband population inversion. Phys. Rev. Lett. 76, 411–414 (1996).

[4] Faist, J.: Quantum Cascade Lasers. Oxford University Press, Oxford (2013).

[5] Haldas, G., Kolek, A., Tralle, I.: Modeling of mid-infrared quantum cascade laser by means of nonequilibrium Green's functions. IEEE J. Quantum Electron. 47, 878–885 (2011).

[6] Kolek, A., Haldas, G., Bugajski, M.: Nonthermal carrier distributions in the subbands of 2- phonon resonance mid-infrared quantum cascade laser. Appl. Phys. Lett. 101, 061110 (2012).

[7] Kolek, A.: Nonequilibrium Green's function formulation of intersubband absorption for nonparabolic single-band effective mass Hamiltonian. Appl. Phys. Lett. 106, 181102 (2015).

[8] Kubis, T., Yeh, C., Vogl, P.: Theory of nonequilibrium quantum transport and energy dissipation in terahertz quantum cascade lasers. Phys. Rev. B 79, 195323 (2009).

[9] Leavitt, R.P.: Empirical two-band model for quantum wells and superlattices in an electric field. Phys. Rev. B 44, 11270–11280 (1991).

[10] Sirtori, C., Capasso, F., Faist, J., Scandolo, S.: Nonparabolicity and a sum rule associated with bound-to-bound and bound-to-continuum intersubband transitions in quantum wells. Phys. Rev. B. 50, 8663–8674 (1994).

[11] Wacker, A.: Gain in quantum cascade lasers and superlattices: a quantum transport theory. Phys. Rev. B. 66, 085326 (2002).

Chapter 8

Effects of Rapid Thermal Annealing on the Optical Properties of Strain-Free Quantum Ring Solar Cells

Jiang Wu[1], Zhiming M. Wang[1,2], Vitaliy G. Dorogan[2], Shibin Li[1], Jihoon Lee[2,3], Yuriy I Mazur[2], Eun Soo Kim[3], Gregory J. Salamo[2]

[1]State Key Laboratory of Electronic Thin Films and Integrated Devices, University of Electronic Science and Technology of China, Chengdu 610054, China
[2]Institute for Nanoscience and Engineering, University of Arkansas, Fayetteville, AR 72701, USA
[3]College of Electronics and Information, Kwangwoon University, Nowon-gu, Seoul 139–701, Republic of Korea

Abstract: Strain-free GaAs/Al0.33Ga0.67As quantum rings are fabricated by droplet epitaxy. Both photoresponse and photoluminescence spectra confirm optical transitions in quantum rings, suggesting that droplet epitaxial nanomaterials are applicable to intermediate band solar cells. The effects of post-growth annealing on the quantum ring solar cells are investigated, and the optical properties of the solar cells with and without thermal treatment are characterized by photoluminescence technique. Rapid thermal annealing treatment has resulted in the significant improvement of material quality, which can be served as a standard process for quantum structure solar cells grown by droplet epitaxy.

Keywords: Photoluminescence, Photovoltaic Cells, Quantum Rings, Rapid Thermal Annealing, Droplet Epitaxy

1. Introduction

Since the proposal of intermediate band concept for high-efficiency solar cell, great efforts have been devoted to intermediate band solar cells (IBSCs). Luque and Martí have theoretically predicted that a single-junction solar cell with an intermediate band can be used to assist multiple spectral band absorption and to obtain ultrahigh efficiency up to 63%[1]. Several approaches have been taken to achieve IBSCs, such as quantum dots (QDs) and impurity bands[2]. Among these approaches, most of the current studies on IBSCs have been focused on QDs, and prototype QDIBSCs have been demonstrated[3][4]. The discrete energy levels of electrons in the QDs form energy bands which can serve as intermediate bands. However, the intermediate band impact on the cell performance is still marginal, mainly due to the high recombination rate in strongly confined QDs and low absorption volume of QDs.

Sablon *et al.* have demonstrated that QDs with built-in charge can suppress the fast recombination and thus prompt electron intersubband transitions in QDs[5]. On the other hand, several groups reported that strain-compensated QDs can be used to increase the number of QD layers and thus the overall absorption volume[6][7]. Recently, strain-free nanostructures grown by droplet epitaxy have been proposed and demonstrated for photovoltaic applications[8][9]. Moreover, strain-free nanostructures have also gained popularity in other optoelectronic devices, such as lasers and photodetectors[10][11]. In order to better understand the optical properties of these unique nanostructures and to fabricate high-perfor- mance optoelectronic devices, it is critical to gain further insight into the optical properties of droplet epitaxial strain-free nanostructures.

In this letter, strain-free quantum ring solar cells were fabricated by droplet epitaxy. Rapid thermal annealing (RTA) is used to improve the optical quality of the solar cells. The optical properties of the quantum ring solar cells before and after RTA treatment are studied. The post-growth annealing of epitaxial nanostructures is considered to be important in optoelectronic device fabrication because the size and shape of nanostructures as well as the band structures can be modified by annealing[12][13]. This letter shows that RTA plays a major role in modifying the electronic structure and in the improvement of material quality.

2. Methods

The GaAs quantum ring sample is grown on a (100) heavily doped p-type GaAs substrate by molecular beam epitaxy technique. A 0.5-μm undoped GaAs buffer layer is grown at 580°C, followed by a 30-nm $Al_{0.33}Ga_{0.67}As$ barrier layer. Subsequently, As valve is fully closed while the substrate is cooled down to 400°C, and a small amount of Ga, corresponding to coverage of 10 monolayer (ML) GaAs on (100) orientation, is supplied to form Ga droplets. During Ga deposition, Si cell is opened in order to dope the nanostructures with Si equivalent to $1 \times 10^{18}cm^{-3}$. The Ga droplets are then irradiated with As4 flux and crystallized into GaAs quantum rings at the same temperature. After quantum ring formation, a thin $Al_{0.33}Ga_{0.67}As$ cap layer (10nm) is deposited over the quantum ring at 400°C. Subsequently, the substrate temperature is raised to 600°C for the deposition of another 20 nm $Al_{0.33}Ga_{0.67}As$. The $GaAs/Al_{0.33}Ga_{0.67}As$ structure is repeated six times to form the stacked multiple quantum ring structures. After the growth of multiple quantum rings, an emitter layer of 150nm n-type GaAs with Si doped to 1×10^{18} cm^{-3} is grown. Finally, the solar cell structure is finished by a 50-nm highly Si-doped GaAs contact layer. In order to make a fair comparison in terms of effective bandgap, a quantum well solar cell used as a reference cell is fabricated with the same growth procedures, except for the quantum well region. The multiple quantum wells with GaAs coverage of 10ML are grown, instead of the fabrication of quantum rings using droplet epitaxy. An uncapped GaAs quantum ring sample is also grown using the same procedures for atomic force microscopy (AFM) measurement. The high-resolution X-ray diffraction reciprocal space mapping (RSM) of the strain-free solar cell sample was analyzed by an X-ray diffractometer (Philips X'pert, PANalytical B.V., Almelo, The Netherlands).

Rapid thermal annealing is performed on four samples in N2 ambient in the temperature range of 700°C to 850°C for 2min. The samples are sandwiched in bare GaAs wafers to prevent GaAs decomposition during high-temperature annealing. The solar cells are fabricated by standard photolithography processing. An electron beam evaporator is used to deposit $Au_{0.88}Ge_{0.12}$/Ni/Au and $A_{u0.9}Zn_{0.1}$ n-type and p-type contacts, respectively. Life-off is used to create the top grid after metal deposition. Continuous wave photoluminescence (PL) measurements are performed using the 532-nm excitation from an Nd:YAG laser with a spot diameter at the sample of 20μm at 10 K. Two excitation power intensities of the laser are used: IL = 0.3W/cm2 and IH = 3,000W/cm^2. The J-V curves of solar cells are measured under an AM 1.5G solar simulator.

3. Results and Discussion

The surface morphology of the uncapped GaAs/$Al_{0.33}Ga_{0.67}As$ quantum ring sample is imaged by an AFM, as shown in **Figure 1**. The image shows quantum ring structures with a density of approximately $2.4 \times 10^9 cm^{-2}$. The inset AFM image shows double quantum rings. **Figure 1** also shows the results obtained for 2D-RSM around the asymmetric 022 reciprocal lattice point (RSM 022 reflection). Strain-free quantum ring solar cell is evidenced by the RSM patterns.

The photoresponse spectrum of the solar cell is measured using a Fourier transform infrared spectrometer interfaced with a preamplifier at 300K without external bias voltage, as shown in **Figure 2(a)**. The spectrum shows four distinct

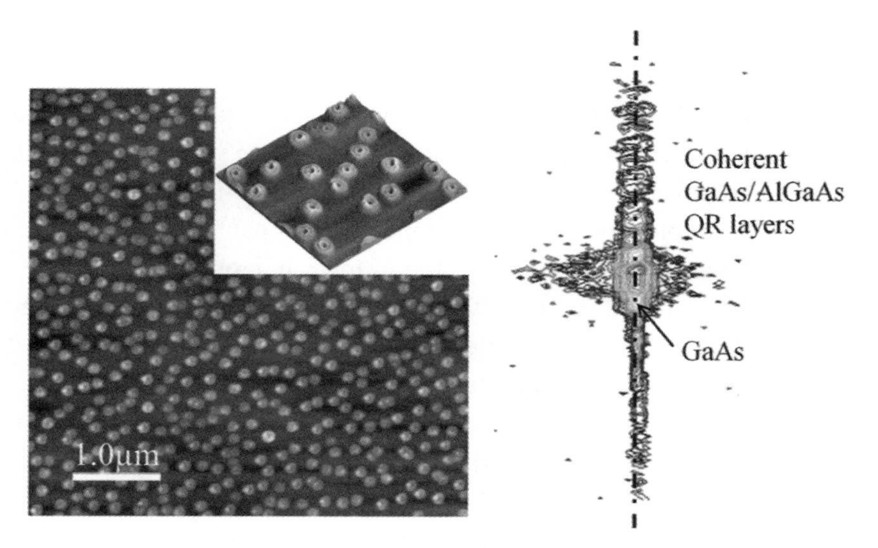

Figure 1. AFM images of surface (left) and reciprocal space map of GaAs/$Al_{0.33}Ga_{0.67}As$ quantum ring solar cell (right).

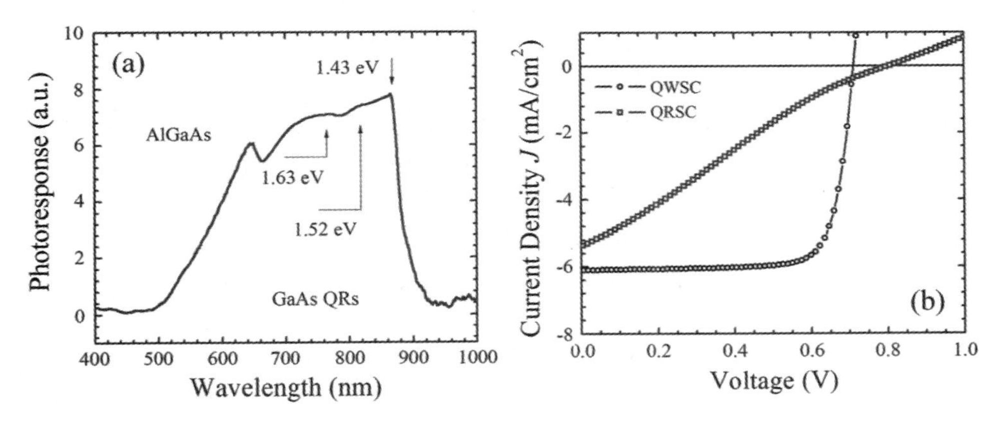

Figure 2. Photoresponse of the quantum ring solar cell and current density voltage characteristics of solar cells. (a) Photoresponse of the quantum ring solar cell at 300K. (b) Current density voltage characteristics of a quantum ring solar cell (QRSC) and a quantum well solar cell (QWSC).

peaks at 645, 760, 817, and 864nm. The photoresponse peak observed around 645nm (1.92eV) is due to interband transitions in the $Al_{0.33}Ga_{0.67}As$ barriers. The broad photoresponse band covering 760nm (1.63eV) and 817nm (1.52eV) can be assigned to the interband transitions through the energy levels in the GaAs quantum rings, while the peak around 864nm (1.43eV) is due to the bulk GaAs. **Figure 2(b)** shows the current density voltage characteristics of a quantum ring solar cell and a quantum well solar cell as reference cells. For the quantum ring solar cell, both the current density and fill factor are low. However, the quantum well solar cell with a similar device structure has a better performance in terms of current density and fill factor. A careful examination can reveal an increase of open-circuit voltage of the quantum ring solar cells. The IBSC is intended to increase the voltage at the expense of some of the sub-bandgap current because some of the intermediate band states are filled with electrons preventing transitions from the valence band to these filled intermediate band states[14]. Here, a plausible explanation is that the quantum ring solar cell, instead of the quantum well solar cell, forms an isolated intermediate band from the conduction band due to three-dimensional confinement and preserves the open-circuit voltage with reduced current. Moreover, since the open-circuit voltage is about the same for both quantum ring and quantum well solar cells, we also attributed the reduction in short-circuit current and fill fact of the quantum ring solar cell to the high series resistance and non-radiative recombination centers. Both quantum ring and quantum well solar cells are fabricated with similar processes, and the possibility for a difference in the contact resistance can be ruled out. Here in this study, the quantum rings and 10nm of AlGaAs (totally 30nm) barrier are fabricated at 400°C, which is lower than the typical growth temperature for GaAs and AlGaAs. The low-temperature growth of quantum rings and barriers is expected to generate various defects and cause degradation of material quality. These defects can act as majority carrier traps which lead to a reduction of carrier concentration and an increase in series resistance.

Post-growth thermal treatments have been used to recover the material quality of quantum structures grown at low temperature. With post-growth annealing, solar cell performance has significantly improved[9]. In order to gain further insight into the properties of the quantum ring solar cells, the PL spectra of the quantum ring solar cell sample before and after rapid thermal annealing are measured and shown in **Figure 3**. At a laser excitation power IL = $0.3W/cm^2$, the PL peak at 1.64eV appears only after post-thermal annealing and the PL spectrum intensity increases distinctly as a function of annealing temperature. This peak can be attributed to the ground energy level transition in the quantum ring, which corresponds to the photoresponse peak at 1.52eV measured at 300K. The PL spectra have shown a blueshift and significant broadening after thermal annealing. The integrated intensity, peak energy, and full width at half maximum of the PL spectra measured to laser excitation IL as a function of the annealing temperature are plotted in **Figure 3(c)**. At high laser excitation IH = $3{,}000W/cm^2$, a second PL peak appears at approximately 1.7eV after annealing, as shown in **Figure 3(b)**. The second peak is assigned to the excited state transitions in the GaAs quantum ring structures which correspond to the photo- response peak at 1.63eV. Similar to the quantum ring ground state transition, the PL spectra experience an emission enhancement as well as a blue shift with increasing annealing temperature [**Figure 3(d)**].

The increase in the PL yield after thermal annealing is due to the considerable improvement of material quality. Post-thermal annealing promotes the depletion of defects generated in GaAs nanostructures as well as the AlGaAs barriers processed at low temperatures. The blueshift and the broadening of the PL spectra after annealing is due to the inter-diffusion of Al and Ga at the GaAs quantum ring and $Al_{0.33}Ga_{0.67}As$ barrier interface. With increasing annealing temperature, the Al and Ga elements become mobilized with diffusion length as a function of annealing

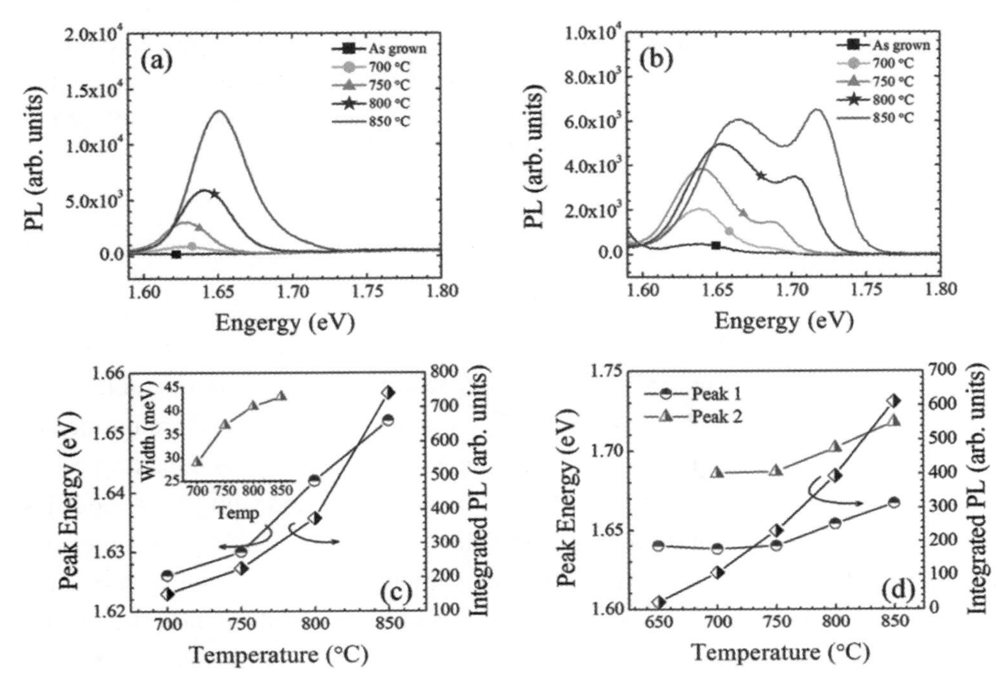

Figure 3. PL spectra of solar cells and PL peak energy and integrated PL intensity. (a) PL spectra of the solar cell samples annealed with different temperatures. The laser excitation power is IL = 0.3W/cm^2. (b) PL spectra of the solar cells annealed with different temperatures. The laser excitation power is IH = 3,000W/cm^2. (c) PL peak energy and integrated PL intensity as a function of annealing temperatures under low excitation power IL. The inset is the PL line width as a function of annealing temperatures. (d) PL peak energy and integrated PL intensity as a function of annealing temperatures under high excitation power IH. The data obtained from the as-grown material is plotted at 650°C.

temperature. As a result, the concentration of Al element is increased in the GaAs quantum ring. The PL line width (PL peak 1.64eV) changes from 29 to 43eV as the annealing temperature increases from 700°C to 850°C [the inset in **Figure 3(c)**]. The PL spectrum broadening is somehow different from the observation for InAs quantum dots. For high-temperature annealing, the size distribution and composition fluctuation of the InAs quantum dots can be improved due to the In-Ga interdiffusion[15][16]. In the case of GaAs quantum ring, the broadening of PL spectra may be explained by the gradient of Al distribution in GaAs quantum ring and barriers introduced by thermal annealing, which may be beneficial for photovoltaic applications. Compared with the In and Ga elements, the diffusion length of Al elements is short and in the range of a few nanometers due to a large Al-As bonding energy[17][18]. Therefore, a gradient of Al distribution results in the GaAs/ AlGaAs interface, instead of the improvement of composition fluctuation. Additionally, the interdiffusion smooths the quantum ring and barrier interface and modifies the quantum ring geometrical shape and further electronic structures.

4. Conclusions

GaAs quantum rings are fabricated by droplet epitaxy growth method. The effects of rapid thermal annealing on optical properties of quantum ring solar cells have been investigated. Thermal annealing promotes interdiffusion through depletion of vacancies and greatly enhances the material quality of quantum rings grown by low-temperature droplet epitaxy. Post-growth annealing also modifies the sharp GaAs/AlGaAs interface, and a gradient interface caused by the annealing leads to broadband optical transitions and thus improves the solar cell performance. These strain-free quantum structures with improved material quality after being treated by rapid thermal annealing may provide an alternative way to fabricate high-efficiency intermediate band solar cells. Further studies on the thermal annealing process are required to optimize quantum structures for intermediate band solar cell applications. A better correlation between morphological change and optical property enhancement during thermal annealing needs to be identified. For example, the three-dimensional quantum confinement has to be preserved while improving the optical properties after annealing.

Abbreviations

AFM: Atomic force microscopy;

IBSC: Intermediate band solar cell;

PL: Photoluminescence;

QD: Quantum dot;

RTA: Rapid thermal annealing.

Competing Interests

The authors declare that they have no competing interests.

Authors' Contributions

JW carried out the sample growth and device fabrication. VGD participated in the PL measurements. JW and SL carried out the XRD, AFM, J-V, and photoresponse measurements. JW, ZMW, SL, JL, and YIM participated in the statistical analysis and drafted the manuscript. JW, ZMW, ESK, and GJS conceived the study and participated in its design and coordination. All authors read and approved the final manuscript.

Acknowledgements

This work was supported in part by the National Science Foundation through EPSCoR grant number EPS1003970, the NRF through grant numbers 2010-0008394 and 2011-0030821, and the National Natural Science Foundation of China through grant numbers NSFC-51272038 and NSFC-61204060.

Source: Wu J, Wang Z M, Dorogan V G, *et al*. Effects of rapid thermal annealing on the optical properties of strain-free quantum ring solar cells [J]. Nanoscale Research Letters, 2013, 8(1):1–5.

References

[1] Luque A, Martí A: Increasing the efficiency of ideal solar cells by photon induced transitions at intermediate levels. Phys Rev Lett 1997, 78(26):5014.

[2] Luque A, Marti A: The intermediate band solar cell: progress toward the realization of an attractive concept. Adv Mater 2010, 22(2):160–174.

[3] López N, Martí A, Luque A, Stanley C, Farmer C, Díaz P: Experimental analysis of the operation of quantum dot intermediate band solar cells. J Solar Energy Eng 2007, 129(3):319.

[4] Lu HF, Mokkapati S, Fu L, Jolley G, Tan HH, Jagadish C: Plasmonic quantum dot solar cells for enhanced infrared response. Appl Phys Lett 2012, 100(10):103505.

[5] Sablon KA, Little JW, Mitin V, Sergeev A, Vagidov N, Reinhardt K: Strong enhancement of solar cell efficiency due to quantum dots with built-in charge. Nano Lett 2011, 11(6):2311–2317.

[6] Alonso-Álvarez D, Taboada AG, Ripalda JM, Alén B, González Y, González L, García JM, Briones F, Martí A, Luque A, Sánchez AM, Molina SI: Carrier recombination effects in strain compensated quantum dot stacks embedded in solar cells. Appl Phys Lett 2008, 93(12):123114.

[7] Zhou D, Sharma G, Thomassen SF, Reenaas TW, Fimland BO: Optimization towards high density quantum dots for intermediate band solar cells grown by molecular beam epitaxy. Appl Phys Lett 2010, 96(6):061913.

[8] Wu J, Shao D, Li Z, Manasreh MO, Kunets VP, Wang ZM, Salamo GJ: Intermediate-band material based on GaAs quantum rings for solar cells. Appl Phys Lett 2009, 95(7):071908.

[9] Wu J, Wang ZM, Dorogan VG, Li S, Zhou Z, Li H, Lee J, Kim ES, Mazur YI, Salamo GJ: Strain-free ring-shaped nanostructures by droplet epitaxy for photovoltaic application. Appl Phys Lett 2012, 101:043904.

[10] Jo M, Mano T, Sakoda K: Lasing in ultra-narrow emission from GaAs quantum dots coupled with a two-dimensional layer. Nanotechnology 2011, 22(33):335201.

[11] Wu J, Shao D, Dorogan VG, Li AZ, Li S, DeCuir EA, Manasreh MO, Wang ZM, Mazur YI, Salamo GJ: Intersublevel infrared photodetector with strain-free GaAs quantum dot pairs grown by high-temperature droplet epitaxy. Nano Lett 2010, 10(4):1512–1516.

[12] Jolley G, McKerracher I, Fu L, Tan HH, Jagadish C: The conduction band absorption spectrum of interdiffused InGaAs/GaAs quantum dot infrared photodetectors. J Appl Phys 2012, 111(12):123719.

[13] Pankratov EL: Optimization of pulse laser annealing to increase sharpness of implanted-junction rectifier in semiconductor heterostructure. Nano-Micro Lett 2010, 2:256–267.

[14] Martí A, Antolín E, Linares PG, Luque A: Understanding experimental characterization of intermediate band solar cells. J Mater Chem 2012, 22:22832–22839.

[15] Hsu TM, Lan YS, Chang W, Yeh NT, Chyi J: Tuning the energy levels of self- assembled InAs quantum dots by rapid thermal annealing. Appl Phys Lett 2000, 76(6):691.

[16] Fu L, McKerracher I, Tan HH, Jagadish C, Vukmirovic N, Harrison P: Effect of GaP strain compensation layers on rapid thermally annealed InGaAs/ GaAs quantum dot infrared photodetectors grown by metal-organic chemical-vapor deposition. Appl Phys Lett 2007, 91(7):073515.

[17] Pierz K, Ma Z, Keyser UF, Haug RJ: Kinetically limited quantum dot formation on AlAs(100) surfaces. J Cryst Growth 2003, 249(3–4):477–482.

[18] Sanguinetti S, Watanabe K, Kuroda T, Minami F, Gotoh Y, Koguchi N: Effects of post-growth annealing on the optical properties of self-assembled GaAs/AlGaAs quantum dots. J Cryst Growth 2002, 242(3–4):321–331.

Chapter 9

Electrical and Optical Properties of Binary CN_x Nanocone Arrays Synthesized by Plasma-Assisted Reaction Deposition

Xujun Liu, Leilei Guan, Xiaoniu Fu, Yu Zhao, Jiada Wu, Ning Xu

Shanghai Engineering Research Center of Ultra-Precision Optical Manufacturing, Department of Optical Science and Engineering, Fudan University, Shanghai, China

Abstract: Light-absorbing and electrically conductive binary CN_x nanocone (CNNC) arrays have been fabricated using a glow discharge plasma-assisted reaction deposition method. The intact CNNCs with amorphous structure and central nickel-filled pipelines could be vertically and neatly grown on nickel-covered substrates according to the catalyst-leading mode. The morphologies and composition of the as-grown CNNC arrays can be well controlled by regulating the methane/nitrogen mixture inlet ratio, and their optical absorption and resistivity strongly depend on their morphologies and composition. Beside large specific surface area, the as-grown CNNC arrays demonstrate high wideband absorption, good conduction, and nice wettability to polymer absorbers.

Keywords: CN_x Nanocone Arrays, Wideband Absorption, Electrical Conduction, Wettability to Polymer Absorbers

1. Introduction

Since the 1990s, there has been an upsurge in interest in the properties and potential uses of carbon-related nanostructures[1]-[3]. These unique nanostructures are attractive for nanotechnology applications in photovoltaic devices and photo- detectors[4]-[8]. Many novel thin film solar cells rely on highly light-absorbing and well electrically conductive electrodes for their successful operation and good capability. For example, dye-sensitized solar cells and polymer organic hybrid solar cells exploit titanium oxide as electrodes[7][8]. But, this material is far from ideal because of poor electrical conduction and limited optical absorption[9][10]. Carbon-related nanostructures, such as carbon nanotubes and graphene, are attractive electrodes and even absorbers for photovoltaic devices and photodetectors owing to strong optical absorptivity and ultrafast charge transport mobility[6][11]. Besides, their large specific surface area could greatly increase the donor/ acceptor interface, which will effectively increase the separation probability of electrons and holes. Compared with carbon nanotubes and graphene, the binary CN_x nano- cones (CNNCs) will have good mechanical stability and better electrical and chemical stabilities due to the incorporation of nitrogen. So far, the experimentally synthesized carbon nitride, except our previous reports of the growth of the CNNC arrays[12], is mainly limited to amorphous or nanosphere CN_x thin films and nanobells with low nitrogen content (about 2%)[13]-[15].

Here, vertically aligned CNNC arrays with high wideband absorption and good electrical conduction were fabricated by an abnormal glow discharge plasma-assisted reaction deposition (GPRD) method which combines highly dense plasma with proper bias enhancement[12]. The methane/nitrogen (CH_4/N_2) mixture feeding gas ratio, which directly affected the contents and activities of the nitrogen-related and carbon-related precursors in the plasmas, was regulated to control the morphologies and composition of the CNNC arrays. The effects of the morphology, composition, and structure of the CNNC arrays on their optical absorption and electrical conduction were studied. The CNNC arrays with intact shape, high optical absorption, high electrical conduction, and nice wettability to polymer are pursued for potential uses as electrodes or even absorbers in photovoltaic devices and photodetectors.

2. Methods

Optically absorptive and electrically conductive CNNC arrays were grown on nickel-covered silicon (100) substrates by means of the GPRD method, as described previously[12][16]. The sample preparation involves two steps. In the first step, nickel catalyst layers were deposited on silicon (100) wafers by a pulsed laser deposition method. About 100-nm thick nickel catalyst layers were deposited on the prepared substrates under a base pressure of 1×10^{-3}Pa for 8min using a Nd: YAG laser to ablate a pure nickel target. The wavelength, pulse energy, and repetition of the Nd: YAG laser were 532nm, 50mJ, and 10Hz, respectively. The distance between the target and substrate was about 4cm. In the second step, the CNNC arrays were grown by the GPRD method. The plasma source generated reactive plasma just above the substrates through the abnormal glow discharge with a CH_4/N_2 mixture inlet under a total pressure of 750Pa. The discharge current, voltage, and time were set to 180mA, 350V, and 40min, respectively. In the CNNC growth, the CH_4/N_2 inlet ratios were varied from 1/80 to 1/5 in order to obtain the CNNC arrays with different morphologies and compositions. The wettability of the CNNC arrays to poly-3-hexylthiophene mixed with phenyl-C61-butyric acid methyl ester (P3HT: PCBM) layer, which is a commonly used polymer absorber in polymer organic hybrid solar cells, has also been examined by spin coating method using different rotational speeds for different polymer thicknesses.

The morphologies of the samples were characterized by field emission scanning electron microscopy (FESEM) and transmission electron microscopy (TEM). The crystallinity and composition of the individual CNNCs were characterized by selected-area electron diffraction (SAED) and energy-dispersive X-ray spectroscopy (EDXS). The optical absorption spectra were measured by an ultraviolet spectrophotometer. Longitudinal resistance of the as-grown CNNC arrays was measured by a platinum-cylindrical-tip contacting method using a Power SourceMeter (Keithley Instruments Inc., Beijing, China), and the resistivity of the as-grown CNNCs was obtained by calculating the measured resistance.

3. Results and Discussion

The FESEM images of the CNNC arrays grown at different CH_4/N_2 feeding gas ratios of 1/80 to 1/10 are shown in **Figure 1**. It is apparent in **Figure 1** that the morphologies and sizes of the as-grown CNNCs are strongly dependent on the CH_4/N_2 ratios. **Figure 1(a)** shows that there are almost no intact CNNCs, but many dispersive hemispherical clusters were clearly discerned when the CH_4/N_2 ratio is 1/80. These CNNCs are in the incomplete-growth stage. As the CH_4/N_2 ratio was increased, the sizes of the as-grown CNNCs were increased and their

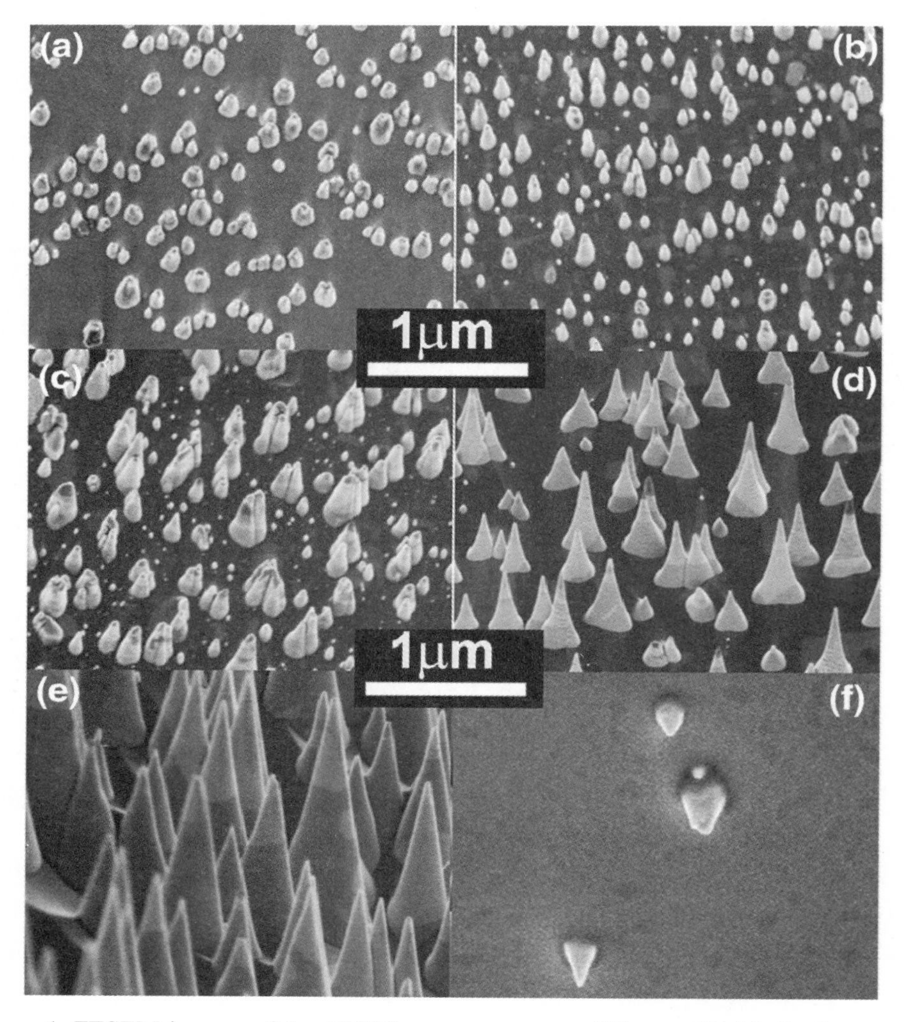

Figure 1. FESEM images of the CNNC arrays grown at different CH_4/N_2 feeding gas ratios. (a) 1/80, (b) 1/40, (c) 1/20, (d) 1/10, (e and f) 1/5. (f) The surface morphologies of the P3HT: PCBM-covered CNNC arrays grown at a CH_4/N_2 feeding gas ratio of 1/5. The samples were prepared on the nickel-covered silicon (100) wafers for 40min, with a discharge current of 180mA and a discharge voltage of 350V.

morphologies were improved [**Figures 1(c)–1(e)**]. It can be seen in the **Figure 1(e)** that the CNNCs grown at the CH_4/N_2 ratio of 1/5 have rather perfect shape, and their average bottom diameter, average height, and identical apex angle are about 400nm, 1,000nm, and 25°, respectively. By comparing the five images [**Figures 1(a)–1(e)**], it could be found that the average height and bottom diameter of the as-grown CNNCs increase quickly, but their distribution density changes in apparently as the CH_4 feeding gas increases. The above phenomena could be explained by that the supersaturation conditions necessary for the nucleation of the CNNCs could be more easily satisfied for a very little CH_4 supply[17]. When the CH_4 supply increases, the CN radicals in the plasma also increase and the N_2^+ or N^+ etching effects become weaker relatively, which will lead to the increment of the growth rate of the CNNCs and their more intact conical shape [**Figure 1(d), Figure 1(e)**].

For novel thin film solar cells, such as polymer inorganic hybrid solar cells, the electrodes made from inorganic nanostructures not only require high optical absorption and good electrical conduction but also nice wettability to absorbers, which is almost the main bottleneck of the development of this kind of solar cells. The wettability of the CNNC arrays to P3HT: PCBM (weight ratio of 1:0.8), which is a commonly used polymer absorber in polymer organic hybrid solar cells, was examined by the spin coating method. **Figure 1(f)** gives the FESEM image of the surface morphology of the P3HT: PCBM-covered CNNC array. It could be seen in **Figure 1(f)** that the P3HT: PCBM layer have fully infiltrated the CNNC arrays, and the several higher CNNC tips protrude from the P3HT: PCBM layer, which indicates that the CNNC arrays have very nice wettability to the P3HT: PCBM absorber layers.

In order to understand the detailed structures and composition of the CNNCs, the TEM, SAED, and EDXS fitted within the TEM were carried out. The TEM images of the two CNNCs grown at the CH_4/N_2 ratios of 1/20 and 1/5 are presented in **Figure 2(a)**, **Figure 2(f)**. The individual CNNCs were directly scraped off from the sample surfaces and transferred onto the copper grids covered by about 10-nm carbon thin films for TEM observation. The grown CNNCs displayed good mechanical stability and strong adhesion to the substrates for the samples need to be forcibly scratched with a steel knife to obtain very few scraped-off CNNCs. **Figure 2(a)**, **Figure 2(f)** shows that there are hollow pipes along the centric axes in the broken CNNCs, and they are completely filled with a kind of black

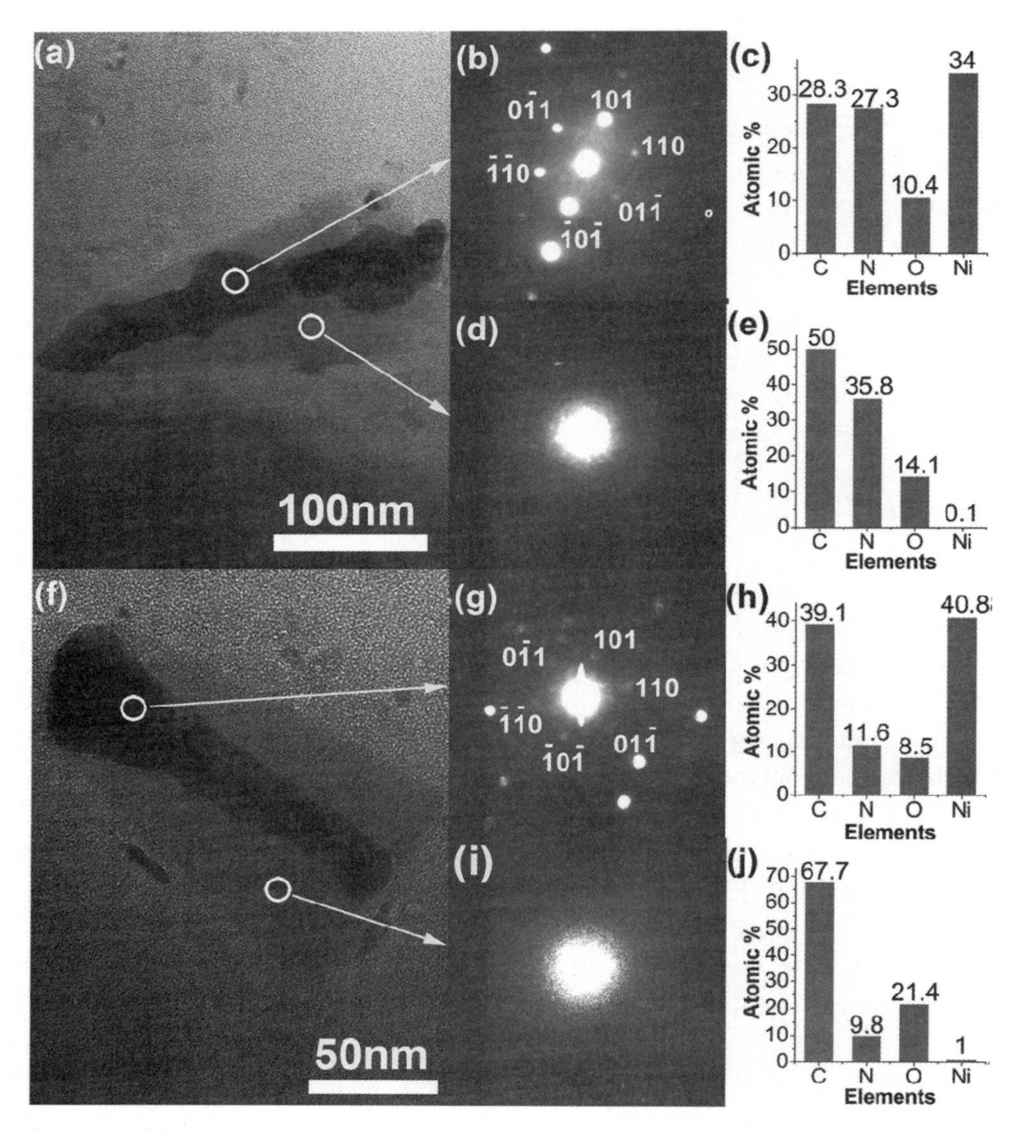

Figure 2. TEM images, SAED patterns, and EDXS analytical histograms. TEM images of the individual CNNCs prepared at the CH_4/N_2 feeding gas ratios of (a) 1/20 and (f) 1/5, SAED patterns taken from the black areas of the central pipes (b and g) and the lateral gray areas (d and i), and EDXS analytical histograms (c, e, h, and j) at the locations corresponding to subgraphs (b, d, g, and i), respectively.

substance, which have obvious contrast with the lateral areas. The SAED patterns demonstrate that the black substance in the central pipes contains crystalline nickel with a face-centered cubic structure [as shown in **Figure 2(b)**, **Figure 2(g)**], and the gray substance in the lateral areas is mainly amorphous [as shown in **Figure 2(d)**, **Figure 2(i)]**. Some diffraction spots can be perceived in **Figure 2(d)**, but it is difficult to distinguish their crystal lattice. The analytical results of the EDXS spectra taken from the locations corresponding to **Figure 2(b)**, **Figure 2(g)** also show that the atomic percentages of nickel at the central black pipes are highest in all ingredients [**Figure 2(c)**, **Figure 2(h)**]. Because the electron beam for X-ray analysis can easily penetrate the CNNC bodies, the partial carbon and nitrogen shown in **Figure 2(c)**, **Figure 2(h)** should come from the CNNC bodies in the front and rear of the central pipes, and the nickel content in the central pipes should be more. In **Figure 2(e)**, **Figure 2(j)**, it could be found that the CNNC bodies at the gray areas are mainly composed of [C] and [N], and the atomic percentages of nickel are below 0.1%. Here, the oxygen is inevitably and should mainly come from the exposure to air for days. After deducting the contribution of the 10-nm carbon thin films on the copper grids (compared with the 50-nm CNNC thickness that the X-ray pass through), the actual atomic ratios of [N]/[C] in the CNNC bodies [given in **Figure 2(e)**, **Figure 2(j)**] can reach about 0.89:1 and 0.18:1, respectively. There may be crystalline C_3N_4 structures at the places adjacent to the central nickel-filled pipes for the actual [N]/[C] which can reach 1.2:1 and 0.4:1 at the CH_4/N_2 ratios of 1/20 and 1/5 (not show here), respectively, significantly higher than elsewhere. But, because the contents of the crystalline C_3N_4 structures near the central pipes are not enough, it is still difficult to distinguish their crystal lattice in the SAED patterns. Because the EDXS is only a semi-quantitative analysis tool, its analysis results usually have some deviation from the actual situation. From the above SAED and EDXS results, it could be certain that the main CNNC bodies are amorphous CN_x, and the [N] content in them synchronously decreases as the CH_4/N_2 ratio increases.

Based on the characterization of morphologies, structures, and composition, the CNNC growth can be outlined as the catalyst-leading growth mode. In this mode, the nickel catalyst layer first melts and fragments into separated hemisphere-like islands under heating of the abnormal glow discharge plasma over the substrate. Then, the incipient CNNCs are formed on the nickel islands due to the deposition of precursors such as CN species, nitrogen atoms, and C_2 species from the discharge plasma[17]. As the CN radicals and other reactive species continue to attach, the heights and lateral diameters of the CNNCs increase simultaneously. Meanwhile, the enclosed molten nickel will be sucked to the top and leave the narrow pipelines in the center of the cone bodies by the capillary effect. The catalyst nickel on the tops will lead to the growth of the CNNCs. As the CNNCs increase in height, the ion streams accelerated by a voltage of 350eV will be focused on the tops by a locally enhanced electric field. The intense ion streams will sputter off the attached species and cut down the diameters of the tops[18]. In this way, the intact CNNC arrays with central pipelines and sharp tips eventually finish the growth. Because the precursors are mainly composed of CN species, nitrogen atoms, and C_2 species[17], the bodies of the as-grown CNNCs are mainly amorphous CN_x other than crystalline C_3N_4 which needs the reaction between atomic C and N without other species involved.

The optical absorption properties of the CNNC arrays are important for their application in optoelectronic devices. The optical absorption spectroscopy results of the CNNC arrays grown at CH_4/N_2 ratios of 1/80 to 1/5 were examined using a UV spectrophotometer in the wavelength range from 200 to 900nm (as shown in **Figure 3**). It could be seen in **Figure 3** that the optical absorption in the wideband of 200 to 900nm increases as the CH_4/N_2 ratio increases. As the CH_4/N_2 ratio increased to 1/5, the absorption of the as-grown CNNC array increased to 78% to 86% in a wideband of 200 to 900nm. By comparing the five absorption spectra, it

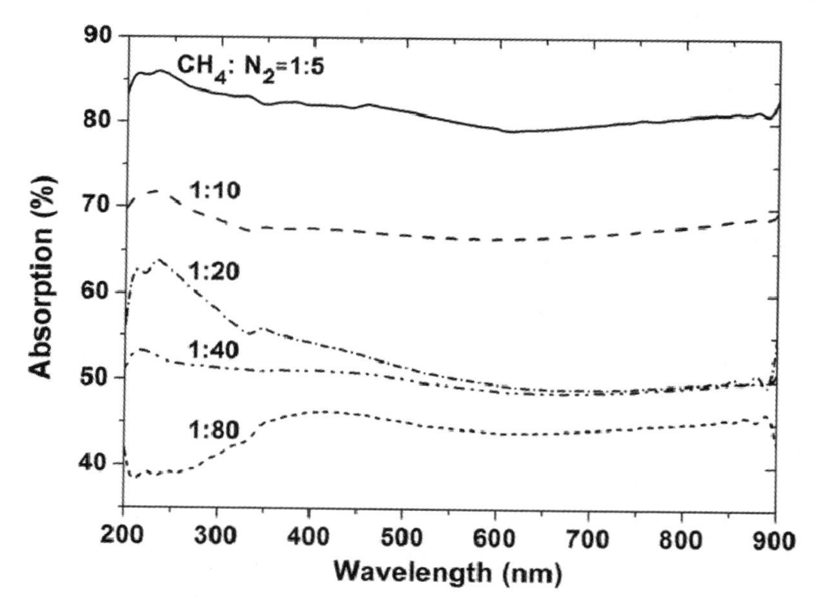

Figure 3. Absorption spectra of the CNNC arrays grown at different CH_4/N_2 feeding gas ratios. The CH_4/N_2 feeding gas ratios were 1/80, 1/40, 1/20, 1/10, and 1/5, respectively.

could be found that the absorption has a larger increment rate when the CH_4/N_2 ratio increases from 1/20 to 1/5. This phenomenon should be mainly caused by the increase of the light refraction and repeated absorption between the CNNCs. At the CH_4/N_2 ratio below 1/20, the light refraction between the small and sparse CNNCs has no apparent effect on the absorption, and the absorption is mainly by base layers. Besides, there is a stronger absorption band between 200 and 400nm for the sample prepared at the CH_4/N_2 ratio of 1/20, but it becomes weak when the CH_4/N_2 ratios are higher or lower. This absorption band may be caused by C_3N_4 phases (the band gaps of the α- and β-C_3N_4 are 3.85 and 3.25eV, respectively) in the as-grown CNNCs[19]. The less C_3N_4 phases in the CNNC arrays grown at the higher CH_4/N_2 ratios make it weaker, while the small and sparse CNNC arrays grown at the lower CH_4/N_2 ratios have no significant absorption (the absorption is mainly by base layers).

For the CNNC arrays used as the electrodes of photovoltaic devices and photodetectors, their electrical properties become very important. Longitudinal resistances of the prepared CNNC arrays were measured by a platinum-cylindrical-tip contacting method. In the method, the top surface of the platinum cylindrical tip with a diameter of 1 mm directly contacted the CNNC arrays. The electrical testing diagram of the CNNC arrays is shown in **Figure 4(a)**, and the TEM micrograph of a CNNC pressed by the platinum cylindrical tip is shown in **Figure 4(b)**. The current-voltage (I-V) curves for the samples prepared at different CH4/N2 ratios of 1/80 to 1/5 are shown in **Figure 4(c)**. All I-V curves are nearly consistent with linear characteristics, and the resistance values in a circular area with a diameter of 1 mm can be obtained by fitting the corresponding slanted lines. According to the distribution density and average size of the CNNCs (estimated through the FESEM and TEM images of the as-prepared samples), the resistivities ρ of the as-grown CNNCs at different CH_4/N_2 ratios can be calculated by the following equation:

$$\rho = \frac{nR}{\int_{h1}^{h2} \frac{dh}{\pi\left(h\tan\frac{\theta}{2}\right)^2}},$$

where R is the resistance value in a circular area with a diameter of 1 mm, n is the

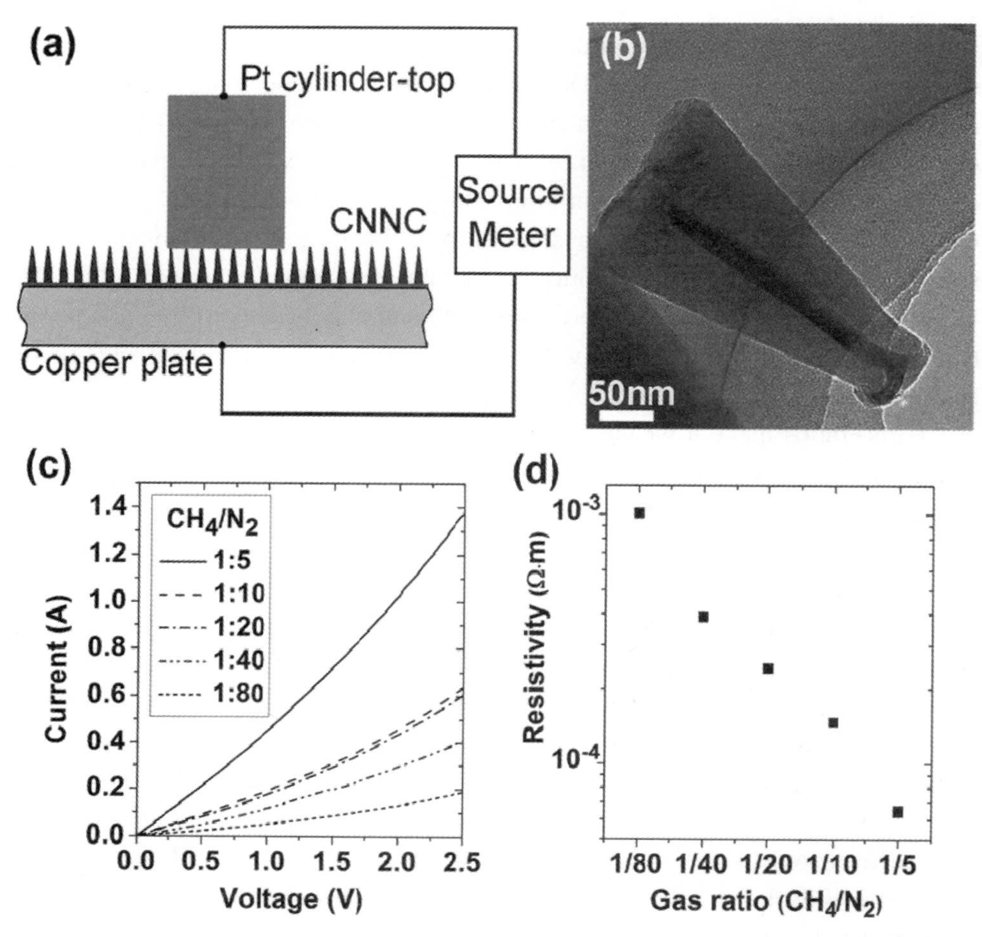

Figure 4. Electrical testing diagram, TEM micrograph, I-V curves, and the corresponding resistivities. (a) Electrical testing diagram of the CNNC arrays; (b) TEM micrograph of a CNNC pressed by the platinum cylindrical tip; (c) and (d) I-V curves and the corresponding resistivities of the samples prepared at CH_4/N_2 feeding gas ratios of 1/80, 1/40, 1/20, 1/10, and 1/5.

number of CNNCs in the area contacted by the platinum cylindrical tip, h_2 is the average height of the nanocones, h_1 is the average loss height caused by the contact with the platinum cylindrical tip, and θ is the cone angle. According to the measured resistance [**Figure 4(c)**], the resistivity of the as-grown CNNCs can be calculated, and the results are shown in **Figure 4(d)**. In the above calculations, the impacts of the Ni-containing substances in the central pipes on the resistance are not considered. Actually, the middle sections of most central pipes (if not all) are empty due to thermal expansion and contraction, and sometimes the central pipes at the tips are also empty by TEM observations (we have not observed the whole central pipes filled by the black substances), *i.e.*, the Ni-containing substances in the central pipes are disconnected. Besides, the resistivity of the Ni-containing substances in the central pipes is uncertain for the atomic percentages of Ni in them are only 30% to 40% or more, and a large part of the ingredients of the Ni-containing substances are CN_x. If there exist central pipes filled with continuous Ni-containing substances and the resistivity of the Ni-containing substances is less than the CN_x bodies, the resistance of the CNNCs may be reduced; if not, the influence of the central pipes on the resistance of the CNNCs will be little. But, it is difficult to establish a model to estimate the extent of the influence due to their discontinuity and uncertain resistivity. Therefore, the resistivity of the CNNCs as a whole is calculated. As shown in **Figure 4(c)**, **Figure 4(d)**, both the resistance and resistivity of the as-grown CNNCs are obviously affected by the CH_4/N_2 ratios. It could be found in **Figure 4(d)** that the resulted resistivity ρ decreases from 1.01×10^{-3} to $6.45 \times 10^{-5} \Omega\cdot m$ as the CH_4/N_2 ratio increases from 1/80 to 1/5, which could be due to the increase of the carbon content in the CNNCs.

4. Conclusions

In summary, the vertically aligned CNNC arrays were synthesized on nickel-covered silicon (100) substrates by the GPRD method. The morphologies and composition of the as-grown CNNC arrays are strongly affected by the CH_4/N_2 feeding gas ratios. The as-grown CNNCs are mainly amorphous CN_x, and the atomic content of nitrogen decreases synchronously as the CH_4/N_2 ratio increases. The CNNC arrays grown at the CH_4/N_2 ratio of 1/5 have rather perfect cone shapes and good wettability to the polymer P3HT: PCBM. The absorption spectra reveal that the optical absorption of the as-grown CNNC arrays increases with increasing CH_4/N_2 ratio and show a very good absorption in a wideband of 200 to 900nm at the CH_4/N_2 ratio of 1/5. The resistivities of the as-prepared samples decrease as the CH_4/N_2 ratios increase and reach about $6.45 \times 10^{-5} \Omega\cdot m$ at the CH_4/N_2 ratio of 1/5, indicating that the as-grown CNNC arrays can have very good conductivity. Due to the large specific surface area, high and wide optical absorption, excellent electrical conduction, and nice wettability (to polymer absorbers) of the as-grown CNNC arrays, such nanocone arrays are supposed to be potential electrodes or even absorbers in the thin film solar cells and photodetectors.

Abbreviations

CH_4/N_2: methane/nitrogen;

CNNC: binary CN_x nanocone;

EDXS: energy-dispersive spectroscopy;

FESEM: field emission scanning electron microscopy;

GPRD: abnormal glow discharge plasma-assisted reaction deposition;

HRTEM: high-resolution transmission electron microscopy;

PLD: pulsed laser deposition;

P3HT: PCBM: poly-3-hexylthiophene mixed with phenyl-C61-butyric acid methyl ester;

SAED: selected-area electron diffraction;

TEM: transmission electron microscopy.

Competing Interests

The authors declare that they have no competing interests.

Authors' Contributions

XL designed and carried out the experiments and wrote the paper. LG, XF, and YZ participated in the experiments. JW participated in the design and the discussion of this study. NX conceived and designed the experiments and revised the paper. All authors read and approved the final manuscript.

Authors' Information

XL, LG, and XF are graduate students major in fabrication of nanometer materials. YZ is an associate professor and MS degree holder specializing in optical devices. JW is a professor and PhD degree holder specializing in optics and nanometer materials. NX is a professor and a PhD degree holder specializing in nanometer materials and devices, especially in nanoscaled super-hard and optoelectronic devices.

Acknowledgements

This work is financially supported by the National Basic Research Program of China (973 Program, Grant No. 2012CB934303) and National Natural Science Foundation of China.

Source: Liu X, Guan L, Fu X, *et al*. Electrical and optical properties of binary CN, x, nanocone arrays synthesized by plasma-assisted reaction deposition [J]. Nanoscale Research Letters, 2014, 9(1):1–7.

References

[1] Iijima S: Helical microtubules of graphitic carbon. Nature 1991, 354:56–58.

[2] Ruoff RS, Lorents DC: Mechanical and thermal properties of carbon nanotubes. Carbon 1995, 33:925–930.

[3] Chen Y, Guo LP, Chen F, Wang EG: Synthesis and characterization of C3N4 crystalline films on silicon. J Phys Condens Matter 1996, 8:L685–L690.

[4] Zhang GY, Jiang X, Wang EG: Tubular graphite cones. Science 2003, 300:472–474.

[5] Wei JQ, Jia Y, Shu QK, Gu ZY, Wang KL, Zhuang DM: Double-walled carbon nanotube solar cells. Nano Lett 2007, 7:2317–2321.

[6] Li XM, Zhu HW, Wang KL, Cao AY, Wei JQ, Li CY: Graphene-on-silicon Schottky junction solar cells. Adv Mater 2010, 22:2743–2748.

[7] Mor GK, Shankar K, Paulose M, Varghese OK, Grimes CA: Use of highly-ordered TiO2 nanotube arrays in dye-sensitized solar cells. Nano Lett 2006, 6:215–218.

[8] Kuwabara T, Nakayama T, Uozumi K, Yamaguchi T, Takahashi K: Highly durable inverted-type organic solar cell using amorphous titanium oxide as electron collection electrode inserted between ITO and organic layer. Sol Energ Mat Sol C 2008, 92:1476–1482.

[9] Tang H, Prasad K, Sanjinès R, Schmid PE, Lévy F: Electrical and optical properties of TiO2 anatase thin films. J Appl Phys 1994, 75:2042–2047.

[10] Hanini F, Bouabellou A, Bouachiba Y, Kermiche F, Taabouche A, Hemissi M, Lakhdari D: Structural, optical and electrical properties of TiO2 thin films synthesized by sol–gel technique. IOSR Journal of Engineering 2013, 3:21–28.

[11] Geim AK: Graphene: status and prospects. Science 2009, 324:1530–1534.

[12] Hu W, Xu XF, Shen YQ, Lai JS, Fu XN, Wu JD, Ying ZF, Xu N: Self-assembled fabrication and characterization of vertically aligned binary CN nanocone arrays. J Electron Mater 2010, 39:381–390.

[13] Zhang GY, Ma XC, Zhong DY, Wang EG: Polymerized carbon nitride nanobells. J Appl Phys 2002, 91:9324–9332.

[14] Yen TY, Chou CP: Growth and characterization of carbon nitride thin films prepared by arc-plasma jet chemical vapor deposition. Appl Phys Lett 1995, 67:2801–2803.

[15] Xu N, Lin H, Pan WJ, Sun J, Wu JD, Ying ZF, Wang PN, Du YC, Li FM: Synthesis of carbon nitride nanocrystals on Co/Ni-covered substrate by nitrogen-atom-beam-assisted pulsed laser ablation. J Mater Res 2003, 18:2552–2555.

[16] Xu N, Du YC, Ying ZF, Ren ZM, Li FM: An arc discharge nitrogen atom source. Rev Sci Instrum 1997, 68:2994–3000.

[17] Hu W, Tang J, Wu JD, Sun J, Shen YQ, Xu N: Characterization of carbon nitride deposition from CH4/N2 glow discharge plasma beams using optical emission spectroscopy. Phys Plasmas 2008, 15:073502–073508.

[18] Levchenko I, Ostrikov K, Long JD, Xu S: Plasma-assisted self-sharpening of platelet-structured single-crystalline carbon nanocones. Appl Phys Lett 2007, 90:113115.

[19] Teter DM, Hemley RJ: Low-compressibility carbon nitrides. Science 1996, 271: 53–55.

Chapter 10

Synthesis, Characterization and Interpretation of Screen-Printed Nanocrystalline CdO Thick Film for Optoelectronic Applications

Rayees Ahmad Zargar[1], Santosh Chackrabarti[1], Manju Arora[2], Aurangzeb Khurram Hafiz[1]

[1]Department of Physics, Jamia Millia Islamia, New Delhi 110025, India
[2]CSIR-National Physical Laboratory, Dr. K.S. Krishnan Marg, New Delhi 110-012, India

Abstract: The transparent conductive oxide CdO thick films are prepared by screen printing method followed by sintering route. The structural, optical and electrical properties of as-grown films are characterized by powder X-ray diffraction (XRD), scanning electron microscopy (SEM), atomic force microscopy (AFM), Fourier transform infrared (FTIR), ultraviolet-visible (UV-VIS) and DC conductivity measurement techniques. XRD, SEM and AFM studies reveal that the film deposited is polycrystalline, single phase and granular in nature. The crystallite size from XRD pattern's most strong (111) peak calculated using Debye-Scherrer's formula is 30nm. IR transmission spectrum exhibits Cd-O stretching and bending mode peaks at 831 and 672cm^{-1}, respectively. Electrical properties were characterized by two-probe measurement in the temperature range between 300 and 400K. Optical transmission spectroscopy and DC conductivity measurements have revealed the semiconducting nature of this film with direct band gap energy 2.53 and 0.29eV activation energy and dc conductivity 4.5×10^{-7} ('Ωcm)$^{-1}$.

Keywords: Semiconductor Compounds, X-Ray Diffraction, Scanning Electron Microscopy (SEM), Morphology of Films, Infrared and Raman Spectra

1. Background

In recent years, transparent conductive oxide (TCO) layers have been drawing considerable attention for their potential applications in photovoltaic devices, gas sensors, photo-transistors and other electronic and optoelectronic devices[1]. Cadmium oxide is one of the providential TCOs possessing high electric conductivity[2][3], transmittance with moderate refractive index. It belongs to a class of n-type semiconductor oxide with a rock salt crystal structure having band gap energy in the interval of 2.2eV–2.8eV. Optically, the films of cadmium oxide act as a selectively transmitting layer being transparent to the visible light and reflective to the thermal infrared radiation. CdO thick films have already been developed by various techniques such as spray pyrolysis[4], DC magnetron sputtering[5][6], sol-gel[7][8], pulsed laser deposition[9], sol-gel technique, etc. In this work, we have employed fast emerging screen printing technique to deposit CdO thick films.

Screen printing technique has been used as a multifaceted method for the fabrication of semiconductor layers in photovoltaic devices, especially II–IV compound semiconductors[10]-[12]. Compared to the other costly methods, screen printing is very simple, quicker, eco-friendly and provides a worthy method for film preparation on large area substrates with maximum utilization.

The aim of the present work is to deposit CdO thick films by screen printing technique and investigate their structural, optical and electrical properties to use them in optoelectronic device applications.

2. Methodology of Materials

Merck-made Analytical Reagent (AR)-grade cadmium oxide (CdO), anhydrous cadmium chloride ($CdCl_2$), ethy-lene glycol ($C_2H_6O_2$, 99.99% purity) are used for synthesis of CdO thick films by screen printing method.

2.1. Synthesis of CdO

For casting CdO thick film on glass substrate, CdO powder (99.999% purity) was thoroughly mixed with anhydrous $CdCl_2$ as adhesive agent and then ground in a mortar with ethylene glycol as binder to form thick paste. Before film deposition, the glass substrates were cleaned with acetone and deionised water and dried at 60°C for 10min in oven. The prepared paste was used in screen printer to deposit CdO thick film on glass substrate. The as-deposited film was heated at 110°C for 2h to remove volatile solvent and for better adherence of film to substrate[13]. This screen-printed film was further annealed in muffle furnace at 500°C for 10min to decompose organic compounds and achieve the desired stoichiometry of film[14].

The thickness of the film was measured by employing a profilometer (Surf-test SJ-301) having the value of 1μm. The steps involved in the preparation of CdO thick films by screen printing method as described above are presented in the following **Figure 1** schematic flow chart.

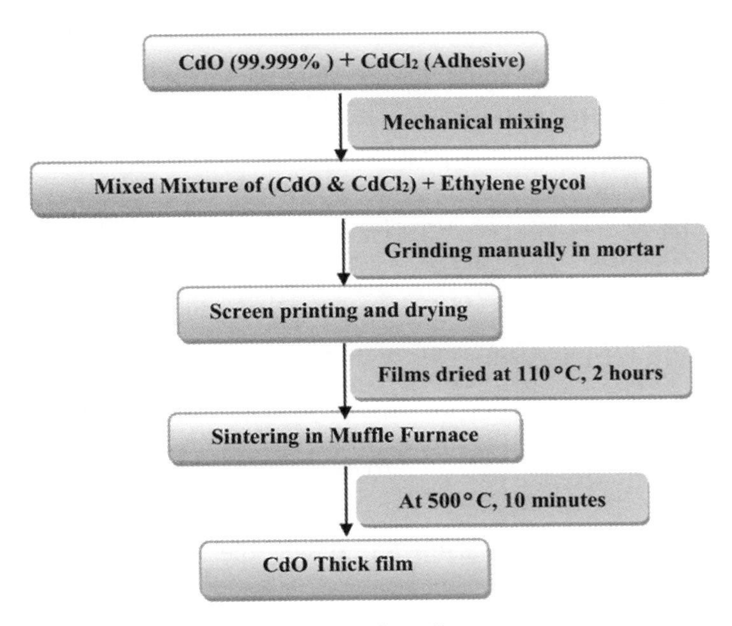

Figure 1. Screen printing procedure diagram.

2.2. Characterization

Advanced Rigaku diffractometer was used to record X-ray diffraction (XRD) pattern in the 2θ range of 20°–60° using Cu-Kα X-ray radiation source. Profilometer (Surftest SJ-301) instrument was used for thickness measurement. The surface morphological information was derived using scanning electron microscope (SEM, Leo-440, UK). Atomic force microscopy (AFM) technique has been used to characterize the topography of CdO thick film. AFM images were recorded on Bruker Multimode 8-AM Atomic Force Microscope in (5μm × 5μm) and (50μm × 50μm) scanning area of the film. IR transmission spectrum was recorded by SHIMAZU-8400S, Japan Fourier transform infrared (FTIR) spectrophotometer in 4000–500cm^{-1} range at ambient temperature with 4cm^{-1} resolution. The optical transmission spectrum was recorded by Hitachi Spectrometer-3900 in the 350–1400nm range. DC conductivity measurement was done using standard two-probe technique.

3. Results and Discussions

3.1. Structure Analysis

The purity and crystallinity of the sintered sample was examined using (XRD). The sharp intense peaks of CdO thick film confirmed good crystalline nature and besides no other impurity peak was seen, it suggested the formation of the single phase of CdO. The peaks originated at (111), (200), (220) reflections are shown in **Figure 2**. The observed 2θ value of diffraction peaks, calculated and standard JCPDS card lattice parameters and crystallite size from Debye-Scherrer's formula[15] are listed in **Table 1**. Debye-Scherrer's formula used for calculation of crystallite size of CdO thick film is given in the following equation:

$$D = \frac{0.9\lambda}{\beta \cos\theta} \tag{1}$$

where D is the grain boundary size (in nm), λ is the X-ray wavelength β is the width (in radians) at half the maximum peak intensity, and θ is the Bragg angle. The crystallite size varies from 29 to 31nm for these three planes.

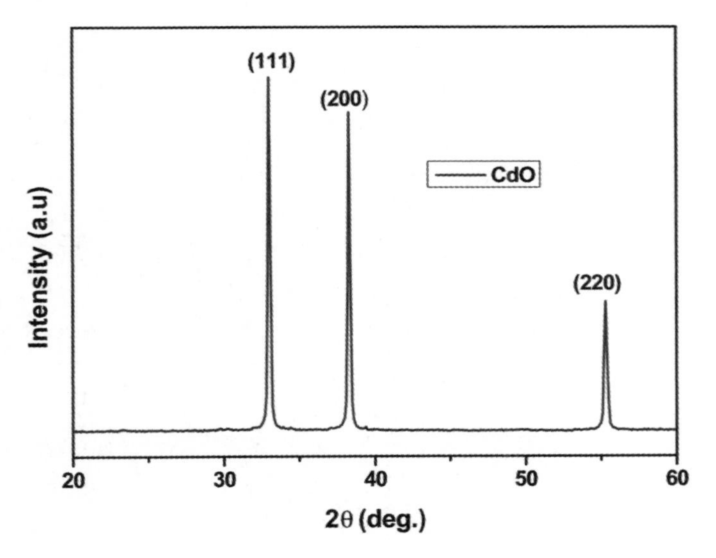

Figure 2. XRD pattern of CdO thick film.

Table 1. XRD parameters: *hkl* plane, 2θ, d values (calculated and JCPDS), and D values of the as-deposited CdO thick film.

hkl Plane	2θ (°)	Calculated d_{hkl} (Å)	JCPDS(05-0640) d_{hkl} ((Å)	D (nm)
111	33.004	2.71188	2.712	30
200	38.304	2.34794	2.349	29
220	55.300	1.65989	1.661	31

The calculated lattice parameters match well with CdO (JCPDS) card No. 05-0640 reported data[16].

3.2. SEM Analysis

Figure 3 shows scanning electron microscopy (SEM) micrograph of deposited thick film scanned by 15kV electron beam at 5000× magnification. The SEM image reveals the polycrystalline, porous morphology with the inter-connected grains present on the film surface. The small crystallites agglomerated to form spindle, dumbbell and cuboidal-shaped particles and fused clusters are also seen in surface morphology of this film. Such structures may provide novel platform for photovoltaic, sensor and other device applications.

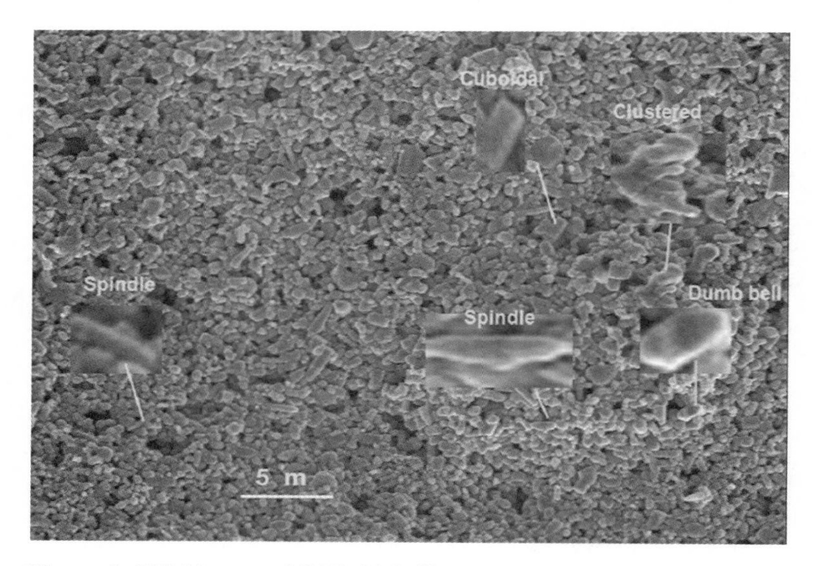

Figure 3. SEM image of CdO thick film.

3.3. AFM Analysis

AFM characterized the 2D and 3D topography of screen-printed CdO thick film, as shown in **Figure 4(a)**, **Figure 4(b)**, respectively. In **Figure 4(a)**, the spherical grains are closely packed, uniformly distributed and amalgamation in some regions within the scanned area. While in **Figure 4(b)**, 3D topography of the film exhibits columnar microstructure perpendicular to the surface. The mean surface roughness was measured to be ~70.00nm. This surface roughness of CdO thick film reduces light reflection, but increases light absorption in the visible region of the solar spectrum in solar cells.

3.4. IR Analysis

IR spectroscopy is a versatile tool to reveal structural details of bonding groups present through fundamental vibrations observed in mid-infrared region. The functional groups, molecular geometry and inter/intra-molecular interactions via hydrogen bonding are considered during IR transmission spectrum analysis[17]. Metal oxides generally give fingerprint absorption bands below $1000cm^{-1}$ arising from inter-atomic stretching and bending vibrations. IR transmittance spectrum of CdO thick film recorded in $4000–600cm^{-1}$ range is presented in **Figure 5**.

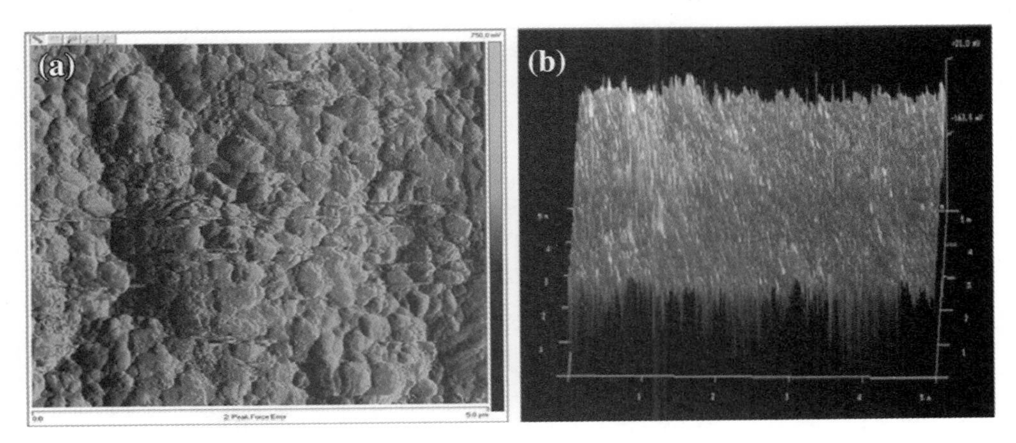

Figure 4. AFM (a) 2D and (b) 3D images of CdO thick film.

The spectrum exhibits broad transmittance band of water stretching mode with maximum at 3431cm^{-1}. The broadness of this mode confirms the pre- sence of hydrogen bonding. C-H antisymmetric and symmetric stretching modes of methylene (CH_2) group are observed at 2930 and 2821cm^{-1}, respectively. The presence of these modes generally arises from the organic compounds present in the atmosphere. In 2500–2000cm^{-1} region, very complex vibrational rotational spectral lines of atmospheric carbon dioxide are obtained. Very strong peaks due to bending modes of CH_2 appeared at 1372cm^{-1}. Cd-OH antisymmetric stretching mode as medium intensity peak is obtained at 1264cm^{-1} while symmetric stretching components of Cd-OH mode appears at 1218 and 1164cm^{-1}. In addition to these peaks, the strong peak of Cd-O stretching mode and weak bending mode are obtained at 831 and 672cm^{-1}, respectively. These studies confirmed the formation of CdO film.

3.5. Optical Properties

Figure 6 shows the optical transmission spectrum of CdO screen-printed film recorded in the 350–1400nm range and the curve shows 80% transmission in the visible region. At 400nm, the transmission is ~10% which starts increasing exponentially with sharp rise in 600 to 850nm region and then at slow pace to 80% at 1400nm. Mostly, all II–VI semiconductors have direct band gaps. The relation between absorption coefficient (α) and incident photon energy ($h\nu$) can be written as[18].

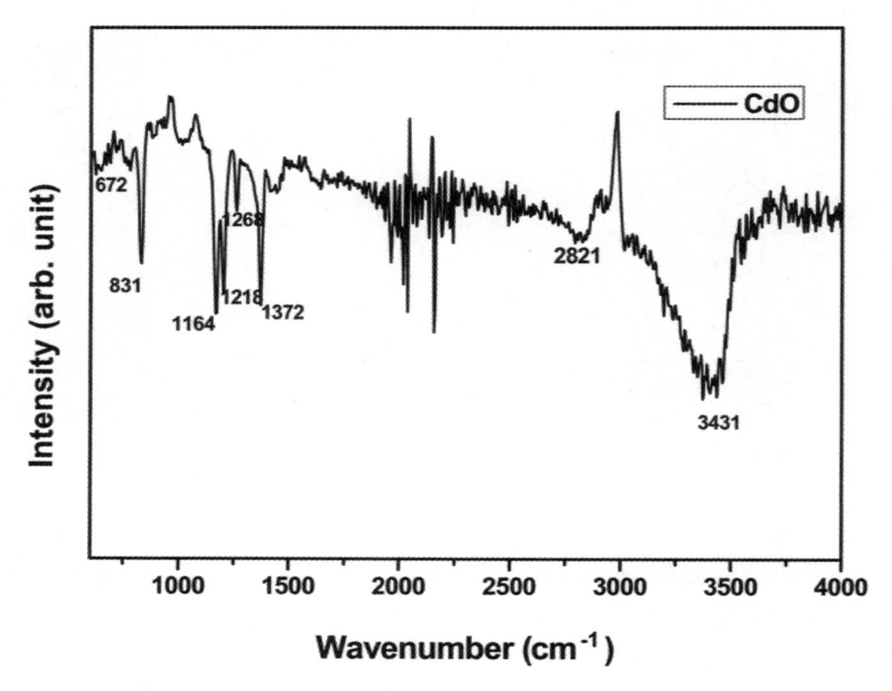

Figure 5. IR transmission spectrum of CdO thick film.

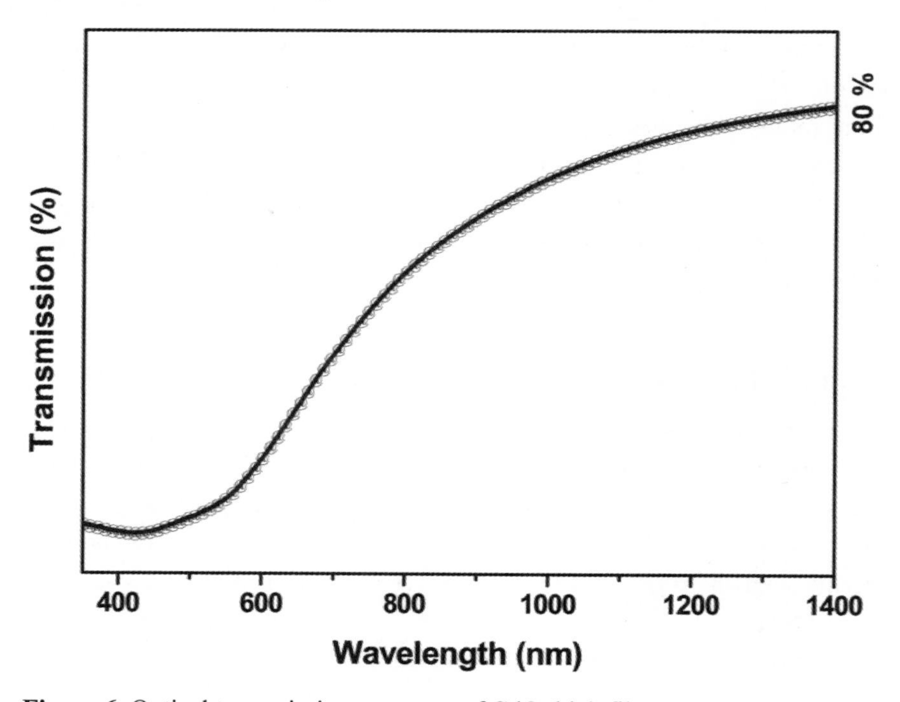

Figure 6. Optical transmission spectrum of CdO thick film.

$$\alpha h\nu = A\left(h\nu - E_g\right)^n \tag{2}$$

where A is constant, E_g is the energy separation between valence and conduction bands. The exponent n depends upon the quantum selection rules for particular material which is equal to 1/2 for direct band gap material. The photon energy ($h\nu$) for Y-axis can be calculated using Equation (3).

$$E = h\nu = h\frac{c}{\lambda} \tag{3}$$

where h is Planck's constant (6.626×10^{-34}J/s), c is speed of light (3×10^8m/s) and λ is wavelength of incident light.

Figure 7 shows the plot between $(\alpha h\nu)^2$ vs. $h\nu$, which is a straight line, indicating the direct transition. The energy band gap is obtained by extrapolating the linear part of the curve to zero absorption coefficient. The band gap is found to be 2.53eV, which is in good agreement with earlier result[19].

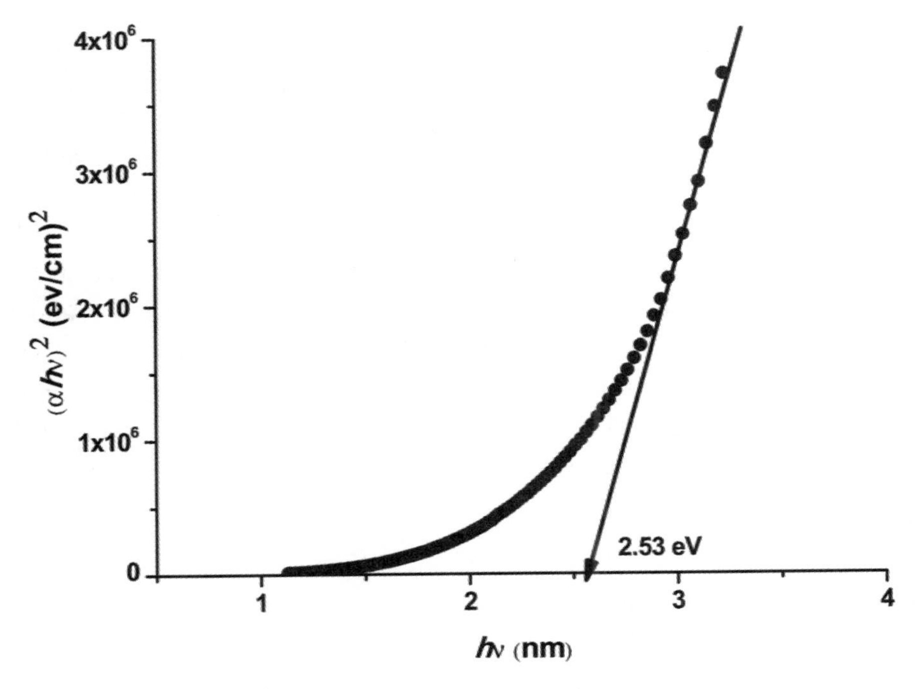

Figure 7. $h\nu$ vs. $(\alpha h\nu)^2$ for determination of CdO thick film band gap energy.

3.6. Electrical Properties

One of the reasons for the application of the CdO thin films in the optoelectronic devices technology is their good electrical conductivity even without any extrinsic doping[20]. The DC electrical conductivity measurements have been carried out in the temperature range 300–400K. The electrical resistivity (ρ) was calculated using the Equation (4)[21].

$$\rho = \frac{VA}{It} \tag{4}$$

where ρ is the resistivity (Ωcm), A is the area, t is the thickness of film (cm), V is the measured voltage and I is the source current (A). To identify the nature of conduction process, activation energy has been calculated using the Arrhenius relation (5):

$$\sigma = \sigma_0 \exp\left(\frac{-\Delta E}{KT}\right) \tag{5}$$

where σ is the conductivity $\left(\frac{1}{\rho}\right)$, σ_0 is the pre-exponential factor, ΔE is the thermal activation energy for generation process, K is Boltzmann's constant and T is the absolute temperature.

In **Figure 8**, graph is plotted between log(σ) and 10^3/T, to understand the charge conduction mechanism in CdO film. The variation of electrical conductivity with temperature indicates the semiconducting nature of the sample because conductivity increases with increase in operating temperature due to rise in electron carriers[22] in the film. The activation energy has been derived from the slope of the graph and it comes out to be 0.29eV which is also in good agreement with the result reported earlier[23].

4. Conclusion

In conclusion, these investigations exhibit the successful utilization of a

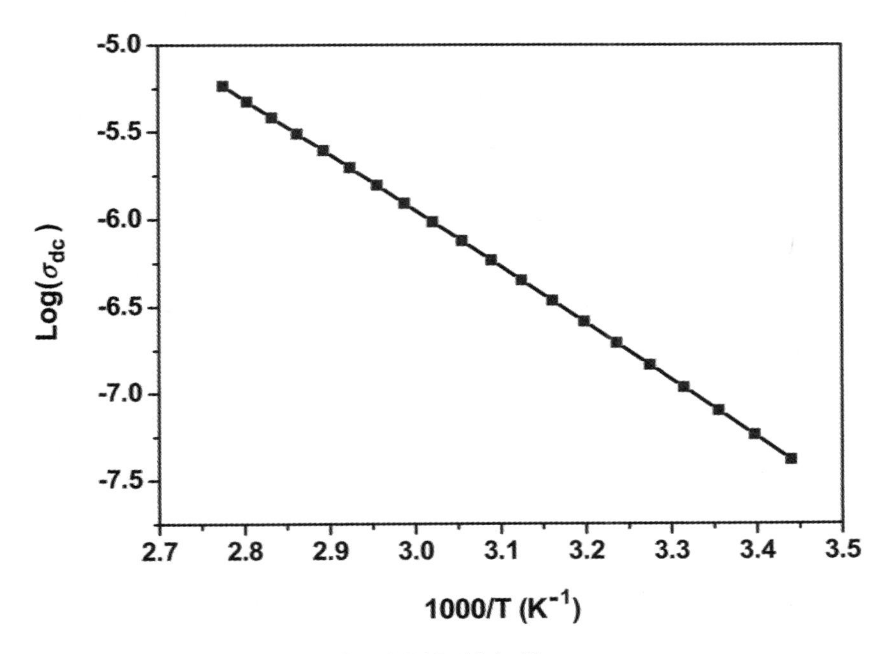

Figure 8. Log (σ_{dc}) vs. 1000/T of CdO thick film.

simple and economical screen printing method for preparing wide band gap semiconductor thick films. The structural, morphological, optical and electrical studies prove the suitability of these films for photovoltaic devices and other electronic applications. The XRD, SEM and AFM studies show that CdO films have polycrystalline nature with cubic structure, spherical grains and porous morphology. The direct energy band gap transition has been found to be 2.53eV. FTIR spectrum reveals the formation of CdO with no trace of any type of impurity. DC conductivity measurement reveals the semiconducting nature of film and gives activation energy value 0.29eV. This exhibits that the conduction process of carried charge is thermally activated. Thus, screen printing is a cost-effective and user-friendly technique and can be used to fabricate polycrystalline thick films having good stability and significant value of activation energy. Such types of films are suitable for solar cells and other optoelectronics devices.

Acknowledgements

RAZ is thankful to the CSIR-NPL, New Delhi for giving permission to use characterization facilities required for analysis.

Compliance with Ethical Standards

Conflict of Interest

The authors declare that they have no competing interests.

Authors' Contributions

RAZ was involved in the acquisition, analysis, and interpretation of data. SC was involved in carried out optical and XRD characterization. MA was involved in analysis of FTIR spectrum of film and drafting the manuscript. AKZ revised the manuscript critically and gave final approval for submission. All authors read and approved the final manuscript.

Source: Zargar R A, Chackrabarti S, Arora M, et al. Synthesis, characterization and interpretation of screen-printed nanocrystalline CdO thick film for optoelectronic applications [J]. International Nano Letters, 2016, 6(2):1–6.

References

[1] Dakhel, A.: Optoelectronic properties of Eu- and H-codoped CdO films. Curr. Appl. Phys. 11, 11–15 (2011).

[2] Yan, M., Lane, M., Kannewurf, C.R., Chang, R.P.H.: Selective-area atomic layer epitaxy growth of ZnO features on soft lithography-patterned substrates. Appl. Phys. Lett. 78, 2342–2344 (2001).

[3] Ghosh, P.K., Das, S., Kundoo, S., Chattopadhyay, K.K.: Effect of fluorine doping on semiconductor to metal-like transition and optical properties of cadmium oxide thin films deposited by sol-gel process. J. Sol-Gel Sci. Technol. 34, 173–179 (2005).

[4] Gandarilla, F., Morales-Acevedo, A., Vigil, O., Hesiquio-Garduo, V., Vaillant, L., Contreras-Puente, V.:Micro-structural characterization of annealed cadmium zinc oxide thin films obtained by spray pyrolysis. Mater. Chem. Phys. 78,840–846 (2003).

[5] Zhou, Q., Ji, Z., Hu, B.B., Chen, C., Zhao, L., Wang, C.: Low resistivity transparent conducting CdO thin films deposited by DC reactive magnetron sputtering at room temperature. Mater. Lett. 61, 531–534 (2007).

[6] Ma, D., Ye, Z., Wang, L., Huang, J., Zhao, B.: Deposition and characteristics of CdO

films with absolutely (200)-preferred orientation. Mater. Lett. 58, 128–131 (2003).

[7] Ghosh, P.K., Das, S., Chattopadhyay, K.K.: Temperature dependent structural and optical properties of nanocrystalline CdO thin films deposited by sol-gel process. J. Nanopart. Res. 7, 219–225 (2005).

[8] Santos-Cruz, J., Torres-Delgado, G., Castanedo-Perez, R., Jimenez-Sandoval, S., Jimenez-Sandoval, O., Zuniga-Romero, C.I., Marquez Marın, J., Zelaya-Angel, O.: Dependence of electrical and optical properties of sol-gel prepared undoped cadmium oxide thin films on annealing temperature. Thin Solid Films 493, 83–87 (2005).

[9] Ismail, R.A., Rasheed, B.G., Salm, E.T., Al-Hadethy, M.: High transmittance-low resistivity cadmium oxide films grown by reactive pulsed laser deposition. J. Mater. Sci. Mater. Electron. 18, 1027–1030 (2007).

[10] Kumar, V., Sharma, D.K., Bansal, M.K., Dwivedi, D.K., Sharma, T.P.: Synthesis and characterization of screen-printed CdS films. Sci. Sinter. 43, 335–341 (2011).

[11] Zargar, R.A., Arora, M., Hafiz, A.K.: Investigation of physical properties of screen printed nanosized ZnO films for optoelectronic applications. Eur. Phys. J Appl. Phys. 70, 10403 (2015).

[12] Zargar, R.A., Khan, S.D., Khan, M.S., Arora, M., Hafiz, A.K.: Synthesis and characterization of screen printed Zn0.97Cu0.03O thick film for semiconductor device applications. Phys. Res. Int. 2014, Article ID 464809, 5 pp. (2014).

[13] Ismail, B., Abaab, M., Rezig, B.: Structural and electrical properties of ZnO films prepared by screen printing technique. Thin Solid Films 383, 92–94 (2001).

[14] Sharma, M., Kumar, S., Sharma, L.M., Sharma, T.P., Husain, M.: CdS sintered films: growth and characteristics. Physica B 348, 15–20 (2004).

[15] Cullity, B.D.: Elements of X-ray Diffractions. Addison-Wesley, Reading (1978).

[16] Lanje, A.S., Ningthoujam, R.S., Sharma, S.J., Pode, R.B.: Luminescence and electrical resistivity properties of cadmium oxide nano particles. Indian J. Pure Appl. Phys. 49, 234–238 (2011).

[17] Nandi, S.K., Chakraborty, S., Bera, M.K., Maiti, C.K.: Structural and optical properties of ZnO films grown on silicon and their application in MOS devices in conjunction with ZrO2 as a gate dielectric. Bull. Mater. Sci. 30, 247–254 (2007).

[18] Tauc, J. (ed.): Amorphous and Liquid Semiconductors, vol. 159. Plenum Press, New York (1974).

[19] Khallaf, H., Chen, C.-T., Chang, L.-B., Lupan, O., Dutta, A., Heinrich, H., Shenoud, A., Chowa, L.: Investigation of chemical bath deposition of CdO thin films using three different complexing agents. Appl. Surf. Sci. 257, 9237–9242 (2011).

[20] Bhosale, C.H., Kambale, A.V., Kokate, A.V., Rajpure, K.Y.: Structural, optical and electrical properties of chemically sprayed CdO thin films. Mater. Sci. Eng. B 122, 67–71 (2005).

[21] Singhal, S., Kaur, J., Namgyal, T., Sharma, R.: Cu-doped ZnO nanoparticles: synthesis, structural and electrical properties. Physica B 407, 1223–1226 (2012).

[22] Ghosh, P.K., Maity, R., Chattopadhyay, K.K.: Electrical and optical properties of highly conducting CdO: F thin film deposited by sol-gel dip coating technique. Sol. Energy Mater. Sol. Cells 81, 279–289 (2004).

[23] Uplane, M.D., Kshrisagar, P.N., Lokhande, B.J., Lokhande, C.D.: Preparation of cadmium oxide films by spray pyrolysis rand its conversation into chalcogenide films. Indian J. Pure Appl. Phys. 37, 616–619 (1999).

Chapter 11

Microstructure and Optical Properties of Ag/ITO/Ag Multilayer Films

Zhaoqi Sun, Maocui Zhang, Qiping Xia, Gang He, Xueping Song

School of Physics & Materials Science, Anhui University, Hefei 230601, China

Abstract: Transflective and highly conductive Ag/ITO/Ag multilayer films were prepared by magnetron sputtering on glass substrates. The microstructure and optical properties of Ag/ITO/Ag multilayer films were systematically investigated by X-ray diffraction, scanning electron microscopy, and ultraviolet-visible spectroscopy. The optical properties of the multilayer films were significantly influenced by the thickness of the Ag surface layer from 3.0 to 12.6nm. The multilayer film of Ag9.3nm/ITO142nm/Ag9.3nm shows the best comprehensive property. It could satisfy the requirement for transflective LCD.

Keywords: Ag/ITO/Ag Multilayer Films, Microstructure, Optical Properties

1. Introduction

Indium tin oxide (ITO) thin films have low electrical resistance and high transmittance in the visible spectrum. Because of their unique photoelectrical properties, they play an important role in optoelectronic devices, such as flat displays, thin-film transistors, solar cells, and so on[1]-[6].

It is well known that transmissive LCD has low contrast ratio in bright light and high power consumption. Reflective LCD has low contrast ratio in weak light, and most of them belong to monochromatic LCD. However, transflective LCD possesses high contrast ratio in bright and weak light as well as low power consumption.

Ag is a noble metal with excellent photoelectrical properties. In addition to good conductivity, it has high reflectivity in the visible range and good chemical stability. Thus, Ag/ITO composite material is the optimizing material to make new transflective LCD. Miedziński reported the electrical properties of Ag/ITO composite films[7]. Choi fabricated ITO/Ag/ITO multilayer films and obtained a high-quality transparent electrode which has a resistance as low as $4\Omega/\Upsilon$ and a high optical transmittance of 90% at 550nm[8]. Bertran prepared Ag/ITO films with a high transmittance (near 80%) in the visible range by RF sputtering and studied their application as transparent electrodes in large-area electrochromic devices[9]. Guillén prepared ITO/Ag/ITO multilayer films with visible transmittance above 90% by sputtering at room temperature and investigated the optical and electrical characteristics of single-layer and multilayer structures. Besides, the transmittance is found to be mainly dependent on the thickness of Ag film[10]. Although much work has paid more attention on the investigation of Ag/ITO/Ag multilayer films, few studies have been carried out to study their photoelectrical properties.

In this study, Ag/ITO/Ag multilayer films with various surface layer thicknesses have been prepared on a glass substrate by direct current (DC) magnetron sputtering. The microstructure and optoelectronic properties of the Ag/ITO/Ag films were investigated using X-ray diffraction (XRD), scanning electron microscopy (SEM), and ultraviolet-visible spectroscopy (UV-vis).

2. Methods

The multilayer films were prepared by an ultrahigh vacuum multifunctional magnetron sputtering equipment (JGP560I, SKY Technology Development Co., Ltd, Shenyang, China). The multilayer films with a sandwich structure were deposited on glass substrates. The Ag layers were deposited by DC magnetron sputtering with a power density of $1.73W/cm^2$, while the ITO coatings were deposited by radio frequency magnetron sputtering with a power density of 2.12W/cm^2. Ceramic ITO targets of In_2O_3:SnO_2 disk (90:10wt.%, 4N) and an Ag metal target (4N) were used for ITO and Ag layer deposition separately. The target-to-substrate distance was 60mm. The base vacuum was 6.0×10^{-4}Pa, and the deposition pressure was 1.0Pa with an argon (4N) flow rate of 45cm. During the deposition, the substrates were kept in room temperature.

The thicknesses of Ag/ITO/Ag multilayer films were controlled by sputtering time. Ag films and ITO films were firstly prepared respectively at a fixed time, and their thicknesses were tested using a surface profiler (XP-1, Ambios Technology, Santa Cruz, CA, USA). The deposition rate of Ag films and ITO films was calculated according to the sputtering time and film thicknesses. Then Ag/ITO/Ag multilayer films could be prepared with different sputtering time.

The multilayer films consist of three layers. The thicknesses of Ag surface layer vary from 3.0 to 12.6nm while the Ag bottom layer keeps the same thickness of 9.3nm. All samples have an ITO interlayer of 142nm. **Table 1** shows the thickness of the multilayer films.

The microstructure was analyzed with a MXP 18AHF X-ray diffractometer (MAC Science, Yokohama, Japan). The X-ray source was Cu$K\alpha$, with an accelerating voltage of 40kV, a current of 100mA, scanning range from 20° to 80°, glancing angle of 2°, scanning step of 0.02°, and scanning speed of 8°/min. The surface morphology of the films was studied with a Hitachi S-4800 type SEM (Hitachi, Tokyo, Japan). The optical properties were tested with a Shimadzu UV-2550 type ultraviolet-visible spectroscope (Shimadzu, Kyoto, Japan). The

Table 1. Microstructure parameters and the average reflectance with 300 to 900 nm of Ag/ITO/Ag multilayer films.

Samples	Compositions	Ag(111) grain size (nm)	Average reflectance R (%)
Ag1/ITO/Ag	Ag 3.0nm/ITO 142nm/Ag 9.3nm	2.477	22.04
Ag2/ITO/Ag	Ag 6.4nm/ITO 142 nm/Ag 9.3nm	5.955	25.07
Ag3/ITO/Ag	Ag 9.3nm/ITO 142 nm/Ag 9.3nm	11.945	28.98
Ag4/ITO/Ag	Ag 12.6nm/ITO 142 nm/Ag 9.3nm	19.885	31.12
ITO	-	-	23.76

scanning range was from 300 to 900nm, scanning step was 1 nm, and slit width was 2nm.

3. Results and Discussion

3.1. Microstructure Analysis

Figure 1 shows the XRD patterns of the Ag, ITO, and Ag/ITO/Ag films. Based on **Figure 1**, it can be noted that broad In_2O_3 (222) diffraction peaks have

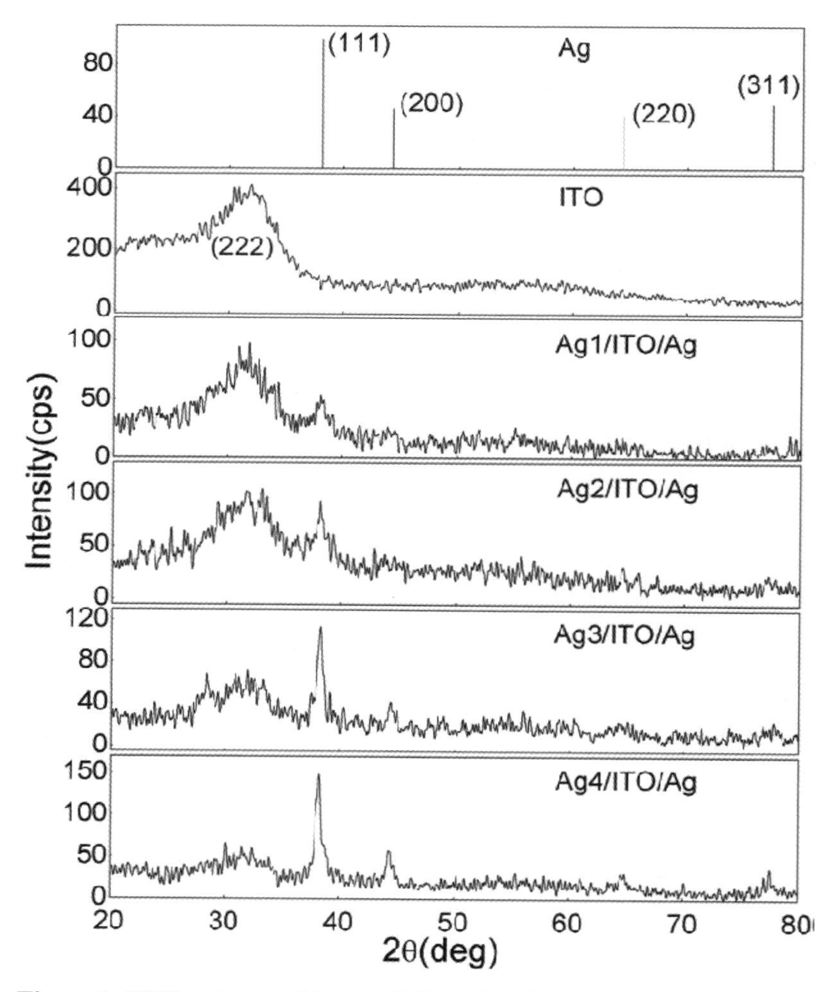

Figure 1. XRD patterns of the Ag, ITO, and Ag/ITO/Ag multilayer films.

been observed in the ITO and Ag/ITO/Ag films. As the thickness of the Ag surface layer increased, the Ag (111) preferred orientation intensified[11][12].

From Scherrer's formula,

$$D = K\lambda/\beta\cos\theta \tag{1}$$

where K is 0.9, λ is 1.54056 Å, β is the full width at half maximum of the diffraction peak, and θ is the diffraction angle. The value of Ag grain size D (111) was calculated, and the results were listed in **Table 1**. With the increase of Ag surface layer thickness from 3.0 to 12.6nm, the Ag grain size of all films increases.

Figure 2 shows the SEM micrographs of single-layer Ag films. According to **Figure 2**, it has been found that the surface morphology of the Ag film is critically dependent on its thickness. As shown in **Figure 2(a)**, the Ag nanoparticles are uniformly distributed in substrate. The Ag film forms in stable nuclei stage. As the thickness of the Ag film increases, randomly connected Ag islands appear, as shown in **Figure 2(b)**.

Figure 3 shows the SEM micrographs of Ag2/ITO/Ag and Ag3/ITO/Ag multilayer films. As shown in **Figure 3(a)**, the Ag nanoparticles are spherical and uniformly distributed in ITO films. The size of Ag nanoparticle is 5 to 60nm. With increasing thickness of the Ag surface layer, randomly connected Ag network also appears, as shown in **Figure 3(b)**.

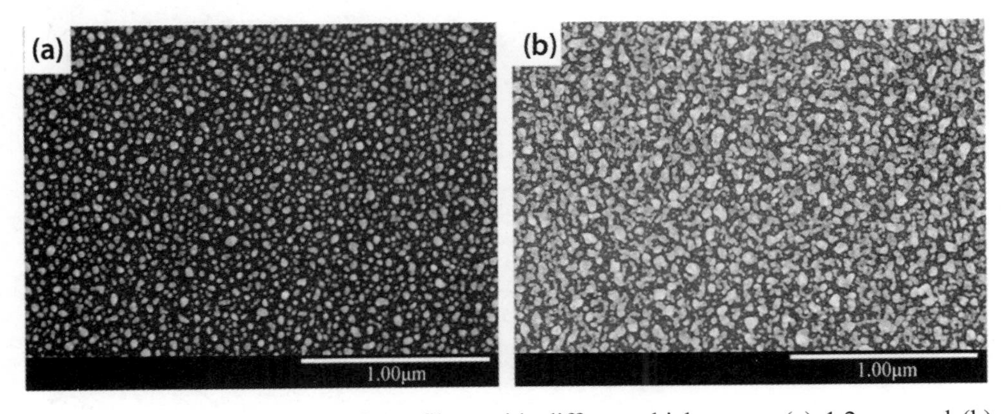

Figure 2. SEM micrographs of Ag films with different thicknesses: (a) 1.2 nm and (b) 1.8nm.

Figure 4 shows a cross-sectional SEM micrograph of Ag3/ITO/Ag multilayer film. The Ag surface layer, ITO interlayer, and Ag bottom layer forming the sandwich structure multilayer film have been observed clearly. From **Figure 4**, it has been seen that the Ag surface layer and bottom layer have a spherical cluster structure, and the interlayer of ITO film has a columnar structure.

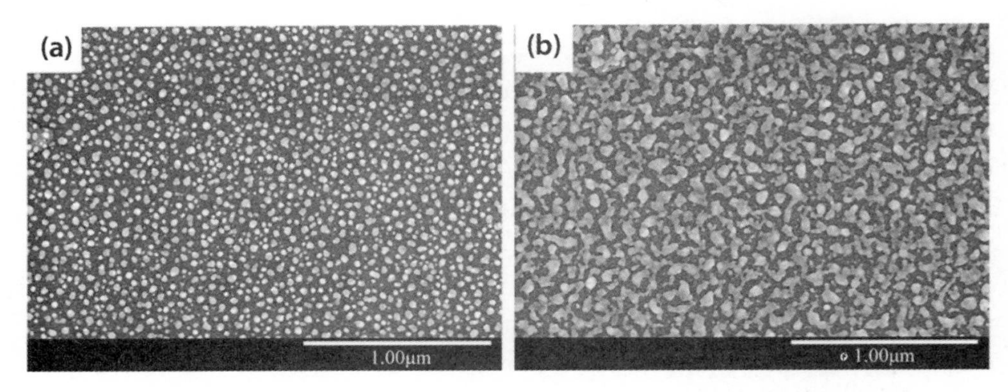

Figure 3. SEM micrographs of Ag/ITO/Ag multilayer films: (a) Ag2/ITO/Ag and (b) Ag3/ITO/Ag.

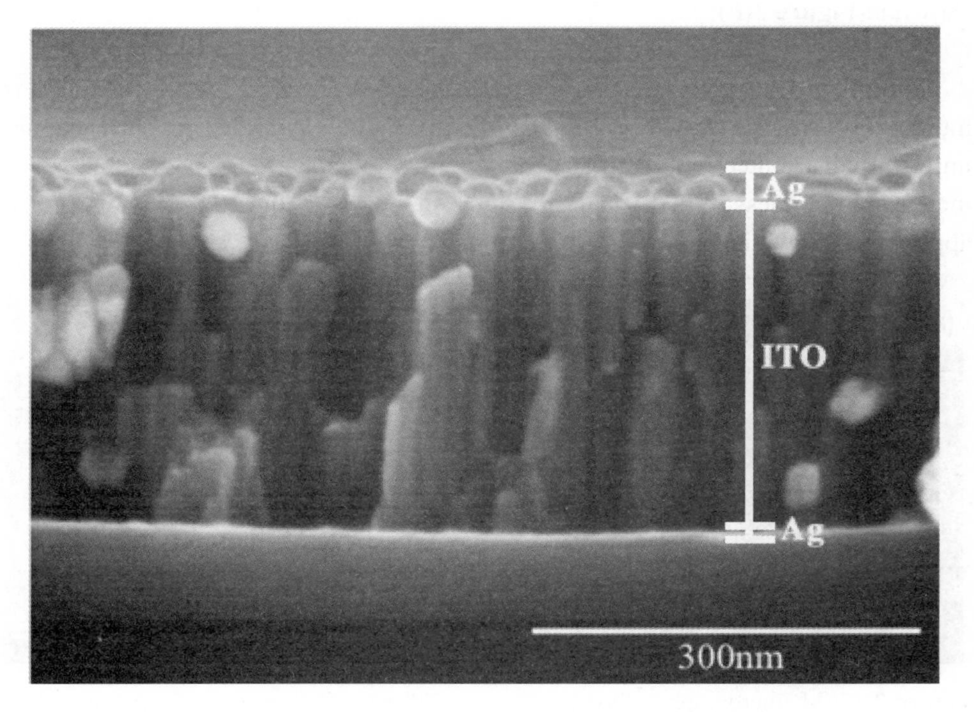

Figure 4. Cross-sectional SEM micrograph of Ag3/ITO/Ag multilayer film.

3.2. Optical Properties

Figure 5 shows the thickness-dependent transmittance spectra of the multilayer films changing wavelength from 300 to 900nm. Compared with the bare ITO, the sandwich structure films have lower optical transmittance. It is suggested that the island structure of the thin Ag surface layer makes its transmittance low due to the large islands and the defects scattering incident light[9][13]. With the increase of Ag surface layer thickness from 3.0 to 12.6nm, the transmittance of the multilayer films decreases, which is caused by the changes of the Ag surface layer first from a stable nuclei stage to randomly connected Ag island stage then to Ag network stage. Besides, Ag1/ITO/Ag, Ag2/ITO/ Ag, and Ag3/ITO/Ag have low optical transmittance at about 500nm. Ag4/ITO/Ag has low optical transmittance at about 450 and 550nm. It is due to the surface plasmon resonance characterization of Ag.

Figure 6 shows the reflectance spectra of the ITO and multilayer films. Based on **Figure 6**, it can be observed that multilayer Ag/ITO/Ag films show higher reflectivity than pure ITO film due to the high reflectivity of Ag. **Table 1** calculated the average reflectance of the bare ITO and multilayer films. When the

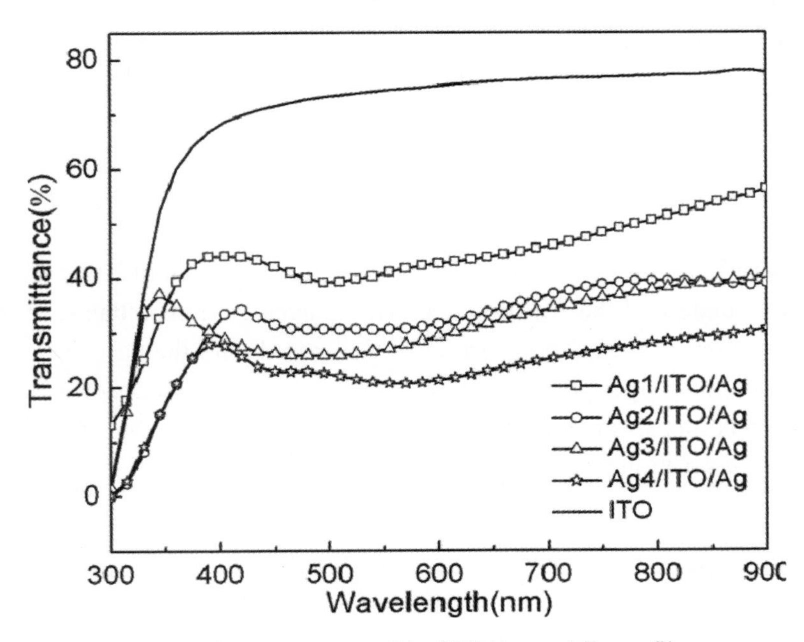

Figure 5. Transmittance spectra of Ag/ITO/Ag multilayer films.

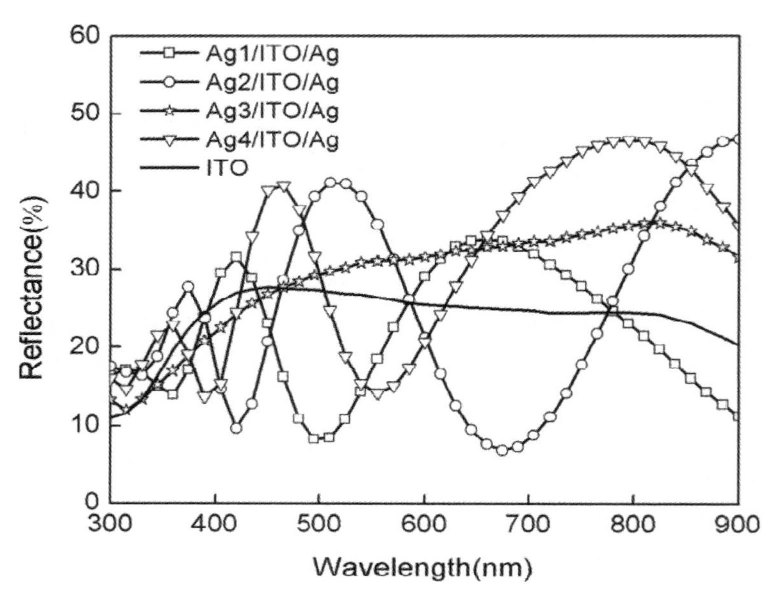

Figure 6. Reflectance spectra of the ITO and Ag/ITO/Ag multilayer films.

thickness of the Ag surface layer increases from 3.0 to 12.6nm, the microstructure and surface morphology of the Ag surface layer changes a lot; the decrease of holes and defects in the films reduces the energy loss of light and the absorption of multilayer film, so the average reflectance of multilayer films increases from 22.04% to 31.12%. Besides, there is an interference phenomenon in the reflectance spectra of Agl/ITO/ Ag, Ag2/ITO/Ag, and Ag4/ITO/Ag; this will lead to uneven reflection and affect the quality of the LCD. The reflectance spectra of Ag3/ITO/Ag are relatively flat and can eliminate the influence of the interference phenomenon.

Figure 7 shows the absorption spectra of the ITO and multilayer films. With increasing thickness of the Ag surface layer, the average transmittance of the multilayer films first increases then decreases. Compared with the bare ITO films, the absorption of multilayer films increases due to the introduction of a double Ag layer. However, the absorption of Agl/ITO/Ag film is close to that of the bare ITO film, and no absorption peaks appeared.

4. Conclusions

Ag/ITO/Ag multilayer films with different thicknesses of the surface Ag

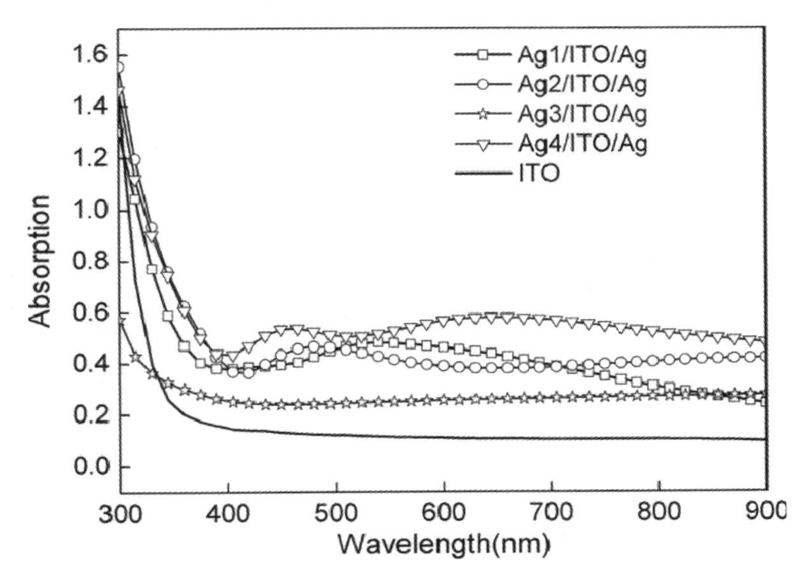

Figure 7. Optical absorption spectra of the ITO and Ag/ITO/Ag multilayer films.

layer were prepared by magnetron sputtering on a glass substrate. Microstructural analysis shows that the multilayer films have a polycrystalline structure. As the thickness of the Ag surface layer increases, the preferred orientation of Ag (111) intensified. With increasing thickness of Ag surface layer, the transmittance spectra and reflectance spectra of Ag/ITO/Ag multilayer films decrease and increase, respectively. Ag3/ITO/Ag multilayer film shows the best comprehensive property and can be used as a potential transflective candidate in future LCD.

Competing Interests

The authors declare that they have no competing interests.

Authors' Contributions

ZQS and QPX prepared the films and tested the surface topography. X-ray diffraction was investigated by XPS and MCZ. The optical properties were measured by GH. The calculations were carried out by ZQS who also wrote the manuscript. Besides, MCZ helped draft the manuscript. All authors read and approved the final manuscript.

Acknowledgements

This work is supported by the National Natural Science Foundation of China (nos. 51072001 and 51272001), National Key Basic Research Program (973 Project) (2013CB632705), and the National Science Research Foundation for Scholars Return from Overseas, Ministry of Education, China. The authors would like to thank Yonglong Zhuang and Zhongqing Lin of the Experimental Technology Center of Anhui University for the electron microscope test and discussion.

Source: Sun Z, Zhang M, Xia Q, et al. Microstructure and optical properties of Ag/ITO/Ag multilayer films [J]. Nanoscale Research Letters, 2012, 8(41):4977–4984.

References

[1] Bhatti MT, Rana AM, Khan AF: Characterization of rf-sputtered indium tin oxide thin films. Mater Chem Phys 2004, 84:126.

[2] Dawar AL, Joshi JC: Semiconducting transparent thin films: their properties and applications. J Mater Sci-Mater M 1984, 19:1.

[3] Meng LJ, Placido F: Annealing effect on ITO thin films prepared by microwave-enhanced dc reactive magnetron sputtering for telecommunication applications. Surf Coat Tech 2003, 166:44.

[4] Deng W, Ohgi T, Nejo H: Development of conductive transparent indium tin oxide (ITO) thin films deposited by direct current (DC) magnetron sputtering for photon-STM applications. Appl Phys A-Mater 2001, 72:595.

[5] Chopra KL, Major S, Pandya DK: Transparent conductors-A status review. Thin Solid Films 1983, 102:1.

[6] Cui HN, Xi SQ: The fabrication of dipped CdS and sputtered ITO thin films for photovoltaic solar cells. Thin Solid Films 1996, 288:325.

[7] Miedziński R, Ebothé J, Kozlowski G, Kasperczyk J, Kityk IV: Laser induced microrelief superstructure of Ag/ITO seed-mediated nanocomposites. Superlattice Microst 2009, 46:637.

[8] Choi KH, Kim JY, Lee YS, Kim HJ: ITO/Ag/ITO multilayer films for the application of a very low resistance transparent electrode. Thin Solid Films 1999, 341:152.

[9] Bertran E, Corbella C, Vives M, Pinyol A, Person C, Porqueras I: RF sputtering deposition of Ag/ITO coatings at room temperature. Solid State Ion 2003, 165:139.

[10] Guillén C, Herrero J: Transparent conductive ITO/Ag/ITO multilayer electrodes deposited by sputtering at room temperature. Opt Commun 2009, 282:574.

[11] Sun X, Huang H, Kwok H: On the initial growth of indium tin oxide on glass. ApplPhys Lett 1996, 68:2663.

[12] Kim DH, Park MR, Lee HJ, Lee GH: Thickness dependence of electrical properties of ITO film deposited on a plastic substrate by RF magnetron sputtering. Appl Surf Sci 2006, 253:409.

[13] Jeong JA, KiKim H: Low resistance and highly transparent ITO–Ag–ITO multilayer electrode using surface plasmon resonance of Ag layer for bulk-heterojunction organic solar cells. J Sol Energ Mat Sol C 1801, 2009:93.

Chapter 12

Modification of Optical and Electrical Properties of Zinc Oxide-Coated Porous Silicon Nanostructures Induced by Swift Heavy Ion

Yogesh Kumar[1,2], Manuel Herrera-Zaldivar[3], Sion Federico Olive-Méndez[2], Fouran Singh[4], Xavier Mathew[5], Vivechana Agarwal[1]

[1]Centro de Investigacion en Ingenieria y Ciencias Aplicadas, UAEM, Av. Univ. 1001, Col. Chamilpa, Cuernavaca, Morelos 62209, México
[2]Centro de Investigacion in Materiales Avanzados, Ave. Miguel de Cervantes 120, Complejo Industrial Chihuahua, Chihuahua 31109, México
[3]Materials Science Group, Centro de Nanocienciay Nanotecnología, UNAM, Ensenada Apdo, Postal 14, Ensenada, Baja California 22800, México
[4]Inter University Accelerator Centre, Aruna Asaf Ali Marg, New Delhi 110067, India
[5]Centro de Investigacion Energia, UNAM, Privada Xochicalco S/N, Temixco, Morelos 62580, México

Abstract: Morphological and optical characteristics of radio frequency-sputtered zinc aluminum oxide over porous silicon (PS) substrates were studied before and after irradiating composite films with 130MeV of nickel ions at different fluences varying from 1×10^{12} to 3×10^{13}ions/cm^2. The effect of irradiation on the composite structure was investigated by scanning electron microscopy, X-ray diffraction (XRD), photoluminescence (PL), and cathodoluminescence spectroscopy. Current-voltage characteristics of ZnO-PS heterojunctions were also measured. As compared to the granular crystallites of zinc oxide layer, Al-doped zinc oxide (ZnO) layer showed a flaky structure. The PL spectrum of the pristine composite structure consists of the emission from the ZnO layer as well as the near-infrared emission from the PS substrate. Due to an increase in the number of deep-level defects, possibly oxygen vacancies after swift ion irradiation, PS-Al-doped ZnO nanocomposites formed with high-porosity PS are shown to demonstrate a broadening in the PL emission band, leading to the white light emission. The broadening effect is found to increase with an increase in the ion fluence and porosity. XRD study revealed the relative resistance of the film against the irradiation, *i.e.*, the irradiation of the structure failed to completely amorphize the structure, suggesting its possible application in optoelectronics and sensing applications under harsh radiation conditions.

Keywords: Porous Silicon, Zinc Aluminum Oxide, Swift Heavy Ions, Photoluminescence, Cathodoluminescence

1. Introduction

Nowadays, efforts are being made to look for suitable types of nanocomposites for optoelectronic applications. Semiconductor nanocrystallites have been considered as the emission source for the next-generation light-emitting diodes due to their electro-optical properties and tunable size[1][2]. Zinc oxide (ZnO) with trivalent elements such as aluminum (Al) is a unique n-type semiconductor and transparent material with a direct bandgap of 3.37eV, along with a large exciton binding energy of 60meV[3]. Al-doped ZnO (AZO) is considered as an important material for its application as a transparent electrode in flat panel displays[4] due to its high conductivity and good transparency. Till now, AZO films with resistivity lower than 1 to $5 \times 10^{-4}\Omega$ cm[5]-[7] and transmittance more than 85% have been attained. On the other hand, porous silicon (PS) has been investigated due to its room-temperature luminescence, and efforts have been focused to obtain an efficient electroluminescent (EL) device[8] based on PS for its possible integration with the present microelectronic industry. Along with various semiconducting and piezoelectric properties suitable for various applications in the optoelectronic industry, the EL efficiency could be increased strongly by filling the pores of PS with AZO. Due to the open structure and large surface area, together with the unique optical properties, PS is a good candidate for a template. It is known that the emission energy of PS increases with a decrease in silicon nano crystallite size covering the entire visible spectrum from red to blue[9]. It has been reported that the blue luminescence band with a relatively fast decay time is observed in the oxidized PS samples. On the other hand, it is easy to get the red emission from the PS, and if it could be added with any other semiconductor with emission in the blue green region, it could be useful to obtain the white light through a simple route for possible applications like the display technology[4][10]. Apart from that, our group recently demonstrated the importance of PS-ZnO composites for sensing application through a control over the spacial distribution of zinc oxide and its transport properties on the porosity of the PS substrate[11].

In the last few years, a considerable amount of progress has been done to enhance the optical properties and other physical characteristics of the ZnO film with different techniques. Among them, energetic ion beams have been employed to modify the electrical, optical, and structural properties of different materials[12]-[15]. Matsunami *et al.*[16] studied the effect of irradiation on the electrical, structural, and optical properties of indium-doped ZnO films and found an enhancement in the electrical conductivity. Sugai *et al.*[17] reported a two-order increase in the conductivity of AZO films after irradiation with Ni and Xe ions. Recently, Singh *et al.*[18] have reported an increase in the ethanol sensing response of irradiated SnO_2 films with ZnO demonstrating a strong resistance to damage caused by ion irradiation. Another work on ZnO-PS nanocomposites[19], where ZnO deposited onto the microporous silicon with sol-gel technique, showed the suppression of X-ray diffraction (XRD) peaks after irradiating it with heavy ions (Au). Hence, the effect of high-energy light ions (such as nickel) could be interesting in studying the stability of the ZnO structure along with its optical properties. In this work, we have investigated the ion irradiation effects on AZO films deposited onto the mesoporous silicon substrate and shown a white light emission from the resulting composites after irradiating with high-energy nickel ions. Swift heavy ion (SHI)-induced morphological and structural changes, in terms of XRD and scanning electron microscopy, have also been studied. The emission after 325-nm excitation from a xenon arc lamp has been compared with the cathodoluminescence (CL) in studying the modifications in the deep-level defects induced by the high-energy radiations. Comparison between the low- and high- porosity mesoporous substrates is also presented. The nanocomposites are found to retain the crystalline structure after irradiation.

2. Methods

PS samples were fabricated by wet electrochemical etching of p^{++}-type Si(100) wafers with a resistivity of 0.01 to 0.05Ω cm and at different current densities of 10 (LP) and 70mA/cm^2 (HP) using a 3:7:1 solution of HF/ethanol/glycerol. The porosity of samples LP (low porosity) and HP (high porosity) was 50% and 70%, respectively. The thickness of both samples was kept to 7μm. After the fabrication, the samples were rinsed with ethanol and dried in pentane.

In order to study the effect of PS on PL and other structural and transport properties, AZO films were deposited by radio frequency magnetron sputtering. A sputtered target with a mixture (2 wt.% Al_2O_3-ZnO) was used, and the PS substrate temperature was kept at 300°C during the deposition of the AZO film with a thickness of 150nm. After deposition, the low- and high- porosity PS-AZO composites were named as ZLP and ZHP, respectively. As-deposited films were later annealed at 700°C for 1h in the tubular furnace in argon atmosphere. The annealed films were irradiated with 130-MeV nickel ions using the 15UD Pelletron Accelerator at the Inter University Accelerator Centre, New Delhi. The samples were mounted on a rectangular-shaped ladder and were irradiated in high vacuum chamber. The focused ion beam was scanned over an area of 1×1cm^2. The films of low porosity (ZLP) were irradiated with fluences of 1×10^{12} and 3×10^{13}ions/cm^2, and the films of high porosity (ZHP) were irradiated with fluences of 3×10^{12} and 1×10^{13}ions/cm^2. The beam current was kept constant at approximately 1.5pnA. The electronic stopping power (energy dissipated in electronic excitations) and nuclear stopping power (energy dissipated in atomic collisions) by such ions in ZnO are around 24.63 and 0.44keV/nm, respectively (calculated using SRIM 2003 simulation code). The modifications in the properties of ZnO films are expected to be mainly due to the electronic excitations, though the contributions of small nuclear stopping power could not be ignored.

The structural properties and the thickness of the PS and nanocomposites were analyzed using a high-resolution field emission scanning electron microscope

(SEM; JSM-7401 F, JEOL Ltd., Akishima-shi, Japan). The orientation and crystallinity of the ZnO crystallites were analyzed using an X-ray diffractometer (X'Pert PRO, PANalytical B.V., Almelo, The Netherlands) using CuKα radiation having a wavelength of 1.54Å. PL properties were studied using a Varian Fluorescence spectrophotometer (Cary Eclipse, Agilent Technologies, Inc., Santa Clara, CA, USA) under the excitation by 325-nm photons using a 500-W xenon lamp. CL spectroscopy was done using JEOL JSM 5300 SEM with an electron beam energy of 15keV. CL measurements were performed at 100K in the UV-visible spectral range using a Hamamatsu R928P photomultiplier (Hamamatsu, Japan). A SPEX 340-E computer-controlled monochromator (Metuchen, NJ, USA) was used for the spectral analysis.

3. Results and Discussion

3.1. X-Ray Diffraction

The crystalline nature of the PS-AZO film was studied through XRD spectra (**Figure 1**) of pristine and irradiated samples at different fluences. The pristine PS-AZO film had a polycrystalline hexagonal wurtzite structure as confirmed by the dominant (002) peak. The ZHP sample, after irradiation at a fluence of 1×10^{13}ions/cm^2, demonstrates a relative decrease (of about 50%) in the peak intensities corresponding to the zinc oxide [*i.e.*, (100), (002), and (101) peaks] and PS (at $2\theta = 55°$), indicating a decrease in the degree of crystallinity of the composite, after irradiation. Apart from this, a slight shift of $2\theta = 0.02°$ in the (002) peak is observed after applying Gaussian fit and is attributed to the release of strain in the crystallites, similar to the effect reported earlier by Rehman *et al.*[18] [shown as inset in **Figure 1(b)**]. Applying Scherrer's formula for the (002) peak, the crystallite size is found to decrease from 23.74 to 20.46nm. The decrease in the crystallite size at a fluence of 1×10^{13}ions/cm^2 could be attributed to the sputtering induced by the SHI irradiation, which is in accordance with the effect observed by some other groups[12][20][21]. The changes in the peak intensities for the samples with the low-porosity PS (10mA/cm^2) substrate show similar but less intense effects after irradiation at a fluence of 3×10^{13}ions/cm^2. A relative dependence of the crystalline quality can be attributed to the substrate porosity dependence of ZnO grain growth[11] and its physical stability. The presence of the low-porosity silicon substrate tends to contribute in the retention of the physical and optical characteristics

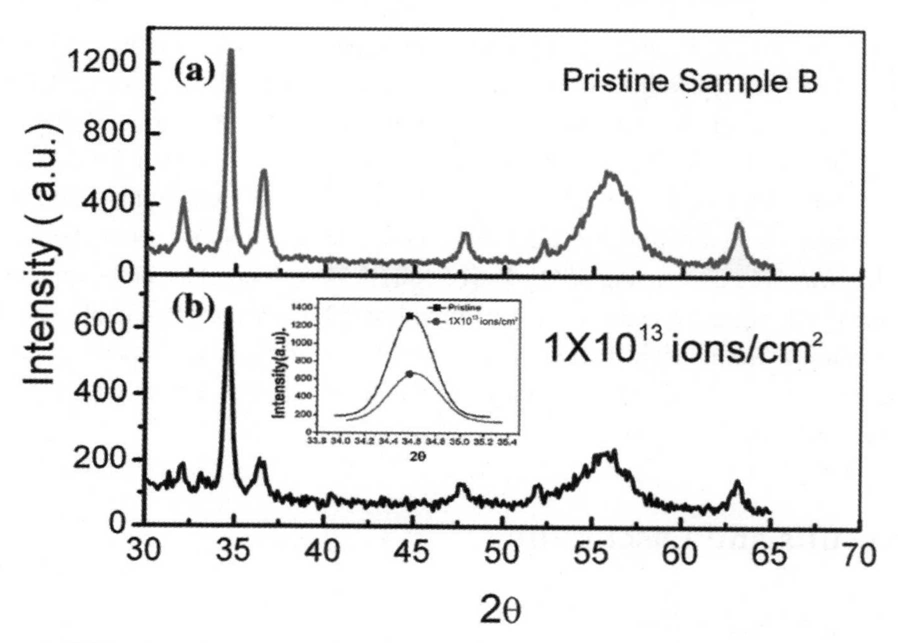

Figure 1. XRD spectra of PS-AZO nanocomposite corresponding to ZHP (PS substrate fabricated at 70mA/cm^2). (a) Pristine and (b) after irradiation at a fluence of 1×10^{13}ions/cm^2. The inset shows the comparison of the (002) peak before and after irradiation.

(shown in the later part of the manuscript). The above-mentioned result can be inferred as an opening for the possible optoelectronic application of PS-AZO films under harsh radiation conditions.

3.2. Measurement of Bandgap by Optical Method

For determining the bandgap of the ZnO film, the absorption coefficient (α) is obtained from transmittance data using the following equation:

$\alpha = -\frac{1}{d}\ln T$, where d is the thickness of the film and T is the optical transmittance.

The bandgap can be estimated using the equation

$$(\alpha h\nu)^2 = A(h\nu - E_g),$$

where h is the Planck constant, ν is the frequency of the incident photon, and A is a constant. Hence, from **Figure 2**, the estimated optical bandgap of the film is 3.35 eV, involving direct electronic transitions[22][23].

3.3. Scanning electron microscopy

In order to check the morphological changes, secondary electron microscope images (**Figure 3**) were analyzed before (pristine) and after irradiation for the ZHP sample. Before irradiation, the films are found to have regular flaky grains spread uniformly over PS. The size of the grains before irradiation appears to be uniform and is in the range of 50 ~ 90nm. However, the irradiated film (at a fluence of 1×10^{13} ions/cm^2) shows apparent changes in the morphology which could be attributed to the inelastic collisions between the electrons of the target material and the high-energy ions. Most of the flaky structure appears as irradiation-induced dark patches [**Figure 3(b)**]. The observed phenomenon after irradiation is similar to the one reported by Rehman *et al.*[20]. The observed changes in the morphology could be co-related to the results of XRD studies.

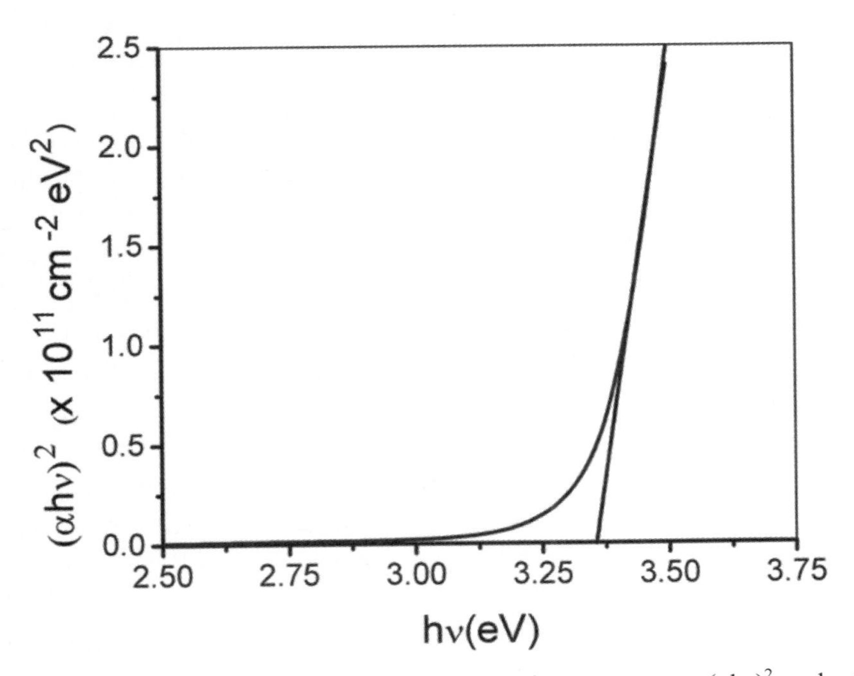

Figure 2. Bandgap calculation of AZO film from UV-vis spectroscopy. $(\alpha h\nu)^2$ vs. hν (photon energy) plot of zinc oxide film deposited over the glass substrate.

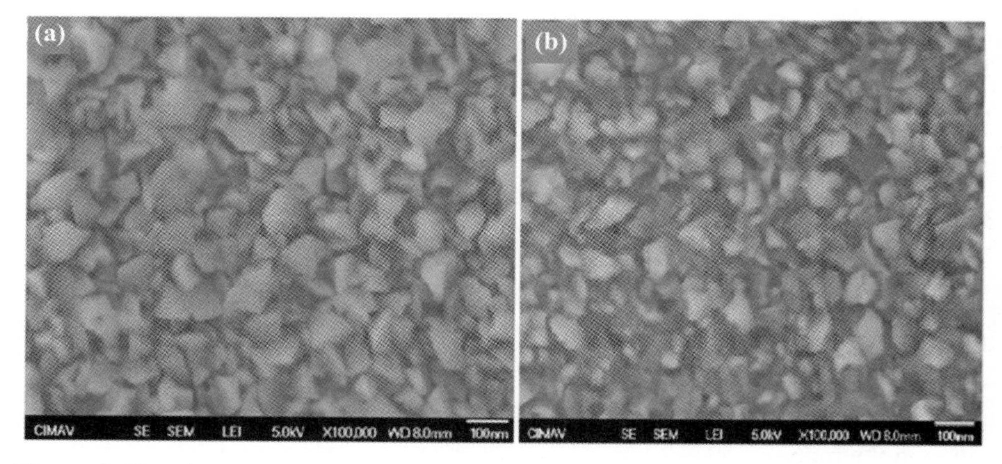

Figure 3. SEM images corresponding to ZHP (PS-AZO). (a) Pristine and (b) after irradiation at a fluence of 1×10^{13} ions/cm^2.

3.4. Photoluminescence Spectroscopy

The photoluminescence (PL) characteristics of the PS-AZO nanocomposite were studied (**Figure 4**) before and after SHI irradiation using nickel ions with a low-energy beam of 130MeV. Figure 4a shows the PL spectrum of the composite formed with the ZLP substrate, with a peak around 3.25eV attributed to the band-edge excitonic transition of ZnO[24][25]. The PL emission peak shows a low-intensity shoulder/tapering end at the left-hand side of the peak which can be attributed to the well-known oxygen vacancies in the zinc oxide layer[26]-[28]. Although the origin of this broad defect emission is controversial, it has been attributed to the singly ionized oxygen vacancies (Vo^{-})[27] and interstitial oxygen (Oi)[27][28]. Particularly, a ZnO emission of 2.5eV has been assigned to the recombination of delocalized electrons close to the conduction band with deeply trapped holes in Vo^{+} centers. **Figure 4(b)** shows PL emission from the PS-AZO nanocomposite with low-porosity substrate (*i.e.*, made with a current density of 10mA/cm^2) after irradiation at a fluence of 1×10^{12}ions/cm^2. No significant change in the defect-related PL intensity around 2.5eV was observed, proving the stability of the structure up to a certain degree of irradiation fluences. On the other hand, an increase in the degree of fluence to 3×10^{13}ions/cm^2 generated a significant increase of the PL intensity corresponding to the defects centered at 2.5eV in **Figure 4(c)**, which is attributed to an increase in the deep-level defects, possibly oxygen vacancies. The formation of defects can be further understood as follows: As the SHI penetrates

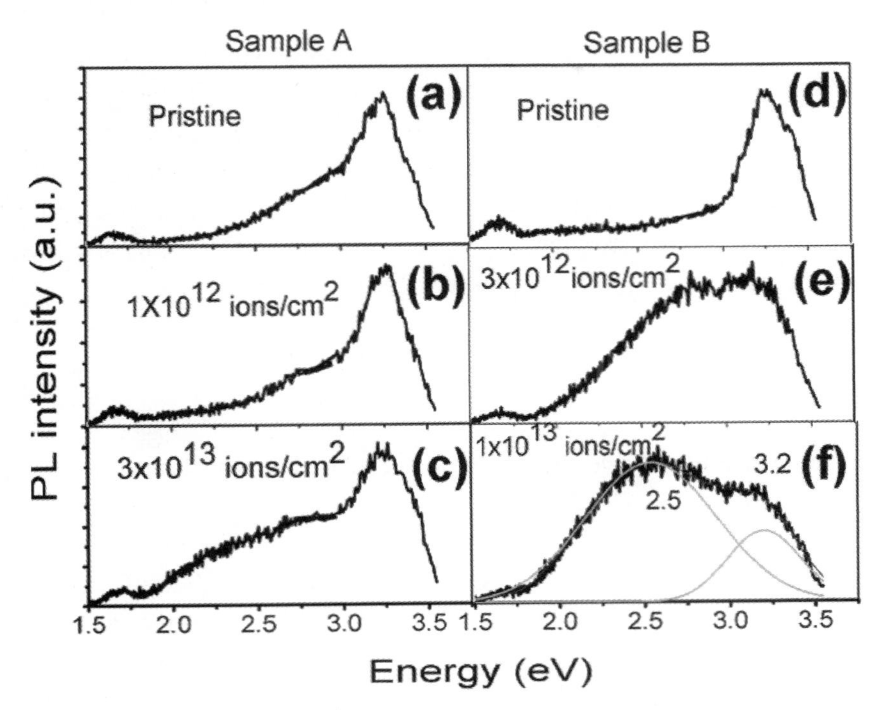

Figure 4. PL spectra of PS-AZO nanocomposites irradiated at different fluences. (a) Pristine, (b) irradiated with 1×10^{12}ions/cm^2, and (c) 3×10^{13}ions/cm^2 correspond to the ZLP sample. (d) Pristine, (e) irradiated with 3×10^{12}ions/cm^2 and (f) 1×10^{13}ions/cm^2 correspond to the ZHP sample.

the solid, inelastic collisions are expected between the ion and the target electrons. In general inelastic collision, it is believed that electronic excitation and ionization of the target atoms plays a dominant role for high-energy heavy ion impact on ZnO. In the case of SHIs, inelastic nuclear collision results in the energy transfer which can yield localized heating/amorphization and a significant increase in the number of defects. On the other hand, due to the fact that in this energy range, electronic stopping is much bigger than nuclear stopping, electronic excitation caused by strong electronic stopping can weaken oxygen bonds, resulting in the formation of oxygen vacancies[29].

Although a similar behavior was observed for the ZHP sample as well [see **Figure 4(e)**, **Figure 4(f)**], an increase of fluence is shown to enhance significantly the defect-related emission centered at 2.5eV. The peaks corresponding to 2.5 to 3.25eV have been observed to merge, resulting in the formation of broad white emission band from 3.25 to 1.50eV. In comparing **Figure 4(e)**, **Figure 4(f)**, a further increase in the level of fluence is found to increase the signal corresponding to the defects (2.5eV) with a simultaneous decrease in the band-edge emission (3.25eV). The intensity of the emission band corresponding to PS (around 1.7eV) is found to decrease slightly with an increase in the level of fluence which is in accordance to the already reported work of Singh *et al.*[19]. The resulting PL spectra are observed to have an almost white emission from the composite structure which can be very useful for the display devices.

3.5. Cathodoluminescence

In order to further investigate the defect-related luminescence mechanisms, CL spectroscopy was performed on all the composite samples. The main advantage of the CL is the spacial resolution, determined by the distribution of the excess carrier in the material and is therefore not limited by the diffraction in the collection and excitation optics, which is very customary in all the far-field techniques. Hence, to enhance our understanding regarding the optical properties of PS-ZnO composites under irradiation conditions, the PL studies have been complimented by CL analysis. The major reason that very few studies reported CL of PS is due to an extremely weak and unstable data. The inset in **Figure 5(c)** shows a strong decrease of the CL emission signal of the PS produced by the electron beam irradiation in SEM. Deposition of AZO layer on the mesoporous silicon layer leads to the stability of the composite structure. As the electron beam can inject into the samples, a strong charge density produces saturation of radiative levels; hence, the luminescence from defects can be explored in detail. **Figure 5(a)** shows CL spectra from the non-irradiated (pristine) sample with a broad emission centered at about 2.6eV, which apparently corresponds to the ZnO defect emission centered at 2.5eV in PL spectra. The low-intensity CL recorded in the spectra is attributed to the low sensibility of our system at room temperature. As compared to the PL spectra from the pristine sample, the ZnO band edge emission was not resolved in CL spectra, possibly due to presence of a high density of defects in the film-substrate interface. **Figure 5(b)** shows the CL spectra from the irradiated ZLP sample, revealing two emissions centered at 3.2 and 2.5eV, associated to the ZnO band edge and defect emission. As commented in the 'Photoluminescence spectroscopy' section, the 2.5eV band is associated to the presence of oxygen vacancies and other native defects[30]-[32]. The CL spectrum of the ZHP sample shows a similar behavior as that of the ZLP sample [**Figure 5(c)**, **Figure 5(d)**], with the

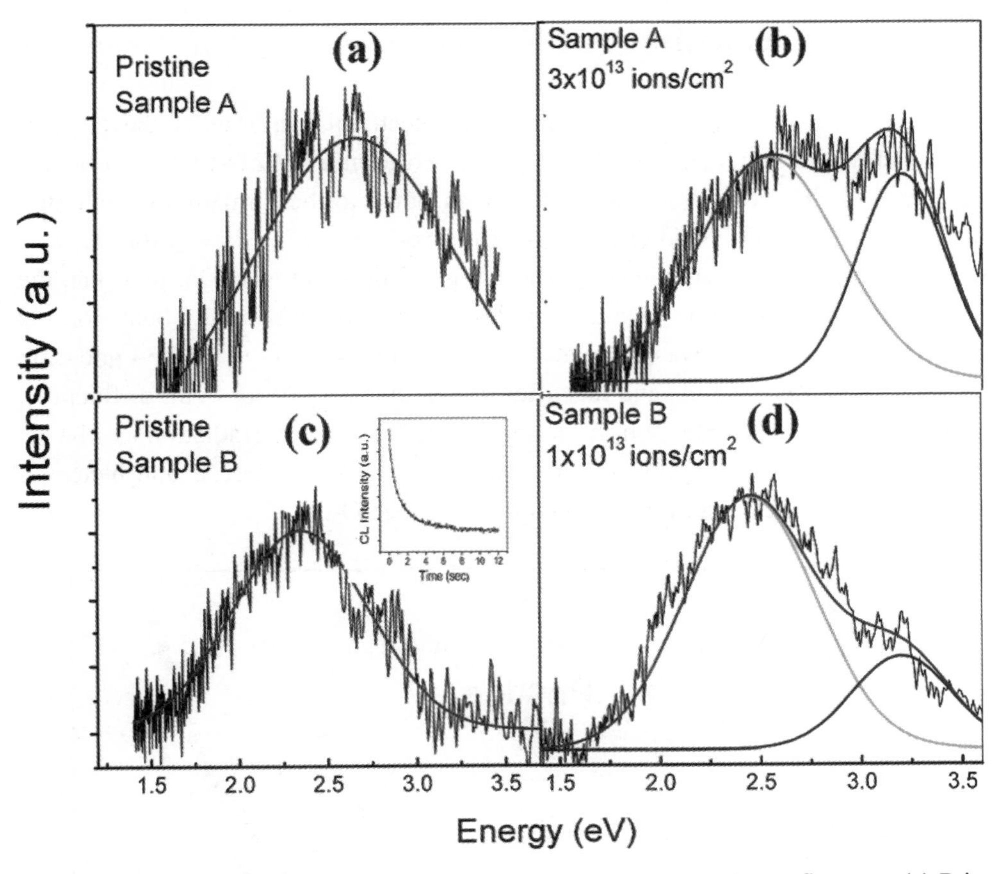

Figure 5. CL spectra of PS-AZO nanocomposites irradiated at different fluences. (a) Pristine ZLP sample. (b) ZLP sample irradiated with 3×10^{13}ions/cm^2. (c) Pristine ZHP sample. (d) ZHP sample irradiated with 1×10^{13}ions/cm^2.

generation of the ZnO band edge (3.2eV) after the SHI irradiation. This effect is not correlated with PL measurements, where the relative intensity of UV emission decreased after irradiation **[Figure 4(c), Figure 4(f)]**. In CL measurements, an increase of the relative intensity of the ZnO band edge emission typically corresponds to an improvement of the crystalline quality, mainly due to annealing effects[30][33]. Possibly, the SHI irradiation generated a high density of point defects mostly at the film's surface, which was recorded in PL spectra with a relative increase of the 2.5eV emission **[Figure 4(c), Figure 4(f)]**, while at the film-substrate interface, the irradiation-generated heating produces annealing of defects (enhancement of crystalline behavior) as shown in the corresponding CL spectra **[Figure 5(b), Figure 5(d)]**.

3.6. Electrical Properties

Figure 6 demonstrates the non-ohmic electrical response of the current-voltage (I-V) measurements performed in a two-terminal AZO-PS-Si configuration (inset of **Figure 6**) which is normally attributed to the formation of Schottky barriers at the ZnO-PS interfaces. It was seen that after irradiation at 1×10^{13}ions/cm^2, the ZHP sample shows little higher forward/reverse current than the pristine sample. In order to quantify the deviation from the Schottky behavior, the rectifying factor (IF/IR) was calculated at 4 V and was found to be 3.34 and 2.62 for the pristine ZHP sample and that after irradiation (1×10^{13}ions/cm^2 of fluence), respectively. A slight reduction in the rectifying factor after irradiation can be attributed to the irradiation-induced thermal annealing of the defects and hence an improvement of the crystallinity at the interface of ZnO-PS.

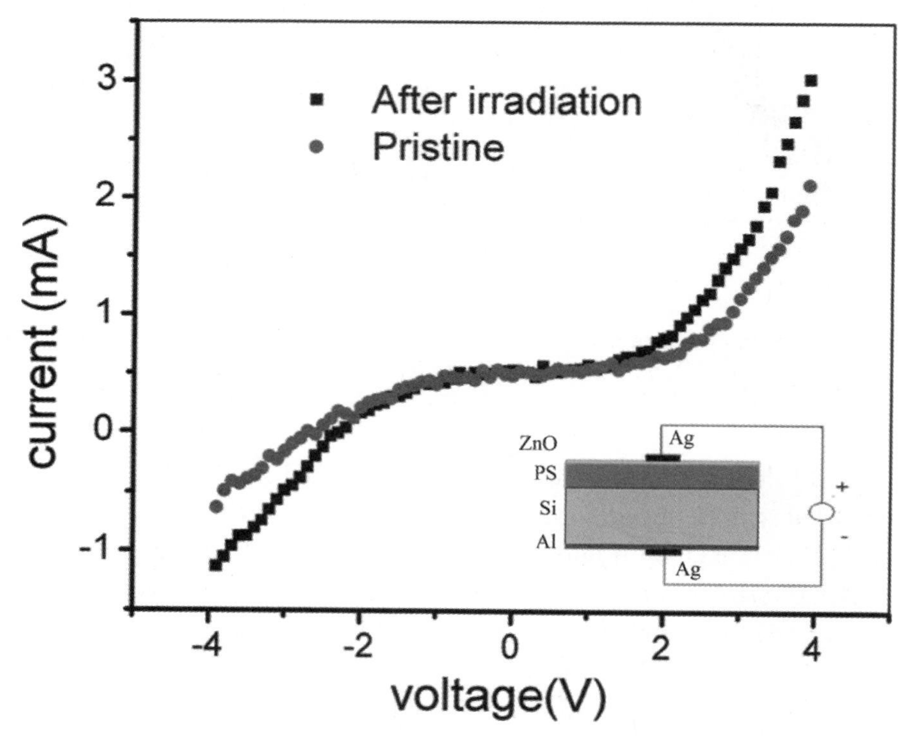

Figure 6. I-V curves of ZnO-PS structure (ZHP sample) in a sandwich configuration, measured at room temperature. The voltage varies from −4 to 4V with a sweep rate of 150mV/s. The inset in the lower part shows the schematic of the configuration used for I-V characteristics. The inset of **Figure 5(c)** shows CL emission spectra as a function of time showing the degradation of the asetched PS layer by high-energy electrons.

4. Conclusions

We report the substrate porosity and fluence-dependent white light emission from RF-sputtered zinc aluminum oxide deposited on PS after SHI irradiation. The structures are shown to partially retain their crystallinity, when irradiated with light Ni ions. Composites are found to have a rectifying behavior which reduces by a factor of 0.78 after irradiation. Its stability under harsh irradiation conditions makes it useful for space applications. Apart from that, the tunability in the optical properties is important for optoelectronic applications.

Competing Interests

The authors declare that they have no competing interests.

Acknowledgements

The authors are thankful to CONACyT, Mexico, and DST, Government of India, for providing the bilateral exchange project (Mexico-India J000.0374 and DST/INT/MEX/RPO-09/2008). VA also acknowledges the partial support given by CONACyT project number 128593. MHZ acknowledges the support of the project CONACyT 102519. YK acknowledges the CONACyT support (179496) for doctoral scholarship and the fruitful discussion with Dr. Karl-Heinz Heinig. We acknowledge the technical support provided by MC Enrique Torres for doing the XRD and Wilber Antunez for SEM at CIMAV-Chihuahua.

Authors' Contributions

YK made the porous silicon samples and performed their characterization before and after irradiation. MHZ participated in performing the cathodoluminescence studies and their corresponding analysis. SFOM participated in acquiring the SEM and XRD facilities. FS performed the SHI irradiation at IUAC, New Delhi. XM participated in the deposition of AZO film on PS and glass substrates. VA conceived the study, participated in its design and coordination, and codrafted the manuscript with YK. All authors read and approved the final manuscript.

Authors' Information

YK is a Ph.D. student registered at CIMAV, Chihuahua, and is doing his research work at CIICAp, UAEM, Mexico. MHZ is an associate professor at CNyN, UNAM and is working on the characterization of semiconductor nanostructures. SFOM is working as a researcher at CIMAV, Chihuahua, in the field of nanostructured materials. FS is a scientist at IUAC, New Delhi. XM is a senior scientist at CIE, UNAM working on thin film semiconductors for photovoltaic applications.VA is working as a professor investigator at CIICAp, UAEM in the field of nanostructured silicon.

Source: Kumar Y, Herrera-Zaldivar M, Olive-Méndez S F, *et al.* Modification of optical and electrical properties of zinc oxide-coated porous silicon nanostructures induced by swift heavy ion [J]. Nanoscale Research Letters, 2012, 7(1):1–7.

References

[1] Pan ZW, Dai ZR, Wang ZL: Nanobelts of semiconducting oxides. Science 2001, 291:1947–1949.

[2] Lin CH, Chiou BS, Chang CH, Lin JD: Preparation and cathodoluminescence of ZnO phosphor. Mat Chem Phys 2003, 77:647–654.

[3] Kashyout A, Soliman M, El Gamal K, Fathy M: Preparation and characterization of nano particles ZnO films for dye-sensitized solar cells. Mater Chem Phys 2005, 90:230–233.

[4] Schuber EF, Kim JK: Solid-state light sources getting smart. Science 2001, 308: 1274–1278.

[5] Nanto H, Minami T, Shooji S, Takata S: Electrical and optical properties of zinc oxide thin films prepared by rf magnetron sputtering for transparent electrode applications. J Appl Phys 1984, 55:1029–1034.

[6] Tang W, Cameron DC: Aluminum-doped zinc oxide transparent conductors deposited by the sol-gel process. Thin Solid Films 1994, 238:83–87.

[7] Kim KH, Park KC, Ma DY: Structural, electrical and optical properties of aluminum doped zinc oxide films prepared by radio frequency magnetron sputtering. J Appl

Phys 1997, 81:7764–7772.

[8] Bsiesy A, Cox TI: Electroluminescence from porous silicon using liquid contacts. In Properties of Porous Silicon. Edited by Canham L. London: INSPEC Publication; 1997:283–289.

[9] Canham LT: Visible photoluminescence from porous silicon. In Properties of Porous Silicon. Edited by Canham L. London: INSPEC Publication; 1997:249–255.

[10] Wang C-F, Ho B, Yi H-H, Li W-B: Structure and photoluminescence properties of ZnS films grown on porous Si substrates. Opt Laser Technol 2011, 43:1453–1457.

[11] Yogesh Kumar, Escorcia Garcia J, Fouran Singh, Olive-Méndez SF, Sivakumar VV, Kanjilal D, Agarwal V: Influence of mesoporous substrate morphology on the structural, optical and electrical properties of RF sputtered ZnO layer deposited over porous silicon nanostructure. Appl Surf Sci 2010, 258:2283–2288.

[12] Zang H, Wang ZG, Peng XP, Song Y, Liu CB, Wei KF, Zhang CH, Yao CF, Ma YZ, Zhou LH, Sheng YB, Gou J: Modification of ZnO films under high energy Xe-ion irradiations. Nucl Instrum Methods Phys Res, Sect B 2008, 266:2863–2867.

[13] Fouran Singh, Kulriya PK, Pivin JC: Origin of swift heavy ion induced stress in textured ZnO thin films: an in situ X-ray diffraction study. Solid State Commun 2010, 150:1751-1754.

[14] Krasheninnikov AV, Banhart F: Engineering of nanostructured carbon materials with electron or ion beams. Nat Mater 2007, 6:723–733.

[15] Krasheninnikov AV, Nordlund K: Ion and electron irradiation-induced effects in nanostructured materials. J Appl Phys 2010, 107:071301–071370.

[16] Matsunami N, Fukushima J, Sakta M, Okayasu S, Sugai H, Kakiuchida H: Electrical property modifications of In-doped ZnO films by ion irradiation. Nucl Instrum Methods Phys Res, Sect B 2010, 268:3071–3075.

[17] Sugai H, Matsunami N, Fukuoka O, Sataka M, Kato T, Okayasu S, Shimura T, Tazawa M: Electrical conductivity increase of Al-doped ZnO films induced by high-energy-heavy ions. Nucl Instrum Methods Phys Res, Sect B 2006, 250:291– 294.

[18] Singh RC, Singh MP, Singh O, Chandi PS, Kumar R: Effect of 100 MeV O7+ ions irradiation on ethanol sensing response of nanostructures of ZnO and SnO2. Appl Phys A 2010, 98:161–166.

[19] Singh RG, Fouran Singh I, Sulania D, Kanjilal K, Sehrawat V, Agarwal RM: Mehra: electronic excitations induced modifications of structural and optical properties of ZnO-porous silicon nanocomposites. Nucl Instrum Meth Phys Res, Sect B 2009, 267:2399–2402.

[20] Rehman S, Singh RG, Pivin JC, Bari W, Singh F: Structural and spectroscopic modifications of nanocrystalline zinc oxide films induced by swift heavy ions. Vaccum 2011, 86:87–90.

[21] Agarwal DC, Avasthi DK, Singh F, Kabiraj D, Kulariya PK, Sulania I, Pivin JC, Chauhan RS: Swift heavy ion induced structural modification of atom beam sputtered ZnO thin film. Surf CoatTechnol 2009, 203:2427–2431.

[22] Pankove JI: Optical Processes in Semiconductors. New York: Dover; 1971.

[23] Smith RA: Semiconductors. Cambridge: Cambridge University Press; 1978.

[24] Pal U, Melendrez R, Chernov V, Flores MB: Thermoluminescence properties of ZnO and ZnO:Yb nanophosphors. Appl Phys Lett 2006, 89:183118-183120.

[25] Key A, Gratzel M: Low cost photovoltaic modules based on dye sensitized nanocrystalline titanium dioxide and carbon powder. Sol Ener Mater Sol Cells 1996, 44:99–117.

[26] Vanheusden K, Seager CH, Warren WL, Tallant DR, Voigt JA: Correlation between photoluminescence and oxygen vacancies in ZnO phosphors. Appl Phys Lett 1996, 68:403–405.

[27] Lin B, Fu Z, Jia Y: Green luminescent center in undoped zinc oxide films deposited on silicon substrates. Appl Phys Lett 2001, 79:943–945.

[28] Xu PS, Sun YM, Shi CS, Xu FQ, Pan HB: The electronic structure and spectral properties of ZnO and its defects. Nucl Instrum Methods Phys Res, Sect B 2003, 199:286–290.

[29] Hemon S, Berthelot A, Durfour C, Domenges B, Paumier E: Irradiation of a tin oxide nanometric powder with swift heavy ions. Nucl Instrum Methods Phys Res, Sect B 2000, 166-167:927–932.

[30] González A, Herrera M, Valenzuela J, Escobedo A, Pal U: CL study of yellow emission in ZnO nanorods annealed in Ar and O2 atmospheres. Superlatt and Microst 2009, 45:421–428.

[31] Dong YF, Tuomisto F, Svensson BG, Kuznetsov AY, Brillson L: Vacancy defect and defect cluster energetics in ion-implanted ZnO. Phys Rev B 2010, 81:81201–81204.

[32] Mosbacker HL, Zgrabik C, Hetzer MJ, Swain A, Look DC, Cantwell G, Zhang J, Song JJ, Brillson LJ: Thermally driven defect formation and blocking layers at metal-ZnO interfaces. Appl Phys Lett 2007, 91:072102–072103.

[33] González A, Herrera M, Valenzuela J, Escobedo A, Pal U: Cathodoluminescence evaluation of defect structure in hydrothermally grown ZnO:Sb nanorods. J Nanoscience and Nanotech 2011, 11:5526–5531.

Chapter 13

Morphological, Compositional, Structural, and Optical Properties of Si-nc Embedded in SiO_x Films

J. Alberto Luna López[1], J. Carrillo López[1], D. E. Vázquez Valerdi[1], G. García Salgado[1], T. Díaz-Becerril[1], A. Ponce Pedraza[2], F. J. Flores Gracia[1]

[1]IC-CIDS Benemérita Universidad Autónoma de Puebla, Ed. 103 C o D, Col. San Manuel, C.P, Puebla, Pue 72570, Mexico
[2]Department of Physics & Astronomy, University of Texas at San Antonio, San Antonio, TX 78249, USA

Abstract: Structural, compositional, morphological, and optical properties of silicon nanocrystal (Si-nc) embedded in a matrix of non-stoichiometric silicon oxide (SiO_x) films were studied. SiO_x films were prepared by hot filament chemical vapor deposition technique in the 900°C to 1,400°C range. Different microscopic and spectroscopic characterization techniques were used. The film composition changes with the growth temperature as Fourier transform infrared spectroscopy, energy dispersive X-ray spectroscopy, and X-ray photoelectron spectroscopy reveal. High-resolution transmission electron microscopy supports the existence of Si-ncs with a diameter from 1 to 6.5nm in the matrix of SiO_x films. The films emit in a wide photoluminescent spectrum, and the maximum peak emission shows a blueshift as the growth temperature decreases. On the other hand, transmittance spectra showed a wavelength shift of the absorption border, indicating an increase in the energy optical bandgap, when the growth temperature decreases. A relationship between composition, Si-nc size, energy bandgap, PL, and surface morphology was obtained. According to these results, we have analyzed the dependence of PL on the composition, structure, and morphology of the Si-ncs embedded in a matrix of non-stoichiometric SiO_x films.

Keywords: Silicon Nanocrystals, High-Resolution TEM, XRD, PL, AFM, HFCVD

1. Introduction

Since the discovery of light emission from porous silicon[1], an intense investigation of materials compatible with silicon technology with excellent optical properties has been under development. Recently, materials containing silicon nanocrystal (Si-nc) have attracted the interest of researchers due to their optical properties. Therefore, a great variety of materials with these characteristics have been studied[2]-[5]. One of these materials is the non-stoichiometric silicon oxide (SiO_x); this material contains Si excess agglomerates to create Si nanoparticles embedded in an oxide matrix. SiO_x shows some special compositional, structural, morphological, and optical properties that vary with the Si excess. In particular, the optical characteristics of SiO_x films can be varied with the growth temperatures. For example, refractive index varies from 1.6 to 2.4 when the growth temperature is changed; also, SiO_x emits visible light. These characteristics have given place to various types of applications such as waveguides, no volatile memories, light radiation, and detection devices[6]-[8]. Furthermore, the fabrication of SiO_x films is completely compatible with complementary metal oxide semiconductor technology, providing an easy way for the optoelectronic integration on silicon. Several techniques have been employed to make thin SiO_x films, such as high-dose Si ion implantation into SiO_2 films[6][9], low pressure chemical vapor deposition[8], sol-gel, hot filament chemical vapor deposition (HFCVD)[10][11], and plasma enhanced chemical vapor deposition[12], in which an improvement of the optical and structural properties as the PL emission and Si excess has been reported. In this work, HFCVD technique was used to deposit SiO_x films on silicon and quartz substrates. This technique allows us to obtain thin SiO_x films with different properties just by varying the source-substrate distance during the deposit; this distance changes the growth temperature. Structural, compositional, morphological, and optical properties of SiO_x films prepared by HFCVD and an analysis of the composition, morphology, Si-nc size, and their relation with the PL emission are presented.

2. Methods

Si-ncs embedded in thin SiO_x films were deposited on quartz and n-type silicon (100) substrates, the silicon substrates with 2,000- to 5,000-Ω cm resistivity in a horizontal hot filament CVD reactor using quartz and porous silicon as the sources. A hot filament at approximately 2,000°C dissociates ultra-high purity molecular hydrogen which flows into the reactor at a 50 sccm rate and produces atomic hydrogen (H). Something worth mentioning in this process is the use of a hydrogen flux on top of the hot filament, which resulted in a remarkable improvement of the optical and structural properties of the SiO_x films deposited. The substrates were carefully cleaned with a metal oxide semiconductor standard cleaning process, and the native oxide was removed with an HF buffer solution before being introduced into the reactor. The heating rate depends on the source-substrate distance (dss). The reactive species (H) forms a volatile precursor (SiO_x) deposited on the silicon substrate and produces Si-ncs embedded in thin SiO_x films. The filament-source distance was kept constant (2mm). The relationship between the filament temperature (approximately 2,000°C) and the variation of the dss of 2, 3, 4, 5, and 6mm provides a change in the growth temperature (Tg) of 1,400°C, 1,300°C, 1,150°C, 1,050°C, and 900°C, which was measured with a thermocouple in each position, respectively. These changes in the dss and Tg, consequently, have modified the silicon excess and defects in the non-stoichiometric SiO_x films. The film refractive index and the film thickness were measured using a null Ellipsometer Gaertner L117 (Gaertner Scientific Co., Chicago, IL, USA) with a laser of He-Ne (632.8nm); the film thickness was measured using a Dek-tak 150 profilometer (Veeco Instruments Inc., Plainview, NY, USA). FTIR spectroscopy measurements were done using a Bruker system model vector 22 (Bruker Instruments, Bellirica, MA, USA). XPS analysis was carried out using a Thermo Fisher spectrometer (Thermo Fisher Scientific, Waltham, MA, USA) with a monochromatic Al radiation XR15 and energy of 15eV. PL response was measured at room temperature using a Horiba Jobin Yvon spectrometer model FluroMax 3 (Edison, NJ, USA) with a pulsed xenon source whose detector has a multiplier tube, which is controlled by computer. The samples were excited using a 250-nm radiation, and the PL response was recorded between 400 and 900nm with a resolution of 1nm. Room-temperature transmittance of the SiO_x films was measured using a UV-vis-

NIR Cary 5000 system (Agi-lent Technologies Inc., Santa Clara, CA, USA). The transmittance signal was collected from 190 to 1,000nm with a resolution of 0.5nm. HRTEM measurements and XEDS were done using a Titan 80-kV to 300-kV model with an energy spread of 0.8eV. HRTEM micrographs were analyzed using Gatan DigitalMicrograph software (Gatan Inc. Pleasanton, CA, USA)[13]. The surface morphology of non-stoichiometric SiO_x films was studied using a scanning probe microscopy of Ambios Technology (Santa Cruz, CA, USA), operated in non-contact mode. A 4×4-μm^2 scanned area was used for each topographic image, and a 460-μm-long single-crystal Si n-type cantilever operated at 12kHz (type MikroMash SPM Probes (San Jose, CA, USA)) was used. Four different scans were done for each sample, showing good reproducibility. AFM images were analyzed using scanning probe image processor software[14].

3. Results and Discussion

The refractive index and thickness of the SiO_x films as a function of the growth temperature are shown in **Figure 1**; the refractive index of the thin SiO_x films changes with Tg. Thicker samples were obtained when the growth temperature was increased from 900˚C to 1,400˚C. A variation in the refractive index from 1.4 to 2.2 was measured when the growth temperature was increased from 1,150˚C to 1,400˚C. This variation has been related to a change of the silicon excess in SiO_x films[12]. Therefore, we can modify the silicon excess in the SiO_x films by changing the growth temperature.

FTIR absorption spectra of thin SiO_x films are shown in **Figure 2**. These spectra show the absorption peaks associated with the rocking (458cm^{-1}) (peak 1), bending (812cm^{-1}) (peak 2), on-phase stretching (1,084cm^{-1}) (peak 4), and out-of-phase stretching (1,150cm^{-1}) (peak 5) vibration modes of the Si-O-Si bonds in SiO_2[7][15][16]. The position of the stretching absorption peak in SiO_x films changes with the growth temperature. Tg produces changes only with regard to the microstructure of the films, where the radiative defects can be activated during the process. The on-phase stretching peak position slightly moves towards a higher wavenumber with a higher growth temperature. On the other hand, the out-of-phase stretching peak position shows similar changes with the growth temperature. The position of peak (4) depends on silicon excess. This is an evidence that the

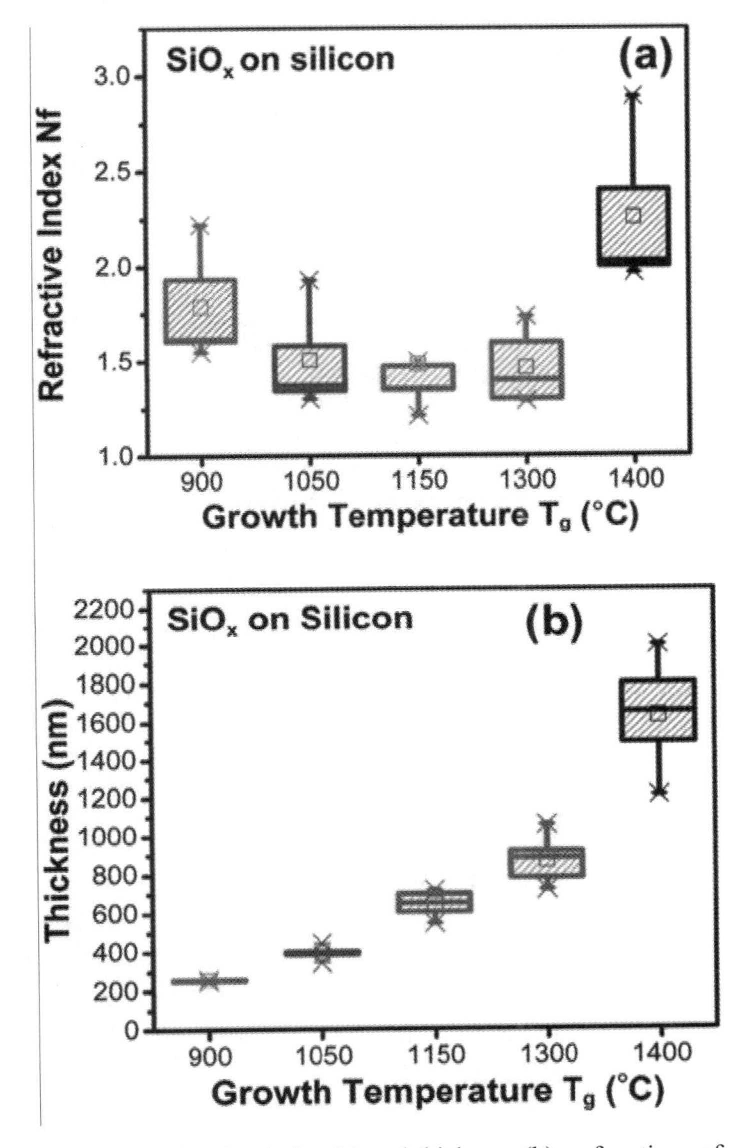

Figure 1. Refractive index (a) and thickness (b) as functions of growth temperature of the SiO_x films.

SiO_x films have a great content of sub-stoichiometric SiO_x ($x < 2$) phase in the as-deposited state. Peaks in the spectra at 883cm^{-1} (peak 3), corresponding to Si-H bending and Si-OH, and the other one located at 2,257cm^{-1} (peak 6), corresponding to Si-H stretching, are observed[17]-[19]; these bonds are present in the films due to hydrogen incorporation during the growth process. Also, a peak centered at

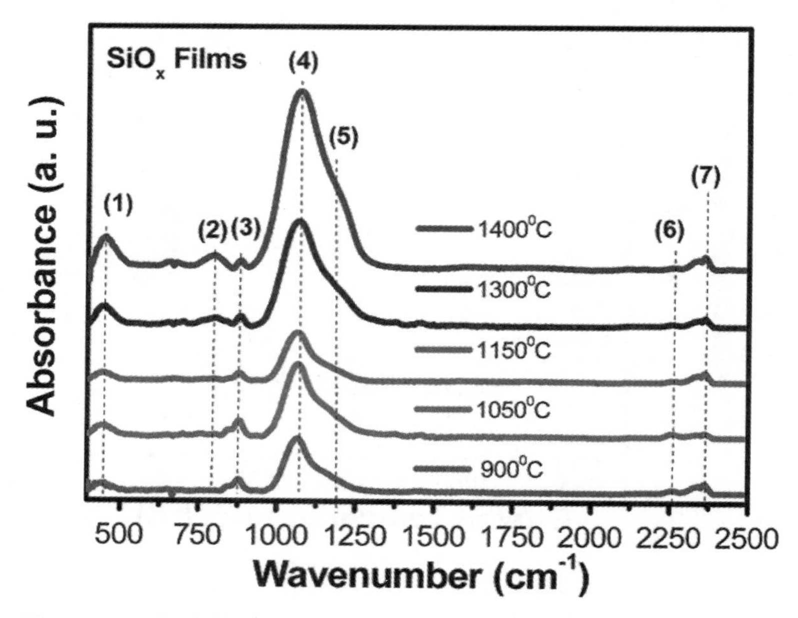

Figure 2. FTIR absorption spectra of the thin SiO_x films for different growth temperature. The numbers (1 to 7) mean different vibration modes, which are described in **Table 1**.

2,352cm^{-1} (peak 7) comes from the CO_2 content in the atmosphere[20]. Furthermore, the peak intensity changes with the growth temperature as shown in **Figure 2**. A relation between peak intensity and thickness is established; to bigger peak intensity, a bigger thickness. In **Figure 1**, we can see that if there is a high growth temperature then the thickness increases. Therefore, the peak intensity is bigger too. Moreover, the oxygen and hydrogen contents change with the growth temperature. The hydrogen and oxygen contents decrease when the growth temperature increases, as can be seen with the behavior of the peaks 3 and 7 and peaks 2, 4, and 5, respectively, as shown in **Figure 2**.

XEDS spectra of the SiO_x films were realized for several growth temperatures; the stoichiometric ratio is determined by the atomic percentages of silicon and oxygen. The peak intensities of oxygen and silicon change with the growth temperature. The peak intensities of silicon are higher when decreasing the growth temperature, and the peak intensities of oxygen decrease when increasing the growth temperature. These variations indicated that the stoichiometry of the SiO_x films changes with the growth temperature. We can see that the oxygen content decreases with the increase of the growth temperature, and the silicon content decreases with the increase of the growth temperature. Then, with higher growth temperature, the silicon content increases and the oxygen content decreases. In **Table 1**, the composition results of the XEDS spectra are listed. **Figure 3** shows the XPS experimental spectra of the Si 2p line and the evolution of the Si 2p line of different SiO_x films. The four oxidation states, as well as the unoxidized state, can be modeled as tetrahedral bonding units, in which a central Si atom is bonded

Table 1. Compositional results (atomic percentages of oxygen (O) and silicon (Si)).

Temperature (°C)	dss (mm)	Atomic percentage		x = O/Si
		O	Si	
1,400	2	30.76	69.22	0.44
1,300	3	61.60	38.38	1.60
1,150	4	56.02	43.96	1.27
1,050	5	64.48	35.50	1.81
900	6	62.49	37.50	1.66

Results of the SiO_x films obtained by XEDS.

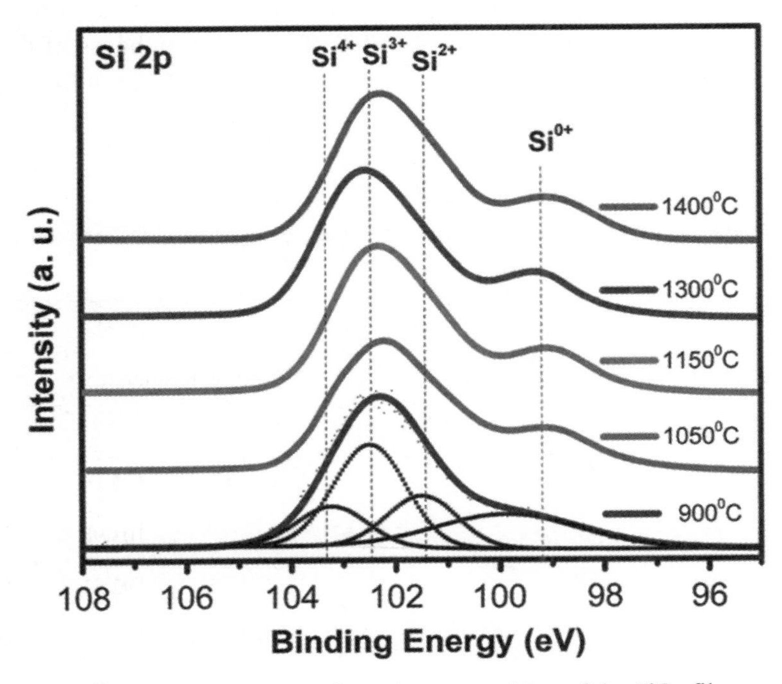

Figure 3. Si 2p XPS spectra show the composition of the SiO_x films.

to (4 − n) Si atoms and n oxygen atoms (Si-Si4 − nOn) with $n = 0$ to 4. Therefore, the 99.5eV peak is associated with elemental silicon. SiO_2 spectra increase the peak energy to 103.3eV, corresponding to n = 4. The Si 2p binding energies are normally about 99 to 103eV. It is widely accepted that the Si 2p photoelectron peak of SiO_x contains five components, corresponding to a non-oxidized state and four different oxidation states of Si[21][22]. The variation of the oxidation states of the SiO_x films leads to peak position's shift, as shown in **Table 2**. A peak at about 99eV accompanied by a peak at about 103eV is present; they can be attributed to Si and SiO_2, respectively, and any variation could be attributed to sub- oxidized silicon[23][24]. The increasing electronegativity of the Si-O bound relative to the Si-Si bond results in a shift to a higher binding energy of the core level electrons in the silicon.

AFM images of the SiO_x films in **Figure 4** are presented. All images exhibit a rough surface. It can be seen that the surface exhibits different characteristics depending on the growth temperature, which influences the size of the grains (roughness), their form, and composition. Average roughness decreases by decreasing the Tg and thickness[8]. The roughness analysis is shown in **Figure 5**. It is observed that the surface roughness of SiO_x films with lower Tg is less than that with higher Tg, except for Tg = 1,150°C. The high roughness of the sample grown at 1,400°C is probably the cause of the index variation with a not clear tendency, and the roughness could be due to the big nanocrystals embedded in the SiO_x films.

Table 2. Oxidation states of the SiO_x films obtained by means of the convolution of the XPS curves.

Oxidation states					
Temperature (°C)	dss (mm)	Peak position (eV)			
		Si^{0+}	Si^{2+}	Si^{3+}	Si^{4+}
1,400	2	99.08	101.06	102.16	102.95
1,300	3	99.67		101.98	103.01
1,150	4	99.08	101.47	102.51	103.24
1,050	5	99.02	101.32	102.28	103.16
900	6	99.90	101.36	102.01	103.34

Figure 4. 3D AFM images of SiO_x films deposited on silicon substrate at different T_g. Scanned area is $4 \times 4\ \mu m^2$.

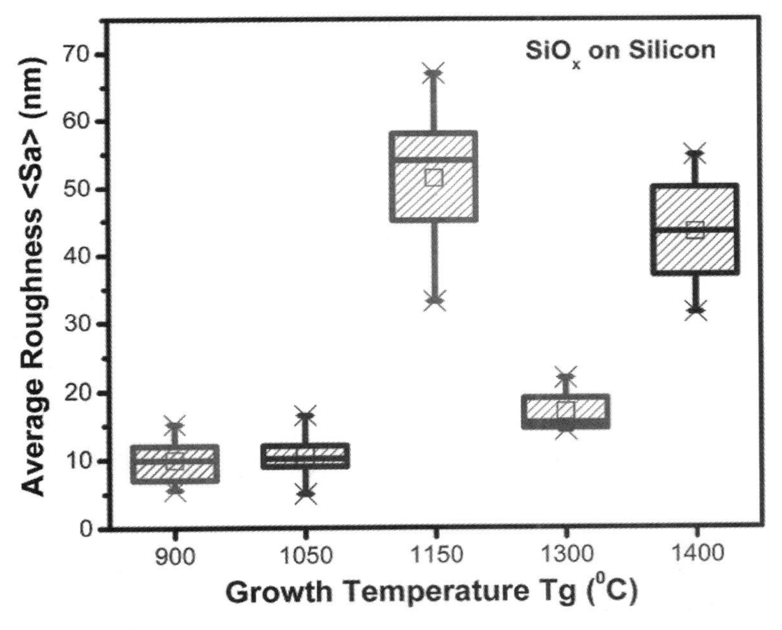

Figure 5. Average roughness (Sa) as a function of Tg for SiO_x films. Scanned area is $4 \times 4\ \mu m^2$.

On the other hand, the structure of SiO_x films was analyzed using the HRTEM technique. **Figure 6** shows the HRTEM images for the SiO_x films, which indicate the presence of Si-ncs embedded in the SiO_x films. Some of them are semi-elliptical, and some other ones have an enlarged shape. The agglomeration process takes place between the nearest Si-nanoclusters forming Si-nc. All micrographs

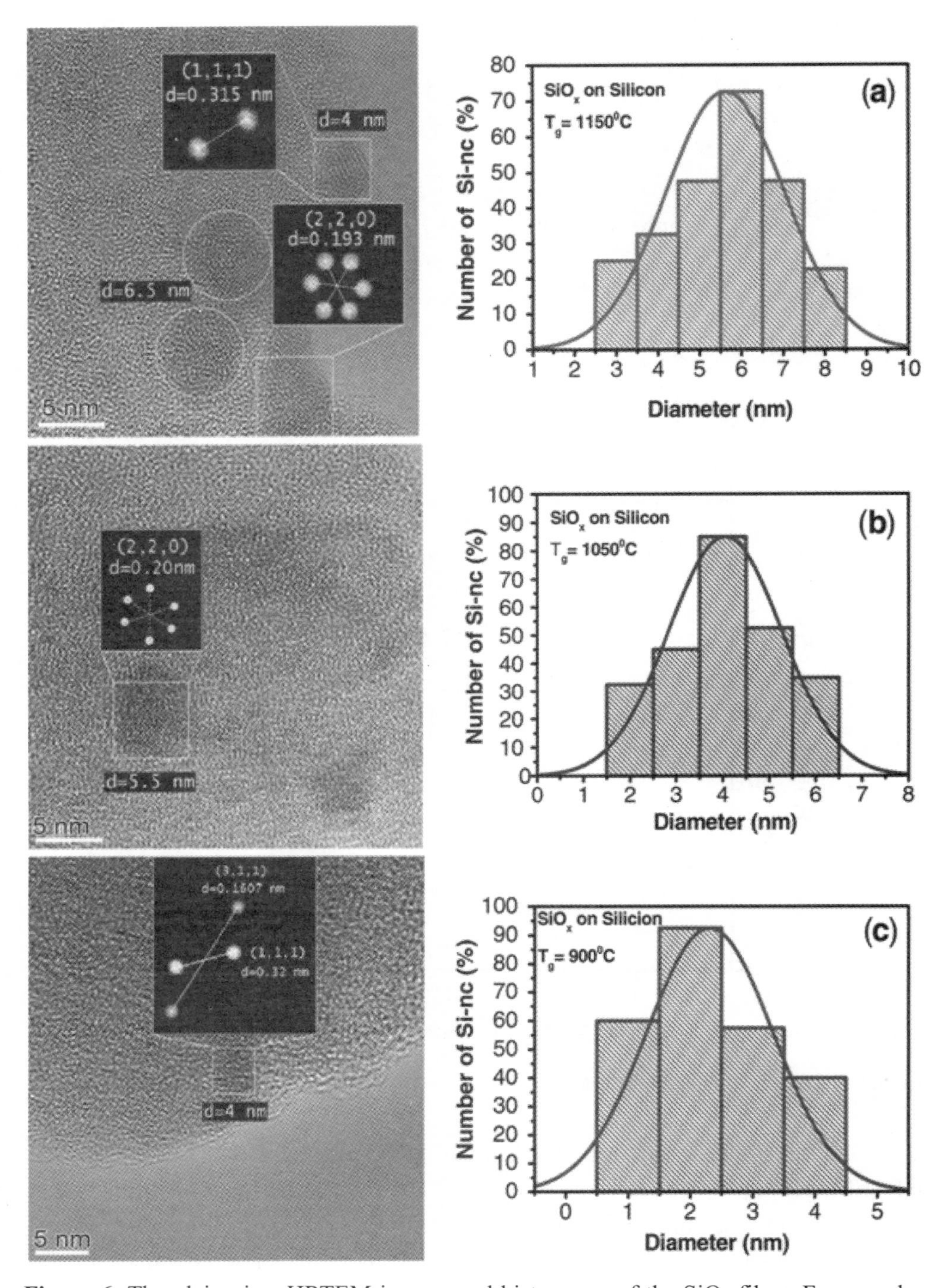

Figure 6. The plain-view HRTEM images and histograms of the SiO_x films. For samples with T_g at (a) 1,150°C, (b) 1,050°C, and (c) 900°C.

show that the SiO_x matrix contains small clusters, which on the basis of the selected area electron diffraction (SAED) analysis can be identified as Si-nc. SAED is referred to as 'selected' because, in the micrograph, it can easily choose which part of the sample we obtain the diffraction pattern; in our case, only on the Si-nc. About ten micrographs were obtained to each sample; with them, a statistical analysis of the distribution of the Si-nc diameter sizes was realized. A great dispersion of diameter sizes is observed; the diameter size goes from 1 up to 9nm, being the average diameter size around 5.5, 4, and 2.5nm for 1,150°C, 1,050°C, and 900°C, respectively, as indicated in the histograms of **Figures 6(a)-6(c)**.

From AFM images, the samples grown with lower Tg look more homogeneous than those grown with higher Tg. As shown in the HRTEM images, the silicon excess agglomerates to create Si-ncs; then, the roughness observed in AFM measurements can be associated with Si-ncs and compounds. In addition, FTIR spectra show a phase separation (Si and SiO_2), which is deduced from the shift of the Si-O stretching vibration mode towards the SiO_2 frequency value, and it is corroborated with both XPS and HRTEM. Therefore, elemental Si, SiO_x, and SiO_2 phases with Tg are present, and depending on Si excess, the roughness, size of Si-nc, oxidation states, and vibration modes of the Si-O-Si bonds, some of these phases could be dominant. This indicates that a direct correlation between the roughness, size of Si-nc, oxidation states, and vibration modes of the Si-O-Si bonds exists. In other words, the roughness is produced by the formation of Si-ncs and oxidation states. The diffusion of excess silicon at high Tg produces Si-ncs in the SiO_x films, *i.e.*, the silicon particles diffuse to create silicon agglomerates around the nucleation sites when the SiO_x is grown at high Tg.

Figure 7 shows the PL response of SiO_x films corresponding to different growth temperatures. At all samples, a wide PL spectrum is observed. At the growth temperatures of 1,150°C, 1,050°C, and 900°C, the PL peaks are at 558, 546, and 534nm, respectively. At the highest growth temperature (1,400°C), the PL has the weakest intensity, and the PL peak has the longest wavelength of 678nm. Moreover, the PL bandwidth and intensity increase for growth temperatures lower than 1,400°C. Therefore, the PL also depends on Tg, size of Si-nc, roughness, oxidation states, and vibration modes of the Si-O-Si bonds, as shown previously.

The optical bandgap of SiO_x films is obtained with transmittance spectra measurements. Transmittance spectra for SiO_x films deposited on quartz are shown

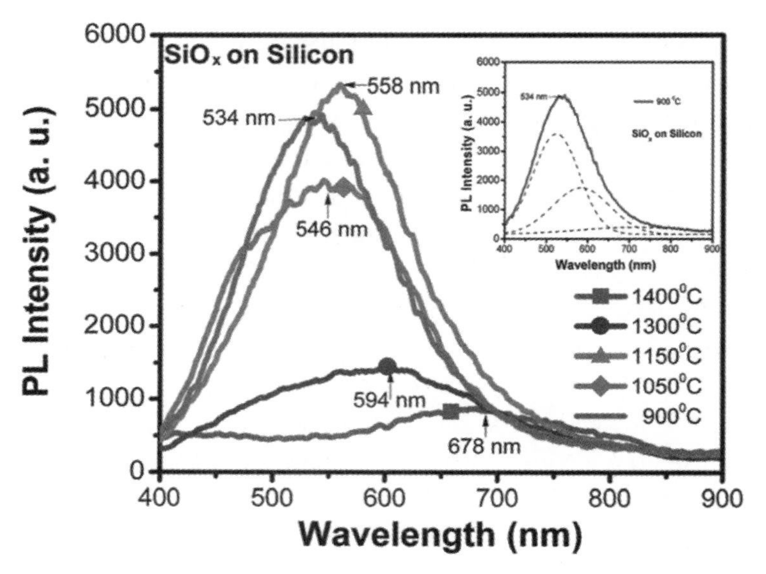

Figure 7. PL spectra of the SiO_x films with different Tg. Inset shows the convolution realized to PL spectra.

in **Figure 8(a)**. The transmittance of all these films is relatively high (>80%) between 600 and 1,000nm, as shown in the figure, and reduces to zero for wavelengths below 600nm. The growing temperature produces a clear change of the curves and a shift towards lower wavelengths related to a silicon excess change of the SiO_x films[11][25][26]. In Tauc's plot, an increase in the energy bandgap (E_g) has been detected when the growth temperature decreases, as shown in the inset of **Figure 8(a)**. The values of the optical bandgap E_g can be estimated from the following equation known as the Tauc plot[27][28]:

$$ahv = A\left(hv - E_g\right)^2$$

where E_g is the optical bandgap corresponding to a particular transition in the film; A, a constant; v, the transmission frequency, which multiplied by the plank constant h we have photon energy hv, and the exponent n characterizes the nature of band transition. The absorption coefficients $\alpha(\lambda)$ were determined from transmission spectra with the following relation:

$$\alpha(\lambda) = \frac{-\ln\left[T(\lambda)\right]}{d}$$

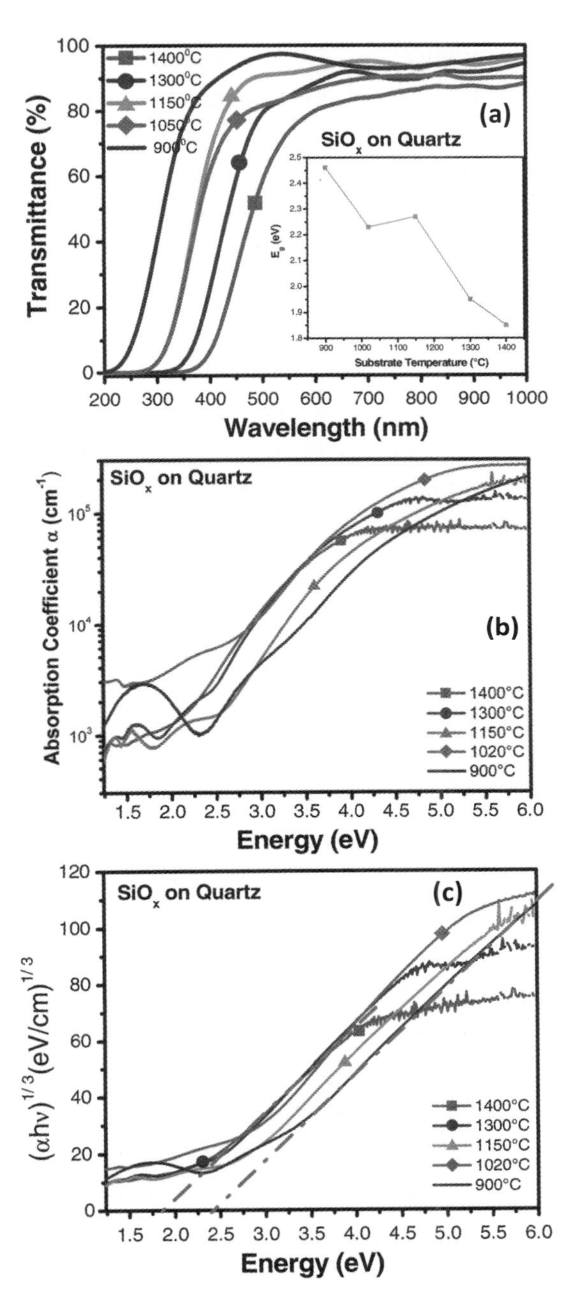

Figure 8. UV-vis transmittance spectra, absorption coefficient versus energy, and $(\alpha h\nu)^{1/3}$ versus energy ($h\nu$). (a) UV-vis transmittance spectra and the inset show the energy optical bandgap from SiO_x films as a function of growth temperature, (b) absorption coefficient versus energy, and (c) $(\alpha h\nu)^{1/3}$ versus energy ($h\nu$). The dashed lines show the linear fit of the straight part.

where $T(\lambda)$ is the transmittance, and d is the thickness of the SiO_x films. α versus $h\nu$ is shown in **Figure 8(b)**. On the other hand, values of n = 1/2 and 3/2 correspond to direct-allowed and direct-forbidden transitions; n = 2 and 3 are related to indirect-allowed and indirect-forbidden transitions, respectively[28]. From a plot $(\alpha h\nu)^{1/n}$ versus $h\nu$, the bandgap can be extrapolated from a straight line to $h\nu = 0$. For all different growth temperatures, the best straight line is observed for $n = 3$ [**Figure 8(c)**], indicating an indirect-forbidden transition mechanism.

The optical properties such as the energy bandgap and the PL bands between (400 to 700nm) are usually some of the important characteristics of these materials. The PL of SiO_x films has been extensively studied in the literature[1]-[25]. Two major mechanisms for PL in this kind of materials are generally accepted: quantum confinement effects in the Si-ncs and defect-related effects, as defects at the Si/SiO_x interface and defects associated with oxygen vacancies in the film. The first mechanism of light emission that we can consider in the SiO_x material is related to some kinds of defects produced during the growth process, as shown in the EDS, XPS, and FTIR spectra, where we have bonding such as neutral charged oxygen vacancies (NOV) (Si-Si bonds), non-bridging oxygen hole center (NBOHC), positively charged oxygen vacancies (E' centers), interstitial oxygen molecules and peroxide radicals[13][14][23][29][30], which can form Si-nps or E' centers. Therefore, the increase of PL with the Tg is due to the activation of some of these radiative defects. In this study, the 550-nm PL band has been associated with silicon excess in the film in the NOV defects and E' centers[2][26] types. These bands appear well defined only if the film has been grown with temperatures within 900°C and 1,150°C. If the film was grown at a higher temperature, the band at 700nm appears with its maximum PL emission.

On the other hand, as a second mechanism of emission, the luminescence peak of SiO_x films shows a blue-shift when the Tg decreases; this behavior is ascribed to quantum confinement effect in the Si-nc. Therefore, in this case the PL spectra are analyzed in terms of a quantum confinement model[1][22]:

$$E_N(eV) = 1,240/\lambda(nm)$$

$$E_N(eV) = 1.12eV + \left(3.73/d^{1.39}\right) \Rightarrow d(nm) = \left[\frac{3.73}{E_N - 1.12}\right]^{\frac{1}{1.39}}$$

which corresponds to the radiative recombination of electron-hole pairs in the Si-nc, where d and EN are the diameter and energy of the Si-nc, respectively, and λ(nm) is the wavelength of the Si-nc emission. **Table 3** shows the theoretical values of average size of Si-nc calculated from the PL spectra, where the size of the Si-nc reduces and the energy band gap increases with decreasing the growth temperature, similar to an effect of quantum confinement. Note that unlike as stated in the literature for the quantum confinement effect, PL spectra are very wide which indicates that two possible mechanisms are involved.

Why is it possible that two mechanisms be involved? When a deconvolution to the PL spectra is made, as shown in the inset of **Figure 7**, different peaks are defined; some of them are related to different kinds of defects, as listed in **Table 4**. Therefore, the high-energy PL peaks are associated with quantum confinement effects in Si-ncs, while the low-energy PL peaks are associated with defects. Such a behavior is described in the schematic diagram of the band structure (**Figure 9**); a similar behavior has been reported previously[31]. This diagram represents the radiative transition giving rise to the emission peaks. For the more energy occurs the generation of electron-hole pairs within the Si-nc core, followed by a thermal relaxation within the conduction band of the Si-nc which in turn suggests the recombination of carriers. In the case of the less energy peaks, phonon relaxation involves more energy because of the transitions between the states of interfacial defects.

The existence of Si-ncs in the SiO_x films was corroborated with the HRTEM measurements. The diffusion of Si excess due to the deposit at high temperature, *i.e.*, when the SiO_x films are being deposited, could produce Si-ncs. The silicon

Table 3. Theoretical values.

Tg (°C)	Gap EN of Si-nc (eV)		Diameter of Si-nc (nm)	
	Silicon	Quartz	Silicon	Quartz
1,400	1.82	1.87	3.3	3.17
1,300	2.08	1.82	2.64	3.3
1,150	2.22	2.07	2.4	2.66
1,050	2.27	2.16	2.32	2.49
900	2.32		2.25	

Values of the E_N and diameter of the Si-nc calculated from the PL spectra as a function of the Tg.

Table 4. Peak position obtained by deconvolution from PL spectra and defect types relationated with the peak position.

Defect types	Peak positions (nm)			Reference		
	1,400°C	1,300°C	1,150°C	1,050°C	900°C	
NOV defects ($O3 \equiv Si\text{-}Si \equiv O_3$)	428	483	458	467		[12]
Centers of defects $E'\delta$				521	523	[12]
E'δ center or oxygen deficiency			553	559		[12][32]
Defect vacancies of oxygen ($O \equiv Si\text{-}Si \equiv O$)		591	600	579	584	[30][33]
Oxide relationated in the interface of Si/SiO_x	675	695				[31][34]
Not identified	796	813				

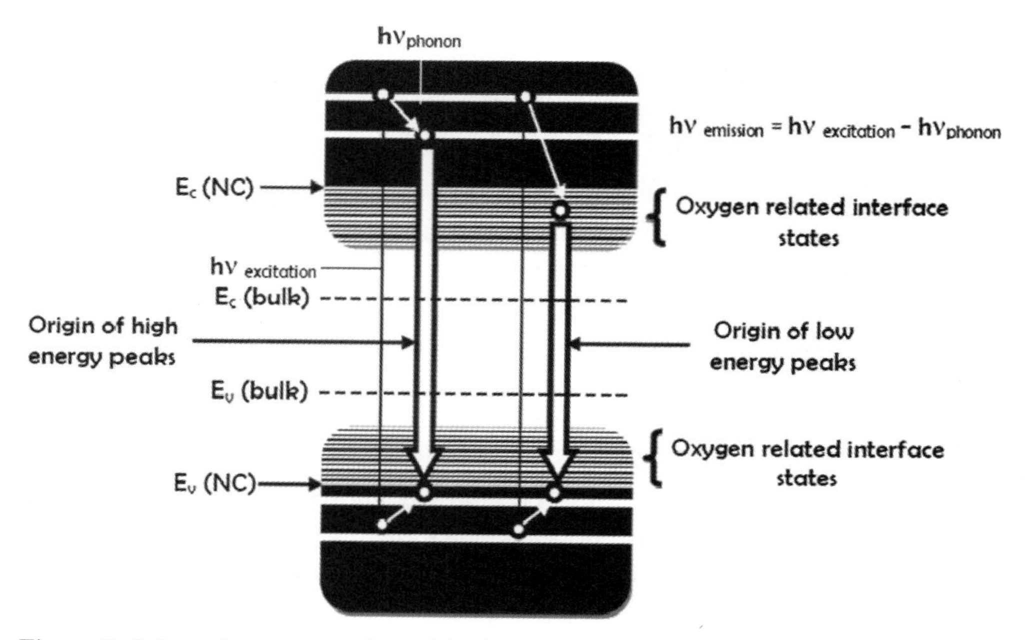

Figure 9. Schematic representation of the band structure and mechanisms responsible for PL from SiO_x films.

particles diffuse themselves to create silicon agglomerates around the nucleation sites. If the Si excess is high enough, the Si agglomerates will be crystallized to form Si-ncs. A decrease in the Si-nc diameter has been detected when the growth temperature reduces. The high growth temperature induces the formation of crystals as the statistical analysis of the crystal size distribution, obtained from the HRTEM images, shown. Therefore, the mean diameter of Si-ncs depends on the growth temperature.

Then, two transition mechanisms are possible as the above results and discussion showed widespread bandgap transitions induced by quantum confinement and interface state transitions associated with defects in the oxide. The widespread transitions in Si-nc may bring about high energy peaks (blueshifted PL peaks), and if this energy decays between defects (NOV, NBOHC, and *E*' centerrelated interface states), it can give place to low energy peaks (redshifted PL peaks). All these data indicate that light emission from the films is due to the Si-ncs embedded in the amorphous SiO_x matrix and defects. Accordingly, we have proposed a combination of mechanisms to explain the photoluminescence in the films.

4. Conclusions

SiO_x films deposited by HFCVD at different growth temperatures were analyzed. These films exhibit an intense PL with a main peak at 550 nm. The strongest PL was obtained for SiO_x films deposited at 1,150°C. Ellipsometry, XEDS, XPS, FTIR, AFM, PL, transmittance, and HRTEM techniques were used to obtain the structural, compositional, and optical properties of the SiO_x films, and they were studied as a function of the growth temperature. Thicker samples were obtained when the growth temperature was increased from 900°C to 1,400°C. A variation in the refractive index from 1.4 to 2.2 was obtained when the growth temperature was increased from 1,150°C to 1,400°C. From AFM images, the samples grown with the lower growth temperatures look more homogeneous than those grown with the higher Tg. As shown in the HRTEM images, the silicon excess agglomerates to create Si-ncs; in this way, the roughness observed through AFM measurements can be associated with Si-ncs and compounds. In addition, FTIR spectra show a phase separation (Si and SiO_2), which is deduced by the shift of the Si-O stretching vibration mode towards the SiO_2 frequency value, which is corroborated with XPS and HRTEM. A clear relation between the surface roughness, size of Si-nc, oxidation states, composition, and the PL property was obtained. Therefore, PL depends strongly on the Tg and properties of the SiO_x films.

Competing Interests

The authors declare that they have no competing interests.

Authors' Contributions

JALL and DEVV participated in the growth of the films, carried out the FTIR, PL, and UV measurements and drafted the manuscript. JCL conducted the ellipsometry and AFM measurements. GGS and TDB conducted the SiO_x growth. APP conducted the HRTEM and XEDS measurements. FJFG coordinated the study. JALL provided the idea and supervised the study. All authors read and approved the final manuscript.

Authors' Information

JALL is currently a researcher and professor in the Science Institute-Center of Investigation in Semiconductors Devices (IC-CIDS) from Autonomous University of Puebla, Mexico. He started to work on electrical and optical characterization of the MOS structures. His research interest is the physics and technology of materials and silicon devices. Moreover, his research interests are, too, the nanotechnology, material characterization, and optoelectronic devices such as sensor, LEDs, and solar cells.

GGS received his PhD in the Electronic and Solid State Department from the Center of Research and Advanced Studies, National Polytechnic Institute, Mexico City in 2003. He started to work on the growth and characterization of non-stoichiometric silicon oxide. His current research interests include metallic oxides obtained by the HFCVD technique, GaN obtained by the metal organic CVD technique and porous silicon gas sensor devices.

FJFG is a coordinator and researcher at the Posgrado en Dispositivos Semiconductores in Benemérita Universidad Autónoma de Puebla, México and was a participant in many important international conferences. Professor FJFG has published many journal articles. His research interests include experiments and models in photoluminescence and quantum confinement in off stoichiometry silicon oxides.

Acknowledgements

This work has been partially supported by CONACyT-154725 and VIEP-

BUAP-2012. The authors acknowledge INAOE and CIQA laboratories for their help in the sample measurements.

Source: López J A L, López J C, Valerdi D E V, et al. Morphological, compositional, structural, and optical properties of Si-nc embedded in SiO_x films [J]. Nanoscale Research Letters, 2012, 7(1):51–64.

References

[1] Canham LT: Silicon quantum wire array fabrication by electrochemical and chemical dissolution of wafers. Appl Phys Lett 1990, 57(10):1046–1048.

[2] Song HZ, Bao XM: Visible photoluminescence from silicon-ion-implanted SiO_2 film and its multiple mechanisms. Phys Rev B 1997, 55(11):6988–6993.

[3] Inokuma T, Wakayama Y, Muramoto T, Auki R, Muramoto Y, Hasegawa S: Optical properties of Si clusters and Si nanocrystallites in high-temperature annealed SiOx films. J Appl Phys 1998, 83(4):2228–2234.

[4] Huisken F, Amans D, Guillois O, Ledoux G, Reynaud, Hofmeister H, Cichos F, Martin J: Nanostructuration with visible-light-emitting silicon nanocrystals. New J Phys 2003, 5(10):10.1–110.

[5] Podhorodeckl A, Misiewicz J, Gourbilleau F, Cardin J, Dufour C: High energy excitation transfer from silicon nanocrystals to neodymium ions insilicon-rich oxide film. Electrochemical and Solid State Letters 2010, 13(3): K26–K28.

[6] DiMaria DJ, Dong DW, Pesavento FL: Enhanced conduction and minimized charge trapping in electrically alterable read-only memories using off-stoichiometric silicon dioxide films. J Appl Phys 1984, 55(8):3000–3019.

[7] Ay F, Aydinly A: Comparative investigation of hydrogen bonding in silicon based PECVD grown dielectrics for optical waveguides. Opt Mater 2004, 26:33–46.

[8] Berman-Mendoza D, Aceves-Mijares M, Berriel-Valdos LR, Yu Z, Falcony C, Domínguez-Horna C, Pedraza J: Design, Fabrication, and Characterization of an Ultraviolet Silicon Sensor. Proc. SPIE 2006, 6024:60242L. http://dx.doi. org/10.1117/12.666968.

[9] Zhu F, Mao ZS, Yan L, Zhang F, Zhong K, Cheng GA: Photoluminescence and radiation effect of Er and Pr implanted silicon-rich silicon oxide thin films. Nuclear Instruments and Methods in Physics Res 2009, 267:3100–3103.

[10] Salazar P, Chávez F, Silva-Andrade F, Illinskii AV, Morales N, Peña-Sierra R: Photoluminescence from amorphous silicon oxide films prepared by HFCVD technique. Modern Physics Letters B 2001, 15(17–19):756–759.

[11] Matsumoto Y, Godavarthi S, Ortega M, Sánchez V, Velumani S, Mallick PS: Size modulation of nanocrystalline silicon embedded in amorphous silicon oxide by Cat-CVD. Thin Solid Films 2011, 519:4498–4501.

[12] Gong-Ru L, Chung-Jung L, Chi-Kuan L, Li-Jen C, Yu-Lun C: Oxygen defect and Si nanocrystal dependent white-light and near-infrared electroluminescence of Si-implanted and plasma-enhanced chemical-vapor deposition-grown Si-rich SiO_2. J Appl Phys 2005, 97:094306-1-094306-8.

[13] Gatan Digital Micrograph Software. http://www.gatan.com/scripting/downloads.php.

[14] Jøgensen JF: Denmark: The Scanning Probe Image Processor (SPIP) Software. http://www.nanoscience.com/products/spip/SPIP.html.

[15] Pai PG, Chao SS, Takagi Y, Lucovsky G: Infrared spectroscopic study of silicon oxide (SiO_x) films produced by plasma enhanced chemical vapour deposition. J Vac Sci Technol, A 1986, 4(3):689–694.

[16] Alayo MI, Pereyra I, Scopel WL, Fantini MCA: On the nitrogen and oxygen incorporation in plasma-enhanced chemical vapor deposition (PECVD) SiO_xNy films. Thin solid Films 2002, 402(1):154–161.

[17] Benmessaoud A: Caracterización de subóxidos de silicio obtenidos por las técnicas de PECVD, PhD thesis: Universidad Autónoma de Barcelona, Departamento de Física; 2001.

[18] McLean FB: A framework for understanding radiation-induced interface states in SiO_2 MOS structures. IEEE Trans Nucl Sci 2001, 27(6):1651–1657.

[19] Shimizu-Iwayama T, Hole DE, Boyd IW: Mechanism of photoluminescence of Si nanocrystals in SiO_2 fabricated by ion implantation: the role of interactions of nanocrystals and oxygen. J Phys Condens Matter 1999, 11 (34):6595–6604.

[20] Smith BC: Fundamentals of Fourier Transform Infrared Spectroscopy. Florida: CRC Press; 1996.

[21] Hayashi S, Tanimoto S, Yamamoto K: Analysis of surface oxides of gas-evaporated Si small particles with infrared spectroscopy, high-resolution electron microscopy, and x-ray photoemission spectroscopy. J Appl Phys 1990, 68(10):5300.

[22] Ma LB, Ji AL, Liu C, Wang YQ, Cao ZX: Low temperature growth of amorphous Si nanoparticles in oxide matrix for efficient visible photoluminescence. J Vac Sci Technol B 2004, 22:2654.

[23] Iacona F, Franzo G, Spinella C: Correlation between luminescence and structural properties of Si nanocrystals. J Appl Phys 2000, 87(3):1295.

[24] Iacona F, Borgiono C, Spinella C: Formation and evolution of luminescent Si nanoclusters produced by thermal annealing of SiO_x films. J Appl Phys 2004, 95(7):3723.

[25] Luna-López JA, Aceves-Mijares M, Malik O, Yu Z, Morales A, Dominguez C, Rickards J: Compositional and structural characterization of silicon nanoparticles em-

bedded in silicon rich oxide. Rev Mex Fis 2007, S53(7):293.

[26] Luna López JA, Morales-Sanchez A, Aceves Mijares M, Yu Z, Dominguez C: Analysis of surface roughness and its relationship with photoluminescence properties of silicon-rich oxide films. J Vac Sci Technol A 2009, 27(1):57.

[27] Pankove JI: Optical Processes in Semiconductors. Englewood Cliffs, New Jersey: Prentice Hall; 1971.

[28] Wang L, Han K, Tao M: Effect of substrate etching on electrical properties of electrochemical deposited CuO. J Electrochem Soc 2007, 154:D91.

[29] Edelberg E, Bergh S, Naone R, Hall M, Aydil E: Luminescence from plasma deposited silicon films. J Appl Phys 1997, 81(5):2410.

[30] Hanaizumi O, Ono K, Ogawa Y: Blue-light emission from sputtered Si: SiO_2 films without annealing. Appl Phys Lett 2003, 82(4):538.

[31] Ray M, Minhaz S, Klie RF, Banerjee K, Ghosh S: Free standing luminescent silicon quantum dots: evidence of quantum confinement and defect related transitions. Nanotechnology 2010, 21(50):505602.

[32] Wehrspohn RB, y Godet C: Visible photoluminescence and its mechanisms from a-SiO_x: H films with different stoichiometry. J Lumin 1999, 80(1):449-453.

[33] Zhu M, Han Y, Wehrspohn RB, Godet C, Etemadi R, Ballutaud: The origin of visible photoluminescence from silicon oxide thin films prepared by dual-plasma chemical vapor deposition. J Appl Phys 1998, 83(10):5386.

[34] Kenjon AJ, Trwoga PF, Pitt CW: The origin of photoluminescence from thin films of silicon-rich silica. J Appl Phys 1996, 79(12):9291.

Chapter 14

One-Dimensional CuO Nanowire: Synthesis, Electrical, and Optoelectronic Devices Application

Lin-Bao Luo[1,2], Xian-He Wang[1], Chao Xie[1], Zhong-Jun Li[1], Rui Lu[1], Xiao-Bao Yang[3], Jian Lu[2,4]

[1]School of Electronic Science and Applied Physics, Hefei University of Technology, Hefei, Anhui 230009, China

[2]Department of Mechanical and Biomedical Engineering, City University of Hong Kong, Kowloon, Hong Kong SAR, China

[3]Department of Physics, South China University of Technology, Guangzhou, Guangdong Province 510641, China

[4]Centre for Advanced Structural Materials, City University of Hong Kong Shenzhen Research Institute, 8 Yuexing 1st Road, Shenzhen Hi-Tech Industrial Park, Shenzhen, China

Abstract: In this work, we presented a surface mechanical attrition treatment (SMAT)-assisted approach to the synthesis of one-dimensional copper oxide nanowires (CuO NWs) for nanodevices applications. The as-prepared CuO NWs have diameter and the length of 50 ~ 200nm and 5 ~ 20μm, respectively, with a preferential growth orientation along $[1\ \bar{1}\ 0]$ direction. Interestingly, nanofield-effect transistor (nanoFET) based on individual CuO NW exhibited typical p-type electrical conduction, with a hole mobility of $0.129 cm^2V^{-1}\ s^{-1}$ and hole concentration of $1.34 \times 10^{18} cm^{-3}$, respectively. According to first-principle calculations, such a *p*-type electrical conduction behavior was related to the oxygen vacancies in CuO

NWs. What is more, the CuO NW device was sensitive to visible light illumination with peak sensitivity at 600nm. The responsivity, conductive gain, and detectivity are estimated to be $2.0 \times 10^2 A\ W^{-1}$, 3.95×10^2 and $6.38 \times 10^{11} cm\ Hz^{1/2}\ W^{-1}$, respectively, which are better than the devices composed of other materials. Further study showed that nanophotodetectors assembled on flexible polyethylene terephthalate (PET) substrate can work under different bending conditions with good reproducibility. The totality of the above results suggests that the present CuO NWs are potential building blocks for assembling high-performance optoelectronic devices.

Keywords: Surface Mechanical Attrition Treatment (SMAT), Semiconductor Nanostructures, The First-Principle Calculation, Metal Oxide, Flexible Photodetector

1. Introduction

Metal oxide semiconductors (e.g. ZnO, [1]TiO_2, [2]NiO, [3]SnO_2[4], and CuO[5]) are one of the most common, most diverse and probably the richest class of materials among the various groups of semiconductors. In the past decade, a number of methods including laser ablation[6][7], thermal oxidation[8][9], solution-phase growth[10], and template-assisted synthesis[11] have been employed to fabricate various one-dimensional metal oxide semiconductor nanostructures, such as nanowires, nanotubes, and nanoribbons[12]. Due to the high surface-volume ratio and quantum-size effect, the resultant nanostructures with improved physical, optical, and electronic properties[13] have been used as building blocks to construct a number of optoelectronic and electronic devices including solar cells[14][15], photo detectors[16][17], gas sensors[18], non-volatile memory devices[19], and so on.

Copper oxide (CuO), as one of the most important metal oxide semiconductors, has been widely used because of its abundance in resources and low cost in synthesis. Low-dimensional CuO nanostructures (zero-dimensional and one-dimensional nanostructures) are used, in particular via simple thermal evaporation method[20], wet chemical method[21]], and metal-assisted growth method[22]. It has been found that the CuO NWs obtained from the above methods normally have good crystallinity and high aspect ratio, which renders them attractive and promising building blocks for fabricating high-performance electronic devices systems[23]. For example, Chang *et al.* reported the growth of CuO NWs on an oxidized Cu wire at 500°C for infrared (IR) photodetection application. The as- obtained high density of CuO NWs on the Cu wire was highly sensitive to IR light illumination (wavelength: 808 nm), with rise-time and fall-time of 15 and 17s, respectively[24]. Zhou *et al.* presented a vertically aligned CuO NWs array-based ultrasensitive sensors for H_2S detection with a detection limit as low as 500 ppb. It was revealed that the high sensitivity was due to the formation of highly conductive CuS layer when H_2S gas was introduced into the detection chamber[25]. Zheng *et al.* developed a simple and effective catalyst system comprised of CuO NWs for CO oxidation. They found that CO oxidation percentage was as high as 85% after Ar or H_2 plasma treatment[5]. In addition to these device applications, it has been observed that highly-aligned CuO NW arrays are good candidates for field emission due to their low turn-on voltages, high current output[26].

Despite of the above research progresses, there is a sparsity of research activity dealing with the transport and optoelectronic property of individual CuO nanostructures[27], which constitutes the basic building blocks of various optoelectronic and electronic devices. Exploration along this direction is highly desirable as it is not only helpful for understanding the electrical property of individual CuO NWs, but also beneficial to the development of high-performance optoelectronic and electronic devices. Herein, we report the synthesis of CuO NWs by heating surface mechanical attrition treatment (SMAT) processed copper foil in tube furnace. The CuO NW is of single crystal with a growth direction of $[1\ \bar{1}\ 0]$. Individual CuO NW-based field-effect transistor displays weak p-type electrical conduction behavior, which was probably due to the O defects, according to the theoretical simulation based on first-principle calculation. Further optoelectronic characterization shows that the CuO NW is sensitive to incident light of 600nm, with high producibility and stability. It is also observed that the photodetector fabricated on flexible polyethylene terephthalate (PET) substrate showed good reproducibility under different bending conditions. The above result suggests that our CuO NWs will have promising potential in future devices applications.

2. Methods

2.1. Synthesis and Structural Characterization of the CuO NWs

In this study, the CuO NWs were fabricated via SMAT-assisted thermal oxidation method. Briefly, copper plates (99.99%) with size of 20 × 20 × 5mm were cleaned by alcohol to remove surface impurities including grease and other organics. The copper plates were then treated by an SMAT process in which millimeter-size steel balls were acoustically driven to bombard the Cu surface randomly and in all directions to generate nanocrystalline Cu[28]. After drying in N_2 atmosphere, the clean samples were heated in a horizontal tube at 500°C in pure O_2 atmosphere (375 Torr) for 2.5h. The morphologies and structure of the as- prepared CuO NWs were characterized by scanning electron microscopy (SEM, FEI Quanta 200 FEG, FEI, Hillsboro, OR, USA), energy-dispersive X-ray spectroscopy (EDS), high-resolution transmission electron microscopy (HRTEM, JEOL JEM-2010 at 200kV, JEOL, Akishima-shi, Tokyo, Japan), X-ray diffraction (XRD, Rigaku D/Max-γB, with Cu Kα radiation, Rigaku Corporation, Tokyo, Japan) and X-ray photoelectron spectroscopy (XPS, ThermoESCALAB250, Thermo Fisher Scientific, Waltham, MA, USA).

2.2. Device Fabrication and Characterization

To evaluate the electrical properties of the CuO NWs, back-gate field-effect transistor (FET) was constructed based on individual CuO NW. Firstly, the as-synthesized CuO NW was dispersed on a SiO_2/p^+-Si substrate by a contact print technique[29], then Cu (4nm)/Au (50nm) source and drain electrodes were defined by photolithography and e-beam evaporation. In order to achieve ohmic contact between the NW and electrodes, the as-fabricated devices were annealed at 200°C for 10min in argon atmosphere at a pressure of 0.33 Torr. In this work, flexible photodetectors on PET substrate were constructed by the same process. Both the electrical and optoelectronic characterization of CuO NW-based devices were carried out by using a semiconductor characterization system (Keithley 4200-SCS, Keithley, Cleveland, OH, USA).

2.3. Theoretical Simulation

The first-principle calculation of $[1\ \bar{1}\ 0]$ CuO NW were based on the density functional theory (DFT) implemented in the Vienna ab initio simulation package method[30][31]. The projector-augmented wave (PAW)[32] and the Perdew-Burke-Ernzerhof GGA (PBE)[33] functionals were employed for the total energy calculations. The cutoff energy was 450eV and the criteria of the forces were set to be 0.01eV/Å for all atoms. An 11 × 11 × 11 k-grid mesh was used for the bulk CuO and a 7 × 1 × 1 mesh for the <1 $\bar{1}$ 0> CuO NW, where the vacuum distance was set to be 10Å to avoid cell-to-cell interactions. To improve the calculations of electronic properties, we used the GGA + U extension to the DFT calculation[34][35], dealing with the Cu 3d electrons for a better description, where U = 7.5eV and J = 1.0eV were adopted.

3. Results and Discussion

The fabrication of the CuO NWs was carried out in a tube furnace in oxygen atmosphere. **Figure 1(a)**, **Figure 1(b)** displays typical SEM images of as-synthe- sized CuO NWs with different magnifications. It is obvious that the product is composed of fiber-like structures with length of about 5 ~ 20μm. The statistical distribution in **Figure 1(c)** shows that the diameters of the CuO NWs are in range of 50 ~ 250nm, with an average value of approximately 120nm. According to the corresponding EDX spectrum of the CuO NWs in **Figure 1(d)**, the product is composed of Cu and O elements with a molar ratio of approximately 51:49, indicative of the presence of CuO, rather than Cu_2O. **Figure 1(f)**, **Figure 1(g)** shows the elemental mapping image of an individual CuO NW, from which one can see that the constituting elements (Cu and O) are uniformly distributed in the nanostructure.

The TEM image in **Figure 2(a)** indicated that the surface of the NWs was free of impurities and contaminants. Further HRTEM image along with the corresponding fast Fourier transform (FFT) pattern in the inset of **Figure 2(a)** shows that the CuO NWs are of single crystal with preferential growth orientation along [1 $\bar{1}$ 0] direction. **Figure 2(c)** displays a typical XRD pattern of the product, in which the peaks labeled with red quadrate can be readily indexed to the monoclinic phase of CuO (JCPDS-80-1916)[4][11][20]. In fact, the presence of CuO was verified by the XPS analysis. As is shown in **Figure 2(d)**, two peaks of Cu 2p are located at 933.8 and 953.8eV which represent the Cu 2p3/2 and Cu 2p1/2, respectively. These signals can be ascribed to the Cu 2p in CuO, in consistence with literature result. What is more, the strong shake-up satellites located at 940.92 and 943.83eV confirm the presence of the Cu (II) valence state[6]. In addition, strong peaks (labeled with blue balls) ascribable to Cu_2O phase were present in the pattern

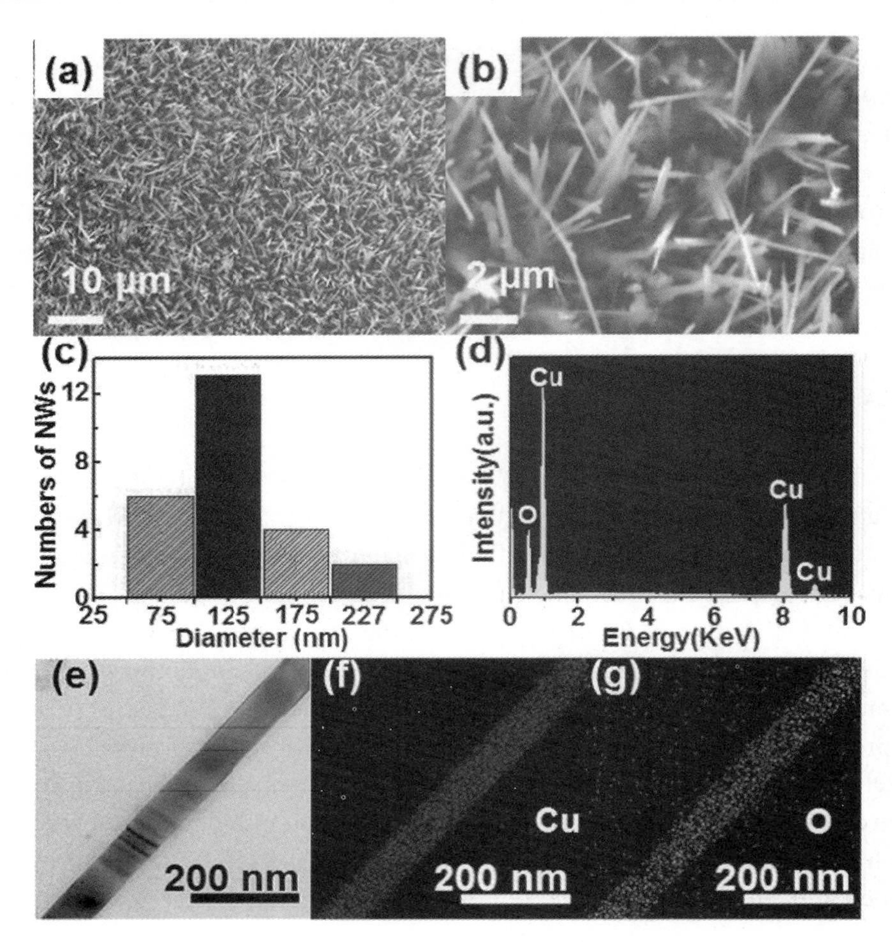

Figure 1. SEM and TEM image, diameter distribution, and EDS spectrum of CuO NWs and Cu/O images. (a, b) Typical SEM image of the CuO NWs at different magnifications; (c) statistical distribution of the diameter of CuO NWs; (d) the EDS spectrum of CuO NWs; (e) TEM image of a single CuO NW; elemental mapping images of Cu (f), and O (g).

(JCPDS-05-0667) as well. We attribute the presence of Cu_2O to the special growth mechanism, as illustrated in **Figure 2(e)**. At initial growth stage, Cu_2O thin film was formed when the copper plate was treated with high temperature in the oxygen atmosphere. According to the previous study, in the underlying Cu_2O layer, the CuO NWs are formed as a result of rapid, short-circuit diffusion of the Cu ions across grain boundaries and/or defects[28][36]. Notably, during this growth pro- cess, the SMAT processing is highly advantageous in that the generation of many dislocations, twins, or stacking faults in the surface of Cu plate will increase the quality, as well as the length of the NWs.

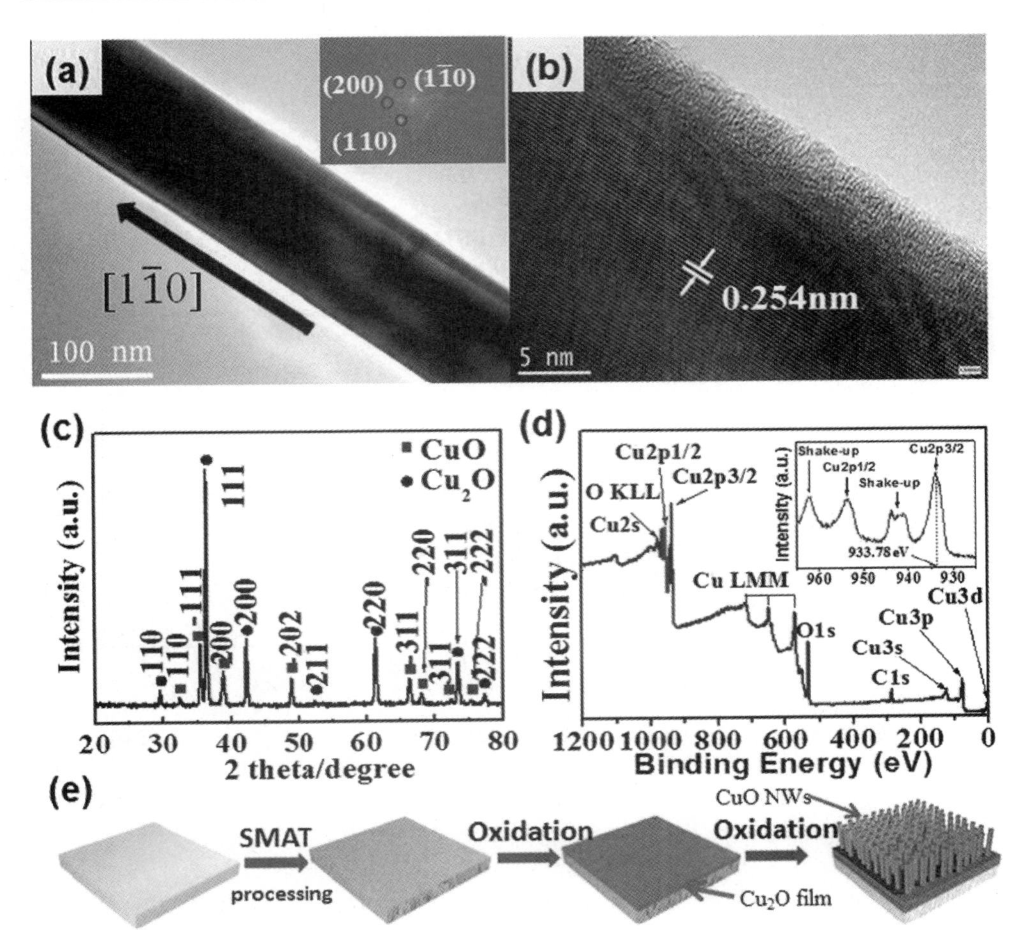

Figure 2. TEM images, XRD pattern, XPS survey spectrum, and schematic illustration of CuO NWs. (a) TEM image of the CuO NW, the inset shows the corresponding FFT pattern; (b) high-resolution TEM image of a CuO NW; (c) the XRD pattern of the CuO NW; (d) XPS survey spectrum of the CuO NW, the inset shows the corresponding high-resolution N1s spectrum; (e) schematic illustration of the formation of CuO NWs.

To study the transport properties of a single CuO NW, back-gate metal-oxide-semiconductor FETs (MOS-FETs) were fabricated on the basis of individual CuO NW. The linear I-V curve in **Figure 3(a)** suggests that Cu (4nm)/Au (50nm) electrode can form good contact with relatively low contact barrier. Electrical study of the single CuO NW-based FET in **Figure 3(c)** exhibits typical p-type conduction behavior. That is, the electrical conduction increases with decreasing gate voltage. By fitting the linear part of the I_{ds}-V_g characteristics, the turn-on threshold voltage (V_T) and transconductance (gm = dI_{ds}/dV_g) are calculated

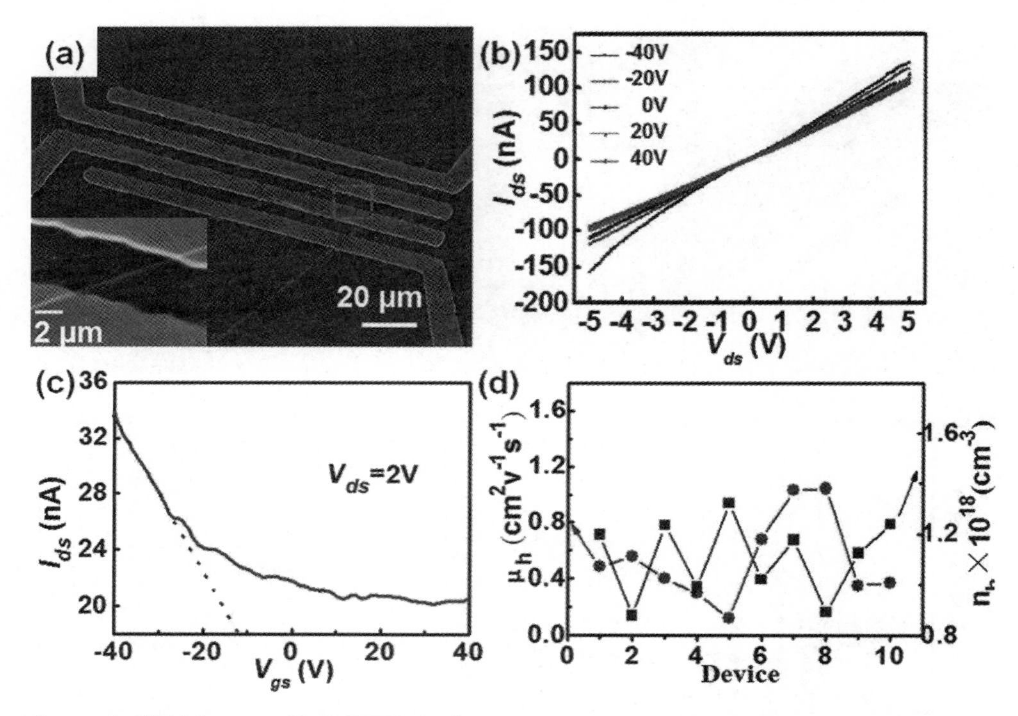

Figure 3. SEM image of MOSFET device, I_{ds}-V_{ds} curves, and characteristics of the nanoFETs. (a) SEM image of the CuO NW MOSFET device, the inset shows the magnified image; (b) I_{ds}-V_{ds} curves at different gating voltages. (c) Transfer characteristics of the nanoFET at V_{ds} = 2V. (d) Hole mobility and concentration of ten representative nanoFETs.

to be −12V and 0.54nS, respectively. To evaluate the property of the CuO NW, two key parameters of hole mobility (μ_h) and concentration (μ_h) were estimated. The hole mobility (μ_h) and concentration (μ_h) can be calculated according to the following equations:

$$\mu_h = g_m \frac{In(4h/d)l}{2\pi\varepsilon_0\varepsilon_{SiO_2}V_{ds}} \quad (1)$$

$$n_h = \frac{\sigma}{q\mu_h} \quad (3)$$

where h, d, and l represent the thickness of oxide layer (300 nm), the NW diameter (125nm), and the channel length (5μm), respectively. ε_{SiO_2} is the dielectric constant of the SiO_2 dielectric layer (approximately 3.9), ε_0 is the permittivity at vacuum, σ is the conductivity of the NW, and q is the charge of an electron. Based on the equation **[Equation (1)]**, the hole mobility is estimated to be $0.134 cm^2V^{-1} s^{-1}$. Such a value is larger than the CuO thin film[37], and CuO NWs synthesized by direct evaporating Cu substrates in oxygen ambient without SMAT process[28], suggesting that the present SMAT-assisted thermal evaporation is an ideal approach to the synthesis of CuO NWs. Furthermore, the hole concentration is calculated to be $1.29 \times 10^{18} cm^{-3}$ according to Equation (2). To obtain a statistical distribution of the CuO NWs, totally ten FETs were analyzed. As displayed in **Figure 3(d)**, the hole mobilities of most CuO NWs are in the range of 0.1 to $1.0 cm2V^{-1} s^{-1}$ with an average value of $0.58 cm^2V^{-1} s^{-1}$. Meanwhile, the hole concentration is in the range of 0.8×10^{18} to $1.4 \times 10^{18} cm^{-3}$ with an average value of $1.13 \times 10^{18} cm^{-3}$. To unveil the physical reason behind the p-type electrical characteristics, we used first-principle calculation to simulate the electronic structures of CuO NW with different surface defects. Firstly, we compared various possible magnetic states and found an anti-ferromagnetic ground state for bulk CuO, in agreement with the previous study[38]. **Figure 4(a)** shows the ground states of anti-ferromag- netic CuO NW, where Cu atoms exhibit local magnetic moments. Obviously, the bands near the Fermi level are not fully filled with electrons and thus the system should exhibit metallic characteristics for an ideal CuO NW. Similar phenomenon was also observed when there is a Cu vacancy in the crystal lattice, in which the corresponding band structures near the Fermi level are partially filled [shown in **Figure 4(b)**]. These results suggest that neither CuO with a Cu vacancy nor ideal CuO without any defect can lead to the observed p-type conduction behavior. However, when an oxygen vacancy is present on the surface of CuO NW, the bands near the Fermi level are fully filled and there are two flat bands at around 0.25 ~ 0.50eV, as is shown in **Figure 4(c)**. In this case, electrons can be readily stimulated and trapped in these flat bands, giving rise to p-type conducting characteristic. As a matter of fact, the presence of huge amount of surface defects was experimentally corroborated by the ESR spectrum as a function of external magnetic field shown in Additional file 1: Figure S1 of the supporting information.

Next, the optoelectronic properties of the individual CuO NW photodetector (nano-PD) were studied. It is obvious that the CuO NW shows an obvious increase in current when the device is exposed to the 600-nm illumination [see **Figure 5(a)**]. Additionally, the CuO NW device can be readily switched between low- and high-conduction states with relatively good reproducibility when the light illumination

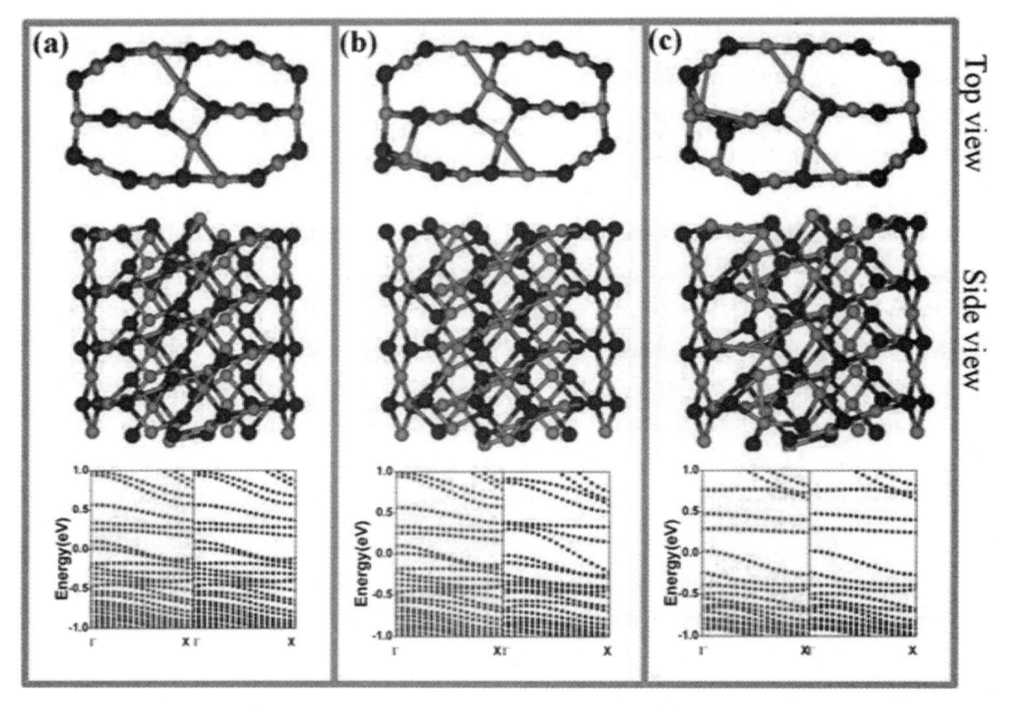

Figure 4. Atomic configurations and band structures of CuO NW with and without defects. (a) Ideal NW; (b) with a Cu vacancy; (c) with an O vacancy. Dark and light balls represent O and Cu atoms, respectively. Red/blue circles represent the band structures from electrons with spin up and down, respectively. The Fermi levels are shifted to zero.

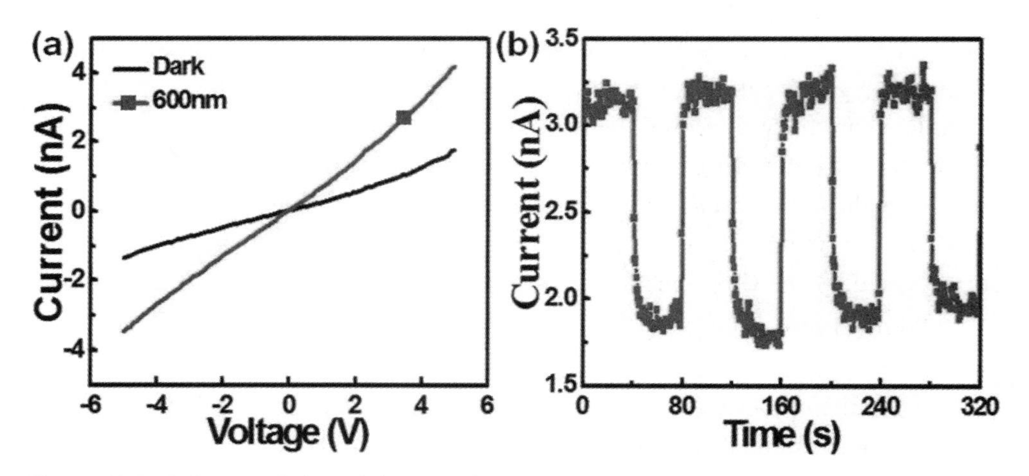

Figure 5. I-V characteristics of the CuO-based photodetector and time response spectra of the device. (a) I-V characteristics of the CuO-based photodetector measured at room temperature with and without light irradiation. (b) Time response spectra of the device when the incident light was manually switched on and off repeatedly (V_{ds} = 5V).

was turned on and off alternatively. In order to quantitatively evaluate the performance of the CuO NW-based device, three key parameters including responsivity (R), gain (G), and detectivity (D^*) were calculated by the following equation:

$$R\left(A\cdot W^{-1}\right)=\frac{I_l-I_d}{P_{opt}}=\eta\left(\frac{q\lambda}{hc}\right)G \tag{3}$$

$$G=N_{el}/N_{ph}=\tau/\tau_{tr} \tag{4}$$

$$D^*=A^{1/2}R/\left(2qI_d\right)^{1/2} \tag{5}$$

where I_l is the photocurrent, Id is the dark current, P_{opt} is the incident light power, η is the quantum efficiency (we assume $\eta = 1$ for simplification), q is the charge of an electron, λ is the incident light wavelength, h is the Planck's constant, and c is the speed of light. The values of P_{opt} q, λ, h, c are $2mW\cdot cm^{-2}$, $1.6 \times 10^{-19}C$, 600nm, $6.626 \times 10^{-34}J\cdot s$, and $2.997 \times 10^8 m/s$, respectively. Based on the equation [Equation (3)], the responsivity (R) of the device is estimated to be $2.0 \times 10^2 A\cdot W-1$ at the voltage of 5V. Physically, the photoconductive gain (G) is defined as the ratio of the number of electrons collected per unit time (N_{el}) and the number of absorbed photons per unit time (N_{ph}), or equal to the ratio of carrier life time (τ) to carrier transit time (τ_{tr}); it can be derived to be 3.95×10^2 according to Equation (3) or Equation (4). By using Equation (5), the detectivity (D^*) is estimated to be $6.38 \times 10^{11} cm\cdot Hz^{1/2}\ W^{-1}$ based on the above value and the active area (A) of $6.25 \times 10^{-9} cm^2$ (effective area that absorbs the incident light). **Table 1** summarizes the key metrics of the current device and other semiconductor nanostructure-based PDs. It is obvious that the R and G are comparable to the device based on pure

Table 1. Summary of the device performances of the CuO-based PD with other PDs based on pure materials.

Materials	R/AW^{-1}	G	$D^*/cm\cdot Hz^{1/2}\cdot W^{-1}$	Reference
CuO NW	2.0×10^2	3.95×10^2	6.38×10^{11}	Our work
CdTe NW	3.6×10^2	5.56×10^2	6.63×10^{10}	[39]
CdSe NW	10 ~ 100	0.05	1.71×10^{11}	[40]
ZnO	Approximately 0.055	Approximately 10^2	7.43×10^{11}	[41]

CdTe NW[39], but much higher than that based on CdSe NW[40] and ZnO NW[41]. We believe this relatively good optoelectronic property is partially associated to the introduction of SMAT prior to oxidation which can improve the quality of the CuO NW during growth process.

Figure 6 plots the UV-vis absorption of the CuO NW and the normalized responsivity of the individual CuO NW-based nano-PD as a function of wavelength. To make the analysis more reliable, we kept the light power identical for all wavelengths during measurements. It is noted that the device exhibits high sensitivity to visible light, with sensitivity peak at 600 nm, in rough consistence with the cutting edge of UV-vis absorption curve (blue line). This agreement is believed to be highly related to the working mechanism of such photoconductive-type photodetector[17][42].

Apart from nano-PDs on hard SiO_2/Si substrate, nano-PDs were also fabricated by selecting flexible PET substrate. **Figure 7(a)** shows the device image under bending condition, from which one can see that the device exhibits excellent flexibility. Remarkably, the CuO NW device also displays obvious sensitivity to 600nm under various bending conditions. **Figure 7(b)** compares the photoresponse of nano-PD after bending the PET substrate to different angles relative to the horizontal level. The Ion/Ioff is 1.35, 1.32, and 1.24 for angles of 0°, 15°, and

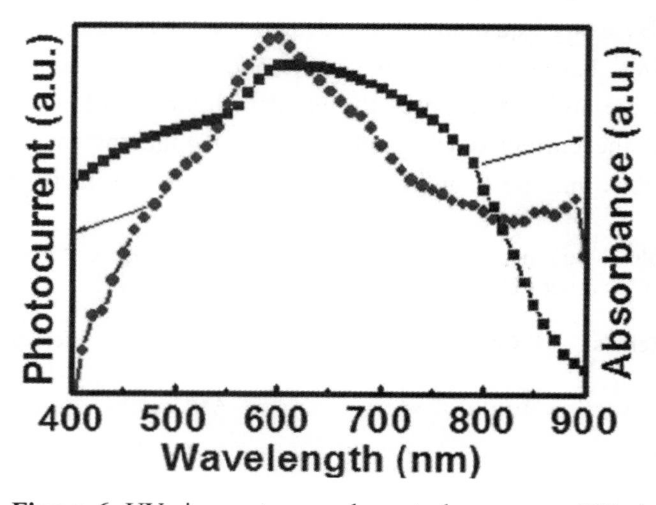

Figure 6. UV-vis spectrum and spectral responses. UV-vis spectrum of CuO NWs (blue line) and spectral response (red line) of individual CuO NW based nano-PD.

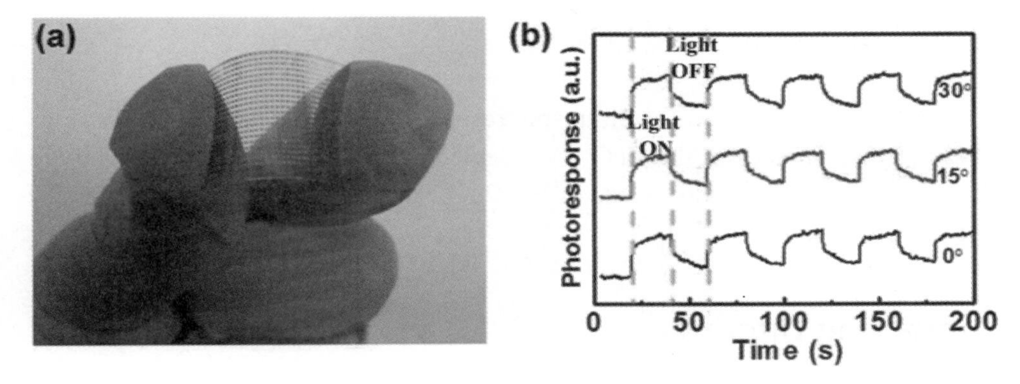

Figure 7. Digital camera picture of and photoresponse of nano-PDs. (a) Digital camera picture of nano-PDs fabricated on the flexible PET substrate; (b) photoresponse of the nano-PD on PET substrate with different bending angles: 0°, 15°, 30°.

30°, respectively, suggesting that the current flexible CuO nano-PDs have great potential for the application in future transparent and flexible optoelectronics. In addition to the flexibility, the present devices are highly transparent in visible light range. As shown in Additional file 1: Figure S2, the transmittance is over 85% in the majority of the visible light range. This characteristic, combined with the flexibility and long-term stability (Additional file 1: Figure S3), makes the current device a good candidate for future optoelectronic device applications.

4. Conclusions

In summary, we have fabricated one-dimensional CuO NW by heating SMAT copper plate in oxygen atmosphere. Electrical field-effect transistor device based on the as-prepared individual CuO NW exhibited typical p-type electrical conduction characteristic, with hole mobility and concentration of $0.134\ cm^2V^{-1}\ s^{-1}$ and $1.29 \times 10^{18} cm^{-3}$, respectively. It is also revealed that the as-synthesized CuO NW was highly sensitive to light irradiation of 600 nm, with a high responsivity and photoconductive gain of $2.0 \times 10^2 AW^{-1}$ and 3.95×10^2, respectively. Further optoelectronic study shows that the photodetector on flexible PET substrate is also highly sensitive to 600-nm wavelength light at different bending conditions. The generality of this study proves that CuO NW obtained via SMAT-assisted thermal evaporation method will have great potential for future high-performance optoelectronic devices application.

Additional File

Additional file 1: The electron spin resonance (ESR) of the CuO NWs (Figure S1). Transmittance of the CuO NWs device on flexible PET substrate (Figure S2). Comparison of photoresponse of the CuO NW before and after 1-month storage (Figure S3).

Competing Interests

The authors declare that they have no competing interests.

Author's Contributions

LBL, XHW, CX, and RL carried out the experiments. ZJL and XBY conducted the theoretical simulation. JL conceived the idea and supervised the whole work. LBL, XBY, and JL drafted the paper. All authors read and approved the final manuscript.

Acknowledgements

This work was supported by the National Key Basic Research Program of the Chinese Ministry of Science and Technology (Grant 2012CB932203), the Natural Science Foundation of China (NSFC, Nos. 51202206, 21101051, 11104080, 61106010), the Croucher Foundation (CityU9500006), and the Fundamental Research Funds for the Central Universities (2012HGCX0003, 2013HGCH0012, 2014HGCH0005).

Source: Luo L B, Wang X H, Xie C, *et al*. One-dimensional CuO nanowire: synthesis, electrical, and optoelectronic devices application[J]. Nanoscale Research Letters, 2014, 9(1):1–8.

References

[1] Wang ZL, Song JH: Piezoelectric nanogenerators based on zinc oxide nanowire ar-

rays. Science 2006, 312:242–246.

[2] Kwon DH, Kim KM, Jang JH, Jeon JM, Lee MH, Kim GH, Li XS, Park GS, Lee B, Han SW, Kim MY, Hwang CS: Atomic structure of conducting nanofilaments in TiO_2 resistive switching memory. Nat Nanotechnol 2010, 5:148–153.

[3] Dar FI, Moonooswamy KR, Es-Souni M: Morphology and property control of NiO nanostructures for supercapacitor applications. Nanoscale Res Lett 2013, 8:363.

[4] Choi SW, Katoch A, Sun GJ, Wu P, Kim SS: NO_2-sensing performance of SnO_2 microrods by functionalization of Ag nanoparticles. J Mater Chem C 2013, 1:2834–2841.

[5] Feng YZ, Zheng XL: Plasma-enhanced catalytic CuO nanowires for CO oxidation. Nano Lett 2010, 10:4762–4766.

[6] Zeng HB, Du XW, Singh SC, Kulinich SA, Yang SK, He JP, Cai WP: Nanomaterials via laser ablation/irradiation in liquid. Adv Funct Mater 2012, 22:1333–1353.

[7] Niu KY, Yang J, Kulinich SA, Sun J, Du XW: Hollow nanoparticles of metal oxides and sulfides: fast preparation via laser ablation in liquid. Langmuir 2010, 26:16652–16657.

[8] Nie B, Hu JG, Luo LB, Xie C, Zeng LH, Lv P, Li FZ, Jie JS, Feng M, Wu CY, Yu YQ, Yu SH: Monolayer graphene film on ZnO nanorod array for high-performance Schottky junction ultraviolet photodetectors. Small 2013, 9:2872–2879.

[9] Luo LB, Liang FX, Jie JS: Sn-catalyzed synthesis of SnO_2 nanowires and their optoelectronic characteristics. Nanotechnology 2011, 22: 485701.

[10] Nguyen P, Ng HT, Yamada T, Smith MK, Li J, Han J, Meyyappan M: Direct integration of metal oxide nanowire in vertical field-effect transistor. Nano Lett 2004, 4:651–657.

[11] Fan HJ, Lee W, Hauschild R, Alexe M, Rhun GL, Scholz R, Dadgar A, Nielsch K, Kalt H, Krost A, Zacharias M, Gösele U: Template-assisted large-scale ordered arrays of ZnO pillars for optical and piezoelectric applications. Small 2006, 2:561–568.

[12] Jie JS, Zhang WJ, Jiang Y, Meng XM, Li YQ, Lee ST: Photoconductive characteristics of single-crystal CdS nanoribbons. Nano Lett 2006, 6:1887–1892.

[13] Miao JS, Hu WD, Guo N, Lu ZY, Zou XM, Liao L, Shi SX, Chen PP, Fan FY, Ho JC: Single InAs nanowire room-temperature near-infrared photodetector. ACS Nano 2014, 8:3628–3635.

[14] Gubbala S, Chakrapani V, Kumar V, Sunkara MK: Band-edge engineered hybrid structures for dye-sensitized solar cells based on SnO_2 nanowires. Adv Funct Mater 2008, 18:2411–2418.

[15] Martinson ABF, Elam JW, Hupp JT, Pellin MJ: ZnO nanotube based dye-sensitized solar cells. Nano Lett 2007, 7:2183–2187.

[16] Law JBK, Thong JTL: Simple fabrication of a ZnO nanowire photodetector with a fast photoresponse time. Appl Phys Lett 2006, 88:133114.

[17] Wang MZ, Liang FX, Nie B, Zeng LH, Zheng LX, Lv P, Xie C, Li YY, Luo LB: TiO_2 nanotube array/monolayer graphene film Schottky junction ultraviolet light photodetectors. Part Part System Ch 2013, 30:630–636.

[18] Chen J, Xu L, Li WY, Gou XL: α-Fe2O3 nanotubes in gas sensor and lithium-ion battery application. Adv Mater 2005, 17:582–586.

[19] Wu CY, Wu YL, Wang WJ, Mao D, Yu YQ, Wang L, Xu J, Hu JG, Luo LB: High performance nonvolatile memory devices based on Cu2-xSe nanowires. Appl Phys Lett 2013, 103:193501.

[20] Jiang XC, Herricks T, Xia YN: CuO nanowires can be synthesized by heating copper substrates in air. Nano Lett 2002, 2:1333–1338.

[21] Ethiraj AS, Kang DJ: Synthesis and characterization of CuO nanowires by a simple wet chemical method. Nanoscale Res Lett 2012, 7:70–74.

[22] Tsai CM, Chen GD, Tseng TC, Lee CY, Huang CT, Tsai WY, Yang WC, Yeh MS, Yew TR: CuO nanowire synthesis catalyzed by a CoWP nanofilter. Acta Mater 2009, 57:1570–1576.

[23] Zhu YW, Yu T, Cheong FC, Xu XJ, Lim CT, Tan VBC, Thong JTL, Sow CH: Large-scale synthesis and field emission properties of vertically oriented CuO nanowire films. Nanotechnology 2005, 16:88–92.

[24] Wang SB, Hsiao CH, Cha Zhu YW, Sow CH, Thong JTL: Enhanced field emission from CuO nanowire arrays by in situ laser irradiation. J Appl Phys 2007, 102:114302.

[25] Chen JJ, Wang K, Hartman L, Zhou WL: H2S detection by vertically aligned CuO nanowire array sensors. J Phys Chem C 2008, 112:16017–16021.

[26] Zhu YW, Sow CH, Thong JTL: Enhanced field emission from CuO nanowire arrays by in situ laser irradiation. J Appl Phys 2007, 102:114302.

[27] Hansen BJ, Kouklin N, Lu G, Lin IK, Chen JH, Zhang X: Transport, analyte detection, and opto-electronic response of p-type CuO nanowires. J Phys Chem C 2010, 114:2440–2447.

[28] Hansen BJ, Chan HL, Lu J, Lu GH, Chen JH: Short-circuit diffusion growth of long bi-crystal CuO nanowires. Chem Phys Lett 2011, 504:41–45.

[29] Wang L, Lu M, Wang XG, Yu YQ, Zhao XZ, Lv P, Song HW, Zhang XW, Luo LB, Wu CY, Zhang Y, Jie JS: Tuning the p-type conductivity of ZnSe nanowires via silver doping for rectifying and photovoltaic device applications. J Mater Chem A 2013, 1:1148–1154.

[30] Kresse G, Furthmüller J: Efficient iterative schemes for ab initio total-energy calculations using a plane-wave basis set. J Phys Rev B 1996, 54:11169–11186.

[31] Kresse G, Joubert D: From ultrasoft pseudopotentials to the projector augmented-wave method. Phys Rev B 1999, 59:1758–1775.

[32] Blöchl PE: Projector augmented-wave method. Phys Rev B 1994, 50:17953–17979.

[33] Perdew JP, Burke K, Ernzerhof M: Generalized gradient approximation made simple. Phys Rev Lett 1997, 78:1396–1396.

[34] Liechtenstein AI, Anisimov VI, Zaanen J: Density-functional theory and strong interactions: orbital ordering in Mott-Hubbard insulators. Phys Rev B 1995, 52:R5467–R5470.

[35] Loschen C, Carrasco J, Neyman KM, Illas F: First-principles LDA + U and GGA + U study of cerium oxides: dependence on the effective Uparameter. Phys Rev B 2007, 75:035115.

[36] Gonçalves AMB, Campos LC, Ferlauto AS, Lacerda RG: On the growth and electrical characterization of CuO nanowires by thermal oxidation. J ApplPhys 2009, 106:034303.

[37] Sanal KC, Vikas LS, Jayaraj MK: Room temperature deposited transparent p-channel CuO thin film transistors. Appl Surf Sci 2014, 297:153–157.

[38] Hu J, Li D, Lu JG, Wu R: Effects on electronic properties of molecule adsorption on CuO surfaces and nanowires. J Phys Chem C 2010, 114:17120–17126.

[39] Luo LB, Huang XL, Wang MZ, Xie C, Wu CY, Hu JG, Wang L, Huang JA: The effect of plasmonic nanoparticles on the optoelectronic characteristics of CdTe nanowires. Small 2014, 10:2645–2652.

[40] Kung SC, Veer WE, Yang F, Donavan KC, Penner RM: 20 micros photocurrent response from lithographically patterned nanocrystalline cadmium selenide nanowires. Nano Lett 2010, 10:1481–1485.

[41] Chang SP, Lu CY, Chang SJ, Chiou YZ, Hsueh TJ, Hsu CL: Electrical and optical characteristics of UV photodetector with interlaced ZnO nanowires. IEEE J Sel Top Quant Electron 2011, 17:990–995.

[42] Luo LB, Yang XB, Liang FX, Jie JS, Li Q, Zhu ZF, Wu CY, Yu YQ, Wang L: Transparent and flexible selenium nanobelt-based visible light photodetector. CrystEngComm 2012, 14:1942–1948.

Chapter 15

Optical Properties of GaP/GaNP Core/Shell Nanowires: A Temperature-Dependent Study

Alexander Dobrovolsky[1], Shula Chen[1], Yanjin Kuang[2], Supanee Sukrittanon[3], Charles W Tu[4], Weimin M Chen[1], Irina A Buyanova[1]

[1]Department of Physics, Chemistry and Biology, Linköping University, Linköping 581 83, Sweden
[2]Department of Physics, University of California, La Jolla, San Diego, California 92093, USA
[3]Graduate Program of Material Science and Engineering, University of California, La Jolla, San Diego, California 92093, USA
[4]Department of Electrical and Computer Engineering, University of California, La Jolla, San Diego, California 92093, USA

Abstract: Recombination processes in GaP/GaNP core/shell nanowires (NWs) grown on Si are studied by employing temperature-dependent continuous wave and time-resolved photoluminescence (PL) spectroscopies. The NWs exhibit bright PL emissions due to radiative carrier recombination in the GaNP shell. Though the radiative efficiency of the NWs is found to decrease with increasing temperature, the PL emission remains intense even at room temperature. Two thermal quenching processes of the PL emission are found to be responsible for the degradation of the PL intensity at elevated temperatures: (a) thermal activation of the localized excitons from the N-related localized states and (b) activation of a competing non-radiative recombination (NRR) process. The activation energy of the latter process is determined as being around 180meV. NRR is also found to cause a significant decrease of carrier lifetime.

Keywords: Nanowires, III-V Semiconductors, Photoluminescence

1. Introduction

GaNP has recently attracted much attention as a promising material for applications in optoelectronic and photonic devices, such as light-emitting diode. The incorporation of N in GaP allows one to tune the band gap energy and also to change the band gap character from an indirect one in GaP to a direct-like one in the GaNP alloys, leading to improvements in light emission efficiency[2][3]. A small lattice mismatch of GaNP to Si also provides a unique opportunity to combine high optical efficiency of the III-V compound semiconductors with the capabilities of mature silicon technologies[4]-[6]. Unfortunately, the properties desired for optoelectronic applications have not been fully utilized due to the degradation of optical quality of GaNP caused by the formation of defects that act as centers of non-radiative recombination (NRR)[7]. The NRR processes often dominate carrier recombination and are largely responsible for a reduced optical efficiency of optoelectronic devices[8].

The growth of semiconductor materials in the form of nanostructures, such as nanowires (NWs), often allows suppression of defect formation and therefore offers a possibility to overcome the limitation imposed by NRR that is inherent to higher dimensional layers/structures. It also provides increased flexibility in structural design, thanks to confinement effects. In fact III-V NWs are currently considered as being among the key material systems for future optoelectronic and photonic devices integrated with Si[9]-[11]. Recently, the epitaxial growth of GaP/ GaNP core/shell NWs on Si (111) has been reported[12]. High optical quality of these structures has been demonstrated based on the observation of intense photoluminescence (PL) emission from a single NW[13]. In spite of the high optical quality, fast PL decay caused by NRR processes in the NWs has been reported. The purpose of this work is to gain a better understanding on the quenching processes of the PL intensity from GaP/GaNP core/shell NWs based on temperature-dependent studies by continuous wave (cw) and also time-resolved PL spectroscopies.

2. Methods

The GaP and GaP/GaNP NW samples were grown by gas source molecular beam epitaxy (MBE) on (111)-oriented Si substrates[12]. Scanning electron microscopy (SEM) showed that NWs are hexagonal in shape (inset in **Figure 1**), indicating that NWs were epitaxially grown following the Si 9 (111) crystal orientation. The NWs are uniform in sizes and have an axial length of about 2.5μm, a total diameter of about 220nm for the GaP/ GaNP NWs, and a typical diameter of approximately 110nm for the GaP NWs. The N content in the GaNP NW shell was estimated[12] to be approximately 0.9% on average from room-temperature (RT) PL data. For a comparison, a 750-nm-thick GaN0.009P0.991 epilayer grown by gas-source MBE on a (001)-oriented GaP substrate was also investigated. PL measurements were carried out in a variable temperature cryostat under optical excitation by the 325-nm line of He-Cd laser, the 532-nm line of a solid state laser or the 633-nm line of a He-Ne laser. The resulting PL was detected by a liquid nitrogen cooled charge coupled device after passing through a grating monochromator. Time-resolved PL was excited by a pulsed Ti/sapphire picosecond laser with a photon wavelength of 375nm and a pulse repetition frequency of 76MHz and was detected using a streak camera system.

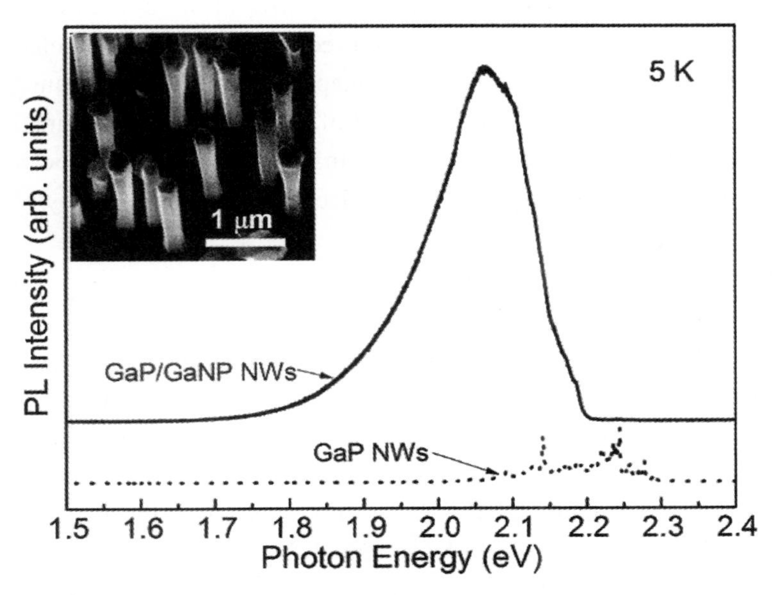

Figure 1. PL spectra from the studied NWs. The inset: an SEM image of the GaP/GaNP NWs.

3. Results and Discussion

Figure 1 shows representative PL spectra measured from the GaP NW (the dotted line, black online) and the GaP/GaNP core/shell NW samples (the solid line, red online) at 5K using the 325-nm line of a solid state laser as an excitation source. The PL emission from the GaP NW is rather weak and is dominated by a series of relatively sharp lines within the 2.05 to 2.32eV spectral range due to the recombination of excitons bound to various residual impurities. Some of the PL lines are very similar to the previously reported emissions due to the recombination of excitons bound to isoelectronic centers involving N impurity, e.g., from an isoelectronic BGa-NP center and its phonon replica[14]. Though the studied GaP NWs are intentionally undoped, the formation of the N-related centers may be caused by contamination of the growth chamber. Further studies aiming to clarify the exact origin of these emissions are currently in progress.

The PL spectra are significantly modified in the GaP/GaNP core/shell NW. First of all, the sharp excitonic lines are replaced by a broad PL band with a rather asymmetric lineshape that peaks at around 2.06eV (**Figure 1**). This emission originates from radiative recombination of excitons trapped at various N-related localized states[13] in the GaNP shell. Secondly, a significant increase of the integrated PL intensity (by about 20 times) is observed which is largely related to the N-induced transition from the indirect bandgap in GaP to a direct bandgap in the GaNP alloy[3]. The observed high efficiency of the radiative recombination in the GaP/GaNP core/shell NW implies that this material system could be potentially promising for applications as efficient nano-sized light emitters.

For practical device applications, it is essential that the high efficiency of radiative recombination is sustained up to RT. Therefore, recombination processes in the studied structures were further examined by employing temperature-dependent PL measurements. In the case of GaP NWs, temperature increase was found to cause a dramatic quenching of the PL intensity so that it falls below the detection limit of the measurement system at measurement temperatures T exceeding 150K. For the GaP/GaNP NW, on the other hand, the PL emission was found to be rather intense at RT even from an individual NW, though significantly weaker than that at 5K. Moreover, thermal quenching is found to be more severe for the high energy PL components which lead to an apparent red shift of the PL maximum position at high T. To get further insights into the mechanisms responsible for the observed thermal quenching, we have analyzed Arrhenius plots of the PL intensity at different detection energies (Edet) as shown in **Figure 2(a)**. The analysis was performed for constant detection energies since (a) the temperature-induced shift of the bandgap energy is significantly suppressed in GaNP alloys[15], and (b) spectral positions of the excitons bound to various deep-level N-related centers do not one-to-one follow the temperature-induced shift of the bandgap energy. This approximation defines error bars of the deduced values as specified below. All experimental data (shown by the symbols in **Figure 2**) can be fitted by

$$I(T) = \frac{I(0)}{1 + C_1 e^{-E_1/kT} + C_2 e^{-E_2/kT}}$$

where $I(T)$ is the temperature-dependent PL intensity, $I(0)$ is value at 4K, E1 and E2 are the activation energies for two different thermal quenching processes, and k is the Boltzman constant [the results of the fitting are shown by the solid lines in **Figure 2(a)**]. The first activation process that occurs within the 30 to 100K

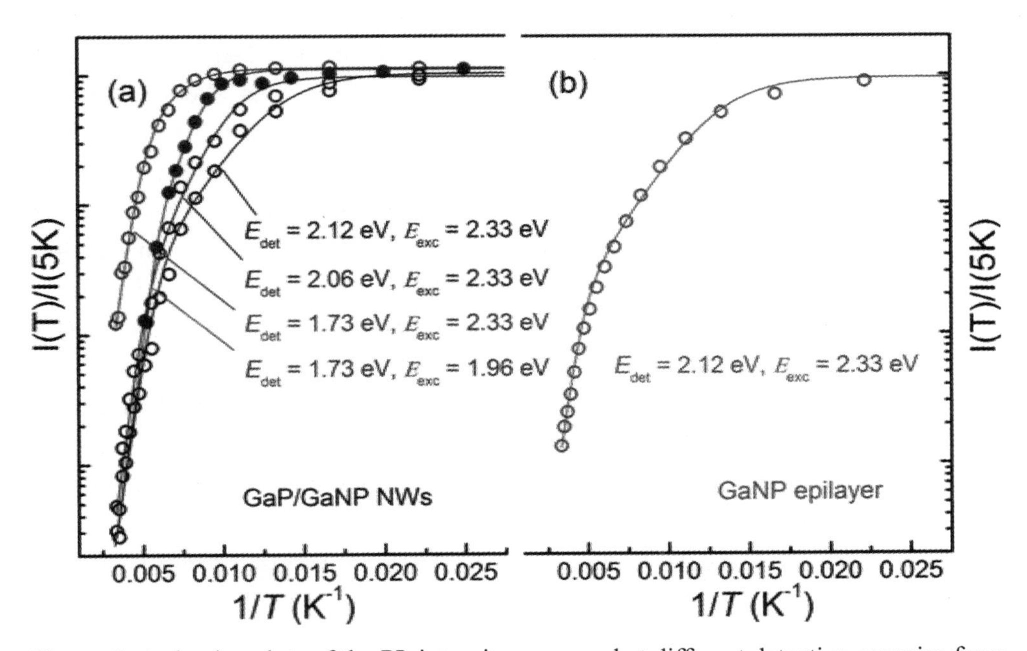

Figure 2. Arrhenius plots of the PL intensity measured at different detection energies from the GaP/GaNP NWs (a) and GaNP epilayer (b).

temperature range is characterized by the activation energy E1 ranging between 40 (at E_{det} = 2.17eV) and 60meV (at E_{det} = 2.06eV). The contribution of this process is most pronounced for high energy PL components that correspond to the radiative recombination at the N-related localized states with their energy levels close to the GaNP band edge. The quenching of the high energy PL components is accompanied by a slight increase in the PL intensity at low E_{det}. Therefore, this process can be attributed to the thermal ionization of the N-related localized states. Such ionization is expected to start from the N-states that are shallower in energy. The thermally activated excitons can then be recaptured by the deeper N states, consistent with our experimental observations. We note that the determined values of E1 do not one-to-one correspond to the "apparent" depth of the involved localized states deduced simply from the distance between E_{det} and the bandgap energy of the GaNP. This is, however, not surprising since such correspondence is only expected for the no-phonon excitonic transitions whereas recombination of excitons at strongly localized states (such as the monitored N states) is usually dominated by phonon-assisted transitions due to strong coupling with phonons.

The second thermal quenching process is characterized by the activation energy E_2 of approximately 180 ± 20meV, which is the same for all detection energies. This process becomes dominant at T > 100K and leads to an overall quenching of the PL intensity irrespective of detection energies. We therefore ascribe it to thermal activation of competing non-radiative recombination which depletes photo-created free carriers and, consequently, causes a decrease in the PL intensity. It is interesting to note that the competing NRR process remains active even when the excitation photon energy (E_{exc}) is tuned to 1.96eV, which is below the GaNP bandgap. Indeed, Arrenius plots of the PL intensity measured at E_{det} = 1.73eV under E_{exc} = 2.33eV [the open circles in **Figure 2(a)**] and E_{exc} = 1.96eV [the dots in **Figure 2(a)**], *i.e.*, under above and below bandgap excitation, respectively, yield the same activation energy E_2. In addition, the PL thermal quenching under below bandgap excitation seems to be even more severe than that recorded under above bandgap excitation. At first glance, this is somewhat surprising as the 1.96eV photons could not directly create free electron–hole pairs and will be absorbed at N-related localized states. However, fast thermal activation of the photo-created carriers from these localized states to band states will again lead to their capture by the NRR centers and therefore quenching of the PL intensity. Moreover, the contribution of the NRR processes is known to decrease at high densities of the photo-created carriers due to partial saturation of the NRR centers which results in a shift of the onset of the PL thermal quenching to higher temperatures. In our case, such regime is likely realized for the above bandgap excitation. This is because of (a) significantly (about 1,000 times) lower excitation power used under below bandgap excitation (restricted by the available excitation source) and (b) a high absorption coefficient for the band-to-band transitions.

The revealed non-radiative recombination processes may occur at surfaces, the GaNP/GaP interface or within bulk regions of GaNP shell. The former two processes are expected to be enhanced in low-dimensional structures with a high surface-to-volume ratio whereas the last process will likely dominate in bulk (or epilayer) samples. Therefore, to further evaluate the origin of the revealed NRR in the studied NW structures, we also investigated the thermal behavior of the PL emission from a reference GaNP epilayer. It is found that thermal quenching of the PL emission in the epilayer can be modeled, within the experimental accuracy, by the same activation energies as those deduced for the NW structure. This is obvious from **Figure 2(b)** where an Arrhenius plot of the PL intensity measured at E_{det} = 2.12eV under E_{exc} = 2.33eV from the epilayer is shown. However, the contribution of the second activation process (defined by the pre-factor C_2 in Equation 1) is found to be larger in the case of the GaNP/GaP NWs. This suggests that the formation of the responsible defects is facilitated in the lower dimensional NWs and that the defects could be at least partly located either at the surface of the GaNP shell or at the GaNP/GaP hetero-interface, consistent with the results of[13].

The activation of the NRR recombination processes at elevated temperatures is also confirmed by the performed time-resolved PL measurements. Typical decay curves of the integrated PL intensity at 5K and RT are shown in **Figure 3**. At 5K, the PL decay is found to be rather slow, *i.e.*, with the decay time τ of the dominant decay component longer than 60ns (the exact value of τ could not be determined from the available data due to the high repetition frequency of the laser pulses). Such slow decay is likely dominated by the radiative lifetime τ_r as it is of the same order of magnitude as previously determined for the radiative transitions within the N-related localized states in the GaNP epilayers[3]. A temperature increase above 100 K causes significant shortening of the PL decay, down to several ns at RT (see the inset in **Figure 3**). The measured decay time contains contributions

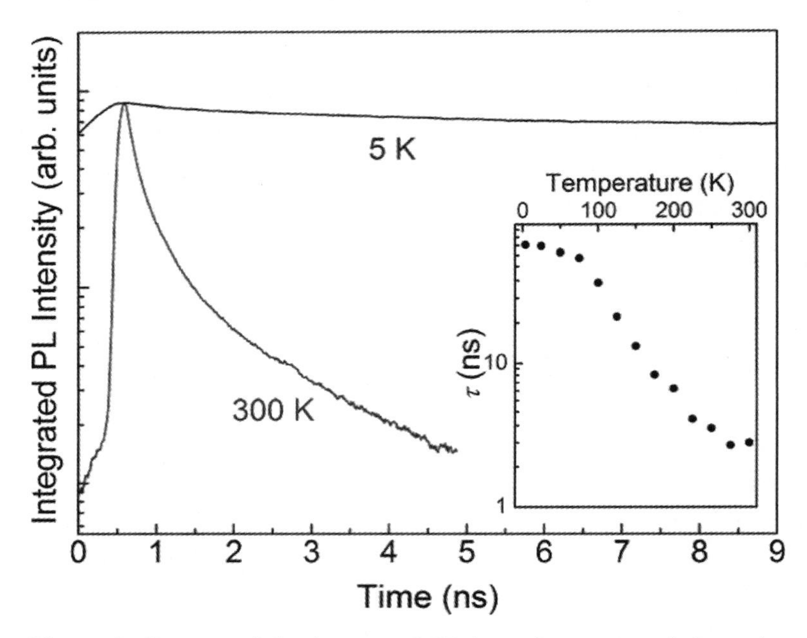

Figure 3. Decays of the integrated PL intensity measured from the GaP/GaNP NWs at 5K and RT.

from both radiative and NRR processes so that $\frac{1}{\tau}=\frac{1}{\tau_{\tau}}+\frac{1}{\tau_{n\tau}}$ where τ_{nr} denotes the non-radiative decay time. Therefore, the observed dramatic shortening of the measured decay time at elevated temperature implies thermal activation of non-radiative carrier recombination, consistent with the results of cw-PL measurements (**Figure 2**).

4. Conclusions

In summary, we have investigated the recombination processes in the GaP NW and GaP/GaNP core/shell NW structures grown on a Si substrate using temperature-dependent cw and time-resolved PL spectroscopies. The GaP/GaNP core/shell NWs are concluded to be a potentially promising material system for applications as efficient nano-sized light emitters that can be integrated with Si. However, the efficiency of radiative recombination in the NWs is found to degrade at elevated temperatures due to the activation of the competing NRR process that also causes shortening of the PL decay time. The thermal activation energy of the

NRR process is determined as being around 180meV.

Competing Interests

The authors declare that they have no competing interests.

Authors' Contributions

AD carried out the experiments and analyzed the data with guidance from IAB and WMC. YK and SS performed the growth of the NWs with guidance from CWT. IAB wrote the final version of the manuscript with contributions from the co-authors. All authors read and approved the final manuscript.

Acknowledgements

Financial support by the Swedish Research Council (grant no. 621-2010-3815) is greatly appreciated. The nanowire growth is supported by the US National Science Foundation under grant nos. DMR-0907652 and DMR-1106369. SS is partially funded by the Royal Government of Thailand Scholarship.

Source: Dobrovolsky A, Chen S, Kuang Y, et al. Optical properties of GaP/GaNP core/shell nanowires: a temperature-dependent study [J]. Nanoscale Research Letters, 2013, 8(1):1–6.

References

[1] Xin HP, Welty RJ, Tu CW: GaN0.011P0.989 red light-emitting diodes directly grown on GaP substrates. Appl Phys Lett 2000, 77:1946–1948.

[2] Shan W, Walukiewicz W, Yu KM, Wu J III, Ager JW, Haller EE, Xin HP, Tu CW: Nature of the fundamental band gap in GaNxP1-x alloys. Appl Phys Lett 2000, 76:3251–3253.

[3] Buyanova IA, Pozina G, Bergman JP, Chen WM, Xin HP, Tu CW: Time-resolved studies of photoluminescence in GaNxP1-x alloys: evidence for indirect–direct band

gap crossover. Appl Phys Lett 2002, 81:52–54.

[4] Furukawa Y, Yonezu H, Ojima K, Samonji K, Fujimoto Y, Momose K, Aiki K: Control of N content of GaPN grown by molecular beam epitaxy and growth of GaPN lattice matched to Si(100) substrate. Jpn J Appl Phys 2002, 41:528–532.

[5] Momose K, Yonezu H, Fujimoto Y, Furukawa Y, Motomura Y, Aiki K: Dislocation-free and lattice-matched Si/GaP1-xNx/Si structure for photo-electronic integrated systems. Appl Phys Lett 2001, 79:4151–4153.

[6] Fujimoto Y, Yonezu H, Utsumi A, Momose K, Furukawa Y: Dislocation-free GaAsyP1-x-yNx/GaP0.98N0.02 quantum-well structure lattice matched to a Si substrate. Appl Phys Lett 2001, 79:1306–1308.

[7] Thinh NQ, Vorona IP, Buyanova IA, Chen WM, Limpijumnong S, Zhang SB, Hong YG, Xin HP, Tu CW, Utsumi A, Furukawa Y, Moon S, Wakahara A, Yonezu H: Properties of Ga-interstitial defects in AlxGa1−xNyP1−y. Phys Rev B 2005, 71:125-209.

[8] Buyanova IA, Chen WM, Tu CW: Recombination processes in N-containing III–V ternary alloys. Solid State Electron 2003, 47:467–475.

[9] Duan X, Huang Y, Cui Y, Wang J, Lieber CM: Indium phosphide nanowires as building blocks for nanoscale electronic and optoelectronic devices. Nature 2001, 409:66–69.

[10] Gudiksen MS, Lauhon LJ, Wang J, Smith DC, Lieber CM: Growth of nanowire superlattice structures for nanoscale photonics and electronics. Nature 2002, 15:617–620.

[11] Mårtensson T, Svensson CPT, Wacaser BA, Larsson MW, Seifert W, Deppert K, Gustafsson A, Wallenberg LR, Samuelson L: Epitaxial III−V nanowires on silicon. Nano Lett 2004, 4:1987–1990.

[12] Kuang YJ, Sukrittanon S, Li H, Tu CW: Growth and photoluminescence of self-catalyzed GaP/GaNP core/shell nanowires on Si(111) by gas source molecular beam epitaxy. Appl Phys Lett 2012, 100:053108.

[13] Dobrovolsky A, Stehr JE, Chen SL, Kuang YJ, Sukrittanon S, Tu CW, Chen WM, Buyanova IA: Mechanism for radiative recombination and defect properties of GaP/GaNP core/shell nanowires. Appl Phys Lett 2012, 101:163106.

[14] Dean PJ, Thomas DG, Frosch CJ: New isoelectronic trap luminescence in gallium phosphide. J Phys C: Solid State Phys 1984, 17:747–762.

[15] Rudko GY, Buyanova IA, Chen WM, Xin HP, Tu CW: Temperature dependence of the GaNxP1−x band gap and effect of band crossover. Appl Phys Lett 2002, 81:2984–2987.

Chapter 16

Optical Properties of Ni and Cu Nanowire Arrays and Ni/Cu Superlattice Nanowire Arrays

Yaya Zhang[1], Wen Xu[1,2], Shaohui Xu[2], Guangtao Fei[2], Yiming Xiao[1], Jiaguang Hu[1,3]

[1]Department of Physics, Yunnan University, Kunming, 650091, China
[2]Key Laboratory of Materials Physics, Institute of Solid State Physics, Chinese Academy of Sciences, Hefei, 230031, China
[3]Department of Math and Physics, Wenshan University, Wenshan, 663000, China

Abstract: In this study, Ni and Cu nanowire arrays and Ni/Cu superlattice nanowire arrays are fabricated using standard techniques such as electrochemical deposition of metals into porous anodic alumina oxide templates having pore diameters of about 50nm. We perform optical measurements on these nanowire array structures. Optical reflectance (OR) of the as-prepared samples is recorded using an imaging spectrometer in the wavelength range from 400 to 2,000nm (*i.e.*, from visible to near-infrared bandwidth). The measurements are carried out at temperatures set to be 4.2, 70, 150, and 200K and at room temperature. We find that the intensity of the OR spectrum for nanowire arrays depends strongly on the temperature. The strongest OR can be observed at about $T = 200$K for all samples in visible regime. The OR spectra for these samples show different features in the visible and near-infrared bandwidths. We discuss the physical mechanisms responsible for these interesting experimental findings. This study is relevant to the application of metal nanowire arrays as optical and optoelectronic devices.

Keywords: Nanowire Array, Optical Properties, Visible and Near-Infrared, Temperature Dependence

1. Introduction

In recent years, quasi one-dimensional (1D) nanostructured materials have received much attention attributed to their interesting physical properties in sharp contrast to the bulk ones and to the potential applications as electronic, magnetic, photonic, and optoelectronic devices[1]-[4]. From a viewpoint of physics, the basic physical properties of nanostructured materials differ significantly from those of bulk materials with the same chemical components. In particular, quantum confinement effects can be observed in the dimensionally reduced nanomaterial systems. Therefore, nanowires have been a major focus of research on nanoscaled materials which can be taken as a fundamental building block of nanotechnology and practical nanodevices. It should be noticed that metal nanowires have displayed unique optical and optoelectronic properties due to surface plasmon resonance (SPR) which is a resonant oscillation of the conducting electrons within the metallic nanostructures. The SPR effect in nanowire structures can cause a tremendous enhancement of the electromagnetic near-field in the immediate vicinity of the particles and can give rise to enhanced scattering and absorption of light radiation. The SPR in metal nanowires and related phenomena (such as the surface-enhanced Raman spectroscopy, nonlinear optic response, plasmonic excitation, to mention but a few) contributes greatly to their promising applications in biosensors, optical devices, and photonic and plasmonic devices[5]-[8]. Moreover, metal nanowire wave guides can excite and emit terahertz (10^{12}Hz or THz) surface plasmon polaritons[9], which can fill the gap of terahertz electronics and optoelectronics. On the other hand, superlattice nanowires have even richer physical properties owing to further quantum confinement of electron motion along the wire direction. They have been proposed as advanced electronic device systems to observe novel effects such as giant magnetoresistance and even high thermoelectric figure of merit[10][11].

Furthermore, with the rapid development of nanotechnology, it is now possible to fabricate nanowire arrays and superlattice nanowire arrays[12][13]. One of the major advantages to apply nanowire arrays and superlattice nanowire arrays as optic and optoelectronic devices is that the optical response of the array structures can be tuned and modulated via varying sample parameters such as the diameter of the wire and the pattern of the array structure. Due to potential applications of the nanowire arrays and superlattice nanowire arrays as optical devices, it is of importance and significance to examine their basic optical properties. In this article, we present a detailed experimental study on the optical properties of three kinds of nanowire array structures such as Ni and Cu nanowire arrays and Ni/Cu superlattice nanowire arrays. We would like to examine how these advanced nanostructured material systems can respond to light radiation, how their optical properties depend on temperature and radiation wavelength, and why the optical properties of the nanowire arrays differ from those observed in bulk materials.

2. Methods

Samples and Measurements

In this study, three kinds of nanowire array structures are fabricated, including Ni arrays, Cu arrays, and Ni/Cu superlattice arrays. Samples are prepared by direct current electrodeposition[14]-[16] of metal into the holes of porous anodic alumina membrane (PAAM) with the pore size of about 50nm. Noteworthy is the diameter of the nanowires used in the investigation, which is about 50nm. The length of the nanowires is about 30μm. The holes of the PAAM are periodically in hexagonal pattern, which can serve as template. The distance between adjacent wires is about 60nm. Because of the confinement of the PAAM material, metal nanowires grow only along the direction of nanopores of the PAAM template and, therefore, form an array structure. In these samples, a layer of Au film (about 200-nm thick) is sputtered onto one side of the PAAM template to serve as the working electrode. The schematic diagram of the Ni or Cu nanowire array in the PAAM template is shown in Figure 1. For the fabrication of the Ni/Cu superlattice arrays, Ni and Cu materials are deposited alternately into the PAAM holes. The details of the sample fabrication were documented in[14]-[16].

For the measurement of optical reflection (OR) spectrum, the incident and emergent light beams are set at an angle of 45° to the sample surface (see **Figure 1**). The measurements are carried out in the visible (400 to 800 nm in wavelength)

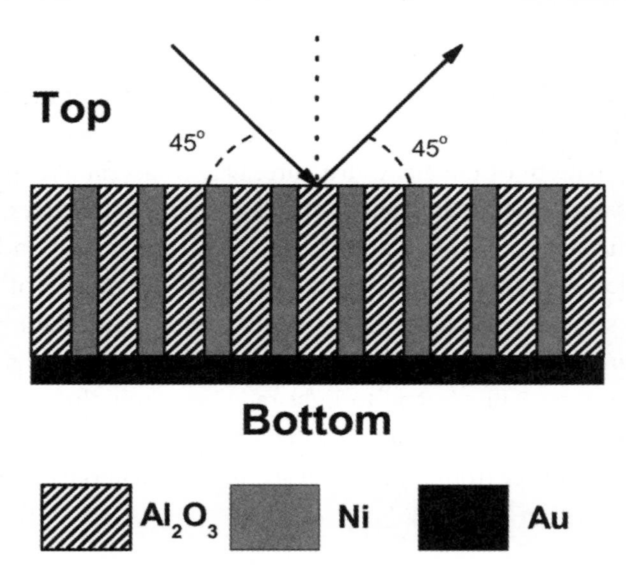

Figure 1. Schematic diagram of Ni or Cu nanowire arrays in porous anodic alumina membrane. In the measurement of optical reflection spectrum, the incident light is set at a 45° angle to the sample surface, and the emergent light beam is also at a 45° angle to the sample surface.

and near-infrared (1 to 2μm in wavelength) bandwidths. The tungsten halogen lamp is taken as a white incident light source for the measurements in the visible bandwidth. The Si carbide rod is employed as broadband infrared incident light source for the measurements in the near-infrared bandwidth. The OR spectrum is recorded using an imaging spectrometer (iHR320 HORIBA Jobin Yvon Inc., Edison, NJ, USA) where the PMT is used for the detection of 400- to 800-nm wavelength regime, and the InGaAs photodetector is employed for the measurement of 1- to 2-μm wavelength regime. For measurements in the visible regime, the temperatures are set at 4.2, 70, 150, and 200K and at room temperature. The change of temperature is achieved in an Oxford cooling system. The measurements in the near-infrared regime are undertaken at room temperature.

3. Results and Discussion

The OR spectra for Ni and Cu nanowire arrays and Ni/Cu superlattice nanowire arrays are shown in **Figure 2** in visible bandwidth for different temperatures at 4.2, 70, 150, 200, and 297K, respectively. As can be seen, the intensity of

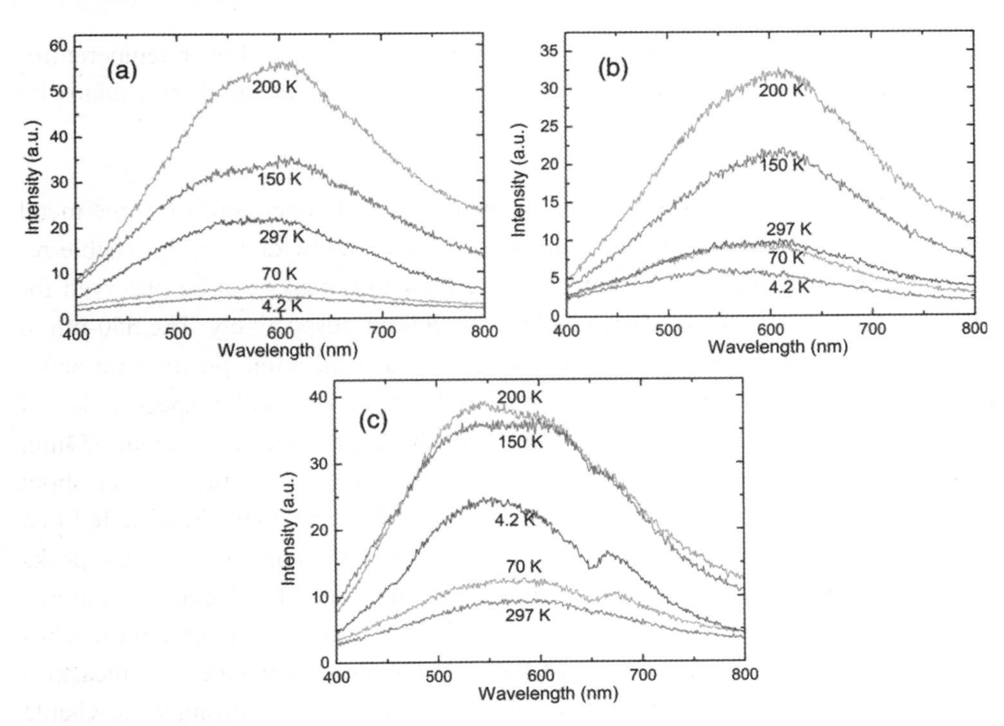

Figure 2. The spectra of optical reflection for nanowire arrays measured at different temperatures of 4.2, 70, 150, 200, and 297K as indicated. The results for a Ni nanowire array (a), a Ni/Cu superlattice nanowire array (b), and a Cu nanowire array (c) are shown.

OR in nanowire array structures depends strongly on temperature. When temperature (T) < 200K, the intensity of OR for a Ni nanowire array sample increases with temperature. When T > 200K, the OR intensity decreases with increasing temperature. The strongest OR can be observed at about 200K. A similar phenomenon can be found for a Ni/Cu superlattice nanowire array sample. In contrast, the OR spectra for Cu nanowire arrays [see **Figure 2(c)**] show different temperature dependence. With increasing temperature, the intensity of OR for a Cu nanowire array first decreases in the 4.2- to 70-K regime, then increases in the 70- to 200-K regime, and decreases again when T > 200K. Again, the strongest OR for Cu nanowire arrays can be observed at about T = 200K. These experimental findings suggest that 200K is an appropriate temperature for the enhancement of optical reflection from Cu, Ni, and Ni/Cu superlattice nanowire array structures. This can provide a basis for further investigation into other optical properties such as optical absorption and emission from metal nanowire arrays in visible regime. We find that when T > 200K, the OR spectrum for Ni/Cu superlattice nanowire array lies between those for Cu and Ni nanowire arrays. However, at lower temperatures (e.g., at 150K), the intensity of the OR spectrum for Ni/Cu superlattice nanowire array is lower than those for Cu and Ni nanowire arrays.

In **Figure 3**, the OR spectra are shown at room temperature for three metal nanowire array samples in visible and near-infrared bandwidths. In the visible regime [see **Figure 3(a)**], two relatively wide reflection peaks can be observed for all samples at about 500 to 650nm and 650 to 700nm, respectively. The 650-nm to 700-nm peaks for the three samples appear at almost the same position (at about 667nm), while the 500-nm to 650-nm ones redshift slightly with respect to that of the incident light source. The peak position of the light source is at about 554nm, whereas the peaks for Cu and Ni/Cu superlattice nanowire arrays are at about 585nm and that for Ni nanowire arrays is at about 600nm. It should be noted that the visible light source provided by the tungsten halogen lamp has two main peaks in the 400- to 800-nm wavelength regime. The intensity of infrared light source given by the Si carbide rod decreases when the radiation wavelength approaches 2μm. The variation of the intensity of the light sources is enhanced via measurement systems. We notice that Ni nanowire arrays reflect more strongly the visible light; Cu nanowire arrays reflect relatively weakly, and the OR spectrum for Ni/Cu superlattice nanowire arrays is just in between them. In the near-infrared range of 1,000 to 2,000nm [see **Figure 3(b)**], the peaks of OR spectra for Cu nanowire

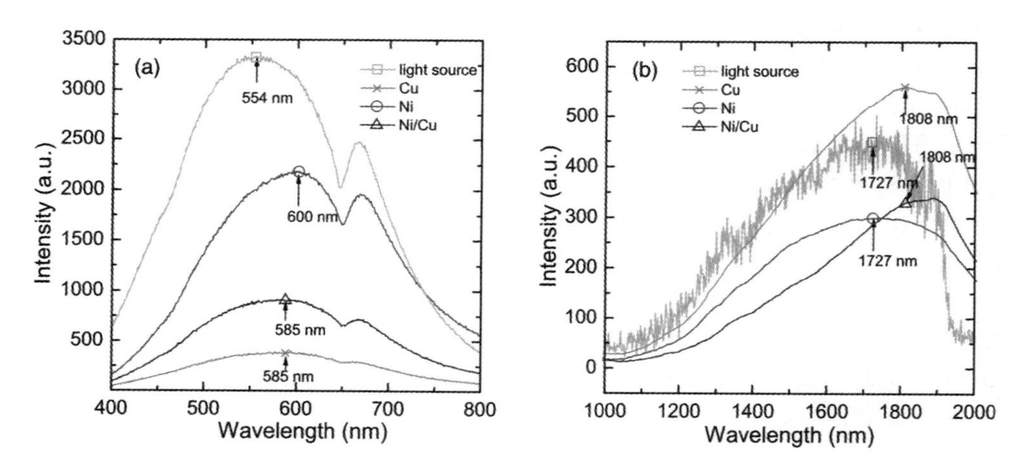

Figure 3. The OR spectra for three kinds of nanowire arrays in visible (a) and near-infrared (b) bandwidths. The measurements are carried out at room temperature. The intensity of the incident light source is shown as a reference. The peak positions are marked to guide the eye.

arrays and Ni/Cu superlattice nanowire arrays are at about 1,808 nm, and Ni nanowire arrays and light source are at about 1,727nm. The OR spectra for nanowire arrays redshift slightly with respect to the spectrum of the light source. In contrast to the visible regime, the Cu nanowire array reflects more strongly the infrared radiation than Ni nanowire array. Interestingly, the OR spectrum for Ni/Cu superlattice nanowire array is below that for Ni nanowire array when radiation wavelength is less than 1,730nm, and it is located in between the OR spectra for Ni and Cu arrays when radiation wavelength is larger than 1,730nm.

It is known that the OR spectrum of a metal nanostructure is determined mainly by surface plasmon modes and corresponding SPR. Our results indicate that the Cu, Ni, and Ni/Cu superlattice nanowire arrays show roughly the same OR spectra when the diameter and the length of the wires are the same. This implies that the features of the SPR in Ni, Cu, and Ni/Cu superlattice nanowire arrays have some similarities. From a fact that a strong optical reflection can weaken optical absorption and transmission, we can predict that Cu nanowire arrays can have stronger (weaker) optical absorption than Ni nanowire arrays in the visible (near-infrared) regime. The strong temperature dependence of the OR spectra for these array structures implies that there exists strong electron-phonon scattering in nanowire array samples. In the presence of light radiation field and phonon scattering, the electrons in an array structure can gain energy from the radiation field and lose energy via emission of phonons and excitation of plasmon and surface plasmon. At relatively low temperatures, the electronphonon interaction is achieved mainly via phonon emission scattering channels, and the strength of the scattering increases with temperature. A strong phonon scattering implies a small electronic conductivity or a weak optical absorption and, thus, a strong OR. This is the main reason why the OR in metal nanowire arrays increases with temperature in the low-temperature regime. At relatively high temperatures, because phonon occupation number increases rapidly with temperature, the electron-phonon interaction is achieved not only through phonon emission, but also through phonon absorption. Phonon absorption can result in a gain of electron energy and in an increase in electronic conductivity. In this case, the effective strength of electron-phonon scattering decreases with increasing temperature, and therefore, the intensity of OR decreases with increasing temperature. It is interesting to note that such a mechanism is responsible to temperature-dependent electronic and optical properties in polar-semiconductor-based electronic systems. For example, it was found that the strongest magnetophonon resonance can be observed at about 180K for GaAs-based bulk and low-dimensional systems. However, we do not know the exact mechanism responsible to the decrease in OR for Cu nanowire arrays with increasing temperature when the temperature is within 4.2 to 70K regime. This may suggest a strong metallic optic conduction in Cu nanowire array samples in this temperature regime.

We note that in a metal nanowire array, the visible (infrared) OR is caused mainly by SPR via interband (intraband) electronic transitions. Due to quantum confinement effect in the nanowire array structure, the surface plasmon and surface plasmon polariton modes induced by interand intraband transitions can have different features. For bulk metals, the interband SPR induced mainly by electronic transition from higher-energy *sp*-band to lower-energy *d*-band determines the color of the metal. At the same time, the intraband SPR within the *sp*- and *d*-bands gives free-carrier optic absorption which leads mainly to a lower-frequency background optic reflection. Because Cu is a better conductor than Ni, Ni normally reflects more strongly the visible light radiation than Cu does. However, for nanowire arrays, the electronic states in different bands are quantized. The intraband electronic transition accompanied by the absorption of photons can be achieved via intersubband transition events which can result in resonant optical absorption when photon energy approaches the energy spacing between two subbands. Thus, intraband optical absorption can be enhanced in nanowire arrays. The results shown in **Figure 3(b)** indicate that the enhancement of intraband optical absorption in Ni nanowire arrays is stronger than that in Cu nanowire arrays. As a result, Cu nanowire arrays reflect more strongly the infrared radiation than Ni arrays do. Because the quantum confinement effect affects mainly the electronic states in different bands in the array structure, the main features of OR due to interband electronic transition does not change very significantly. This is why Ni nanowire arrays can reflect more strongly the visible radiation than Cu arrays can, as shown in **Figure 3(a)** and similar to the case for bulk materials.

Moreover, our results show that in the visible regime and when $T > 200$K, the OR spectrum for Ni/Cu superlattice nanowire arrays lies between those for Cu and Ni nanowire arrays. However, at relatively lower temperatures (e.g., at 150K), the intensity of the OR spectrum for Ni/Cu superlattice nanowire array is lower than those for Cu and Ni nanowire arrays. We believe that this may have resulted from different features of the phonon modes and electron-phonon scattering in nanowire and superlattice nanowire structures. In superlattice nanowire systems formed by different host materials, the phonon modes can be quantized and the conducting electrons are confined along the wire direction. The quantized phonon modes can weaken the electron-phonon scattering because a scattering event requires momentum and energy conservation. On the other hand, the localized electrons can interact more strongly with phonons. Our results suggest that when $T > 200$K, the former case is dominant, and when T"- 150K, the latter effect is stronger.

4. Conclusions

In this study, Cu, Ni, and Ni/Cu nanowire arrays have been fabricated using state-of-the-art nanotechnology. The optical measurements on these nanowire arrays have been carried out in visible and near-infrared bandwidths for different temperatures. We have found that the optical reflection spectra of these samples depend strongly on temperature and on radiation wavelength. In particular, (1) the strongest OR in the visible regime can be observed at about 200K for all samples, and (2) the OR for Cu nanowire arrays show a different dependence on temperature and radiation wavelength from that for Ni nanowire arrays. These results indicate that the surface plasmon resonances induced by inter and intraband electronic transitions, the electron-phonon interaction, and the quantum confinement effect can play important roles in affecting optical properties of the metal nanowire array structure. We hope that the interesting experimental findings from this study can provide an in-depth understanding of optical properties of Cu and Ni nanowire arrays and Cu/Ni superlattice nanowire arrays and can provide a physical base for the application of metal nanowire arrays as advanced optical and optoelectronic devices.

Competing Interests

The authors declare that they have no competing interests.

Authors' Contributions

WX proposed the research work, coordinated the collaboration, and carried out the analyses of experimental results. YYZ designed the experiment and experimental setup, carried out the measurements, and drafted the manuscript. SHX and GTF fabricated the nanowire and superlattice nanowire array samples. YMX and JGH participated in experimental measurements, results and discussion, and analyses. All authors read and approved the final manuscript.

Acknowledgements

This work was supported by the National Natural Science Foundation of China (grant no. 10974206), Department of Science and Technology of Yunnan Province, and by the Chinese Academy of Sciences.

Source: Zhang Y, Wen X, Xu S, et al. Optical properties of Ni and Cu nanowire arrays and Ni/Cu superlattice nanowire arrays[J]. Nanoscale Research Letters, 2012, 7(1):1–6.

References

[1] Lew K-K, Redwing JM: Growth characteristics of silicon nanowires synthesized by vapor-liquid-solid growth in nanoporous alumina templates. *J Crystal Growth* 2003, 254:14.

[2] Cui Y, Lieber CM: Functional nanoscale electronic devices assembled using silicon nanowire building blocks. *Science* 2001, 291:851.

[3] Bok H-M, Shuford KL, Kim S, Kim SK, Park S: Multiple surface plasmon modes for a colloidal solution of nanoporous gold nanorods and their comparison to smooth gold nanorods. *Nano Lett* 2008, 8:2265–2270.

[4] Phillips J: Evaluation of the fundamental properties of quantum dot infrared detectors. *J Appl Phys* 2002, 91:4590–4594.

[5] Germain V, Brioude A, Ingert D, Pileni MP: Silver nanodisks: size selection via centrifugation and optical properties. *J Chem Phys* 2005, 122:124707.

[6] Du YB, Shi LF, He TC, Sun XW, Mo YJ: SERS enhancement dependence on the diameter and aspect ratio of silver-nanowire array fabricated by anodic aluminium oxide template. *Appl Surf Sci* 2008, 255:1901.

[7] Ryua S-W, Kima C-H, Hana J-W, Kima C-J, Jungb C, Parkb HG, Choia Y-K: Gold nanoparticle embedded silicon nanowire biosensor for applications of label-free

DNA detection. *Biosens Bioelectron* 2010, 25:2182–2185.

[8] Zaraska L, Sulka GD, Jaskuła M: Porous anodic alumina membranes formed by anodization of AA1050 alloy as templates for fabrication of metallic nanowire arrays. *Surf Coat Technol* 2010, 205:2432–2437.

[9] Deibel JA, Wang K, Escarra M, Berndsen N, Mittleman DM: The excitation and emission of terahertz surface plasmon polaritons on metal wire waveguides. *C R Physique* 2008, 9:215–231.

[10] Piraux L, George JM, Despres JF, Leroy C, Ferain E, Legras R, Ounadjela K, Fert A: Giant magnetoresistance in magnetic multilayered nanowires. *Appl Phys Lett* 1994, 65:2484.

[11] Blondel A, Meier JP, Doudin B, Ansermet JP: Giant magnetoresistance of nanowires of multilayers. *Appl Phys Lett* 1994, 65:3019.

[12] Wu Y, Fan R, Yang P: Block-by-block growth of single-crystalline Si/SiGe superlattice nanowires. *Nano Lett* 2002, 2:83.

[13] Gudiksen MS, Lauhon LJ, Wang J, Smith DC, Lieber CM: Growth of nanowire superlattice structures for nanoscale photonics and electronics. *Nature* 2002, 415:617.

[14] Wang XW, Fei GT, Xu XJ, Jin Z, Zhang LD: Size-dependent orientation growth of large-area ordered Ni nanowire arrays. *J Phys Chem B* 2005, 109:24326–24330.

[15] Xu SH, Fei GT, Zhu XG, Wang B, Wu B, Zhang LD: A facile and universal way to fabricate superlattice nanowire arrays. *Nanotechnology* 2011, 22:265602.

[16] Zhou WF, Fei GT, Li XF, Xu SH, Chen L, Wu B, Zhang LD: In situ X-ray diffraction study on the orientation-dependent thermal expansion of Cu nanowires. *J Phys Chem C* 2009, 113:9568.

Chapter 17

Optoelectronic Spin Memories of Electrons in Semiconductors

M. Idrish Miah[1,2]

[1]Department of Physics, University of Chittagong, Chittagong 4331, Bangladesh
[2]Queensland Micro-and Nanotechnology Centre, Griffith University, Nathan, Brisbane, QLD 4111, Australia

Abstract: We optically generate electron spins in semiconductors and apply an external magnetic field perpendicularly to them. Time-resolved photoluminescence measurements, pumped with a circularly polarized light, are performed to study the spin polarization and spin memory times in the semiconducting host. The measured spin polarization is found to be an exponential decay with the time delay of the probe. It is also found that the spin memory times, extracted from the polarization decays, enhance with the strength of the external magnetic field. However, at higher fields, the memory times get saturated to sub-ls because of the coupling for interacting electrons with the local nuclear field.

Keywords: Optical Materials, Semiconductors , Magnetic Properties

1. Introduction

Spintronics is a relatively new and emerging field in solid-state physics where, rather than the charge of the electron, its spin plays the dominant role. The study of the spin property of electrons in solid-state systems and its exploitation for future technological applications is the main task of spintronics (Prinz 1998; Dyakonov and Khaetskii 2008; Ziese and Thornton 2001; Awschalom et al. 2002). However, for the successful incorporation of spins into the currently existing semiconductor technology, one has to resolve technical issues such as efficient generation/injection of spins and their optimal control or spin dynamics (Awschalom et al. 2002).

Generation/injection of spins (or spin polarization) usually means creating a non-equilibrium spin population. This has been achieved either by optical method (using a circularly polarized light excitation) or by an electrical means by magnetic semiconductors, or ferromagnetic contacts (Prinz 1998; Awschalom et al. 2002). Although electrical spin injection is desirable, this technique is found not to be efficient and has resulted in low spin injection effects due to the conductivity mismatch. However, the spin generation by the optical methods has been successful (Prinz 1998) and the high spin polarization of conductor band electrons in semiconductor heterostructures has been obtained (Endo et al. 2000).

Despite substantial progress in optical spin generation, a further hurdle still remains in the spin transport which is the lack of a proper understanding of spin dynamics and control in semiconductor-based heterostructures (Gotoh et al. 1998; Cortez et al. 2002; Wang et al. 2007). Spin dynamics in GaAs-based semiconductors and their low-dimensional systems has been studied by time-resolved polarization of photoluminescence (PL) (Seymour and Alfano 1980; Wagner et al. 1993; Endo et al. 2000; Gotoh et al. 1998; Cortez et al. 2002). The PL spin polarization was used to directly measure the spin-flip times which correspond to the spin memory times (ss) in the samples. The measured values range from 100 ps to 20 ns in p-doped GaAs and related materials (Seymour and Alfano 1980; Wagner et al. 1993; Endo et al. 2000), and 1–15 ns in InGaAs quantum wells and n-type InAs/GaAs quantum dots (Gotoh et al. 1998; Cortez et al. 2002). Here, in the present investigation, electron spins in lightly n-doped GaAs are generated optically in the presence of an external magnetic field. The detection of them is performed using a time-resolved pump-probe excitonic PL spin polarization measurement.

2. Experimental Details

Samples were GaAs layers in an AlGaAs heterostructure, grown by molecular beam epitaxy, and were lightly n-doped with a doping density of $3 \times 10^{15}cm^{-3}$. A circularly polarized light from a modulated Ti-sapphire laser, operated at 30mW and 810nm, was used to generate spin-polarized electrons in the samples. The laser intensity was modulated on/off with an acousto-optic modulator (AOM) to obtain light pulses as short as 15ns. The AOM was controlled by an Interface Technology via digital word generator (DWG), which delivered voltage pulses to the AOM's pulse shaping input. The DWG was controlled by a pulse controller, and was triggered by a 20kHz photo-elastic modulator (PEM) in the PL detection path. The PEM operated as a sinusoidally oscillating quarter-wave plate, which was combined with a linear polarizer to make a circularly polarization analyzer. The PEM additionally triggered the two channels of the counter so that the two (σ^+ right and σ^-, left) circular polarizations could be separately recorded. For the application of the magnetic field (B) in the Voigt configuration, an Oxford superconducting magnet was used. Experiments were performed at low temperature. The sample temperature was measured, and it was 4.2K. A SPEX 1680 double grating spectrometer, with a photomultiplier tube (PMT) and photon counter/ quantum sensor (LI-190 SA, LI-COR Inc.), was used to collect and measure the PL. Important characteristics of the present experiment include using excitonic PL polarization to monitor the spin polarization of doped electrons, initializing the electronic polarization with a lengthy pump light pulse, and reading out the electronic polarization with a shorter probe light pulse (probe), where the spin polarization relaxes during the dark period between pump and probe pulses. The degree of PL circular polarization (P) was calculated by defining it as the ratio of the difference of the PL emission intensities of the right and left circularly polarized PL to their sum, i.e.,

$$P = \frac{I_{PL+} - I_{PL-}}{I_{PL+} + I_{PL-}}$$

where I_{PL+} (I_{PL-}) is the *PL* emission intensity of the right (left) circularly polarized *PL*. The *PL* emission intensities were measured in the left and right circular polarizations under the right circularly polarized excitation.

3. Results and Discussion

The pump pulse generated spin-polarized electrons, while a probe pulse read out their polarization. **Figure 1** shows typical PL emission intensities measured using pulses with the same and opposite circular polarizations. There is a difference between the different spin polarization conditions, which is caused by spin-dependent phase-space filling (Wang et al. 2007). From the measured data, the PL integrated intensity was used for the calculations of the PL polarization.

In order to explore conditions for pump and probe pulses, we also studied a dependence of the circular polarization on the length of a single pulse for a constant pump power density. It was found that the polarization increases with the increase in the pulse length, but for the higher pulse lengths, it gets saturated. The free exciton line is polarized to a degree that depends on the pulse length. At a small pulse length, the polarization is small and depends on the spin relaxation, electron density, and generation rate, whereas at larger pulse lengths, the polarization saturates while the power density is held constant. The number of photo-generated electrons must be comparable to the number of doped electrons in order for an appreciable polarization to be set by the light pulse. The results suggest a measurement using a probe pulse smaller than 40ns and a pump larger than 220ns. The present pump-probe measurements were performed with a delay ($\Delta\tau$) using 250ns pump pulses and 25ns probe pulses. In the measurement conditions, the quantum sensor ensured that PL was only collected from the probe pulse.

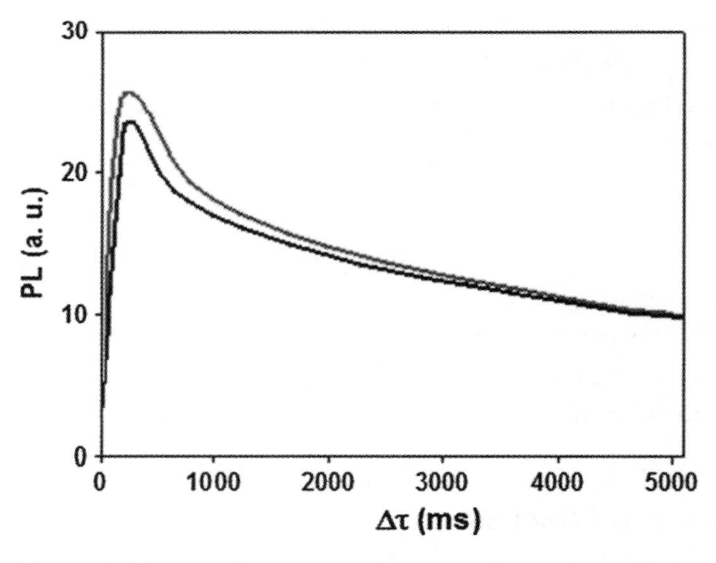

Figure 1. PL intensities measured using optical pulses with the same (upper) and opposite (lower) circular polarizations

Figure 2 shows the spin polarization (mean data point) as a function of $\Delta\tau$ for different strengths of the magnetic field. As obvious, the PL polarization decays with $\Delta\tau$. The observed PL polarization decay corresponds to electron spin relaxation in the sample for the transverse magnetic field. For a long delay, there is still a small polarization value. This is because the probe pulse cannot measure the system without affecting it, as it is circularly polarized (Miah 2011a). However, the probe must be weak enough that one can observe the existing polarization, but strong enough so that it produces enough photoluminescence yield to detect rigorously. For this reason, for example, for less-doped samples, weaker probe beams are required, but there might be a difficulty for detecting the signals. For shorter delays, a larger polarization was seen, demonstrating that polarization persisted into the dark period between pulses. The polarization decayed exponentially between the short- and long-time limits in accordance with the decay law. The experimental data are well fitted by the exponential decay. Exponential fits to the data give the spin-flip times, corresponding to the spin memory times for the respective fields.

The dependence of the spin memory times on the magnetic field is shown in **Figure 3**. As can be seen, the spin memory time increases with the increase in the

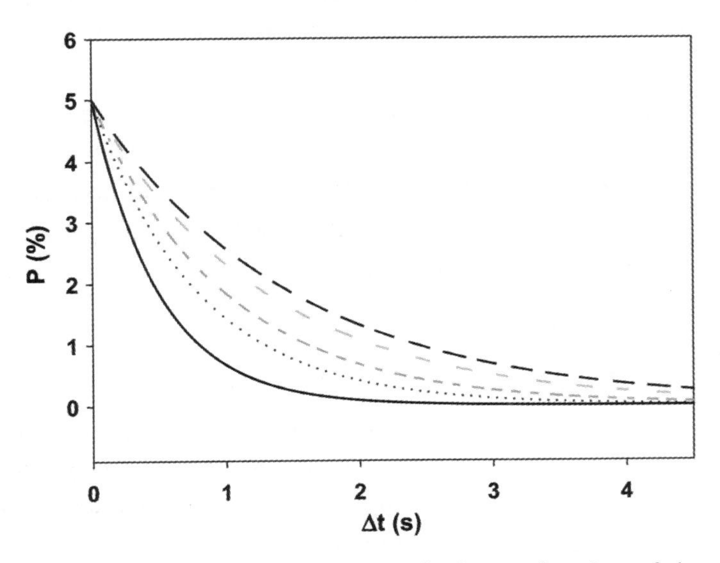

Figure 2. Spin polarization (mean data) as a function of the pump- probe delay for different magnetic field strengths: 0.5T (solid line), 0.9T (dotted line), 1.6T (short dashed line), 3T (dot-dashed line), and 5T (long dashed line).

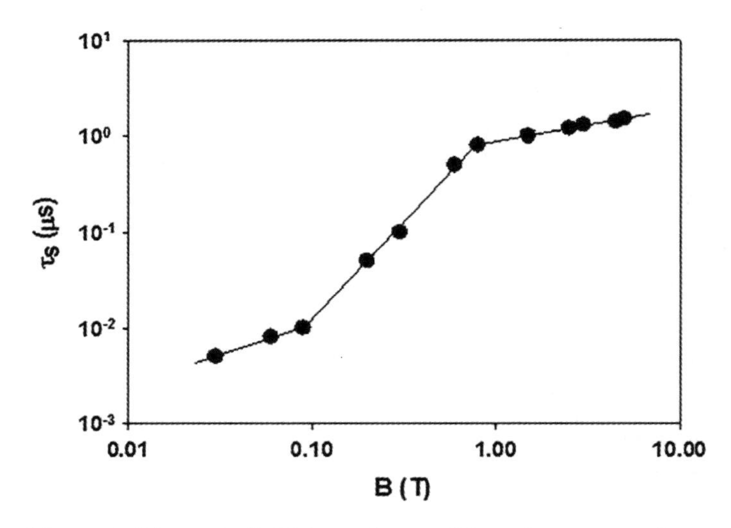

Figure 3. Magnetic field dependence of the spin memory time.

strength of the magnetic field up to about 8T. However, at higher magnetic fields, it got saturation. In the intermediate field strengths, there is also another feature where a rapid increase of the memory time is seen. This regime corresponds to interacting electrons at slightly higher fields, because there is a natural distribution of donor separations, which can lead to more- and less-localized electrons (Pikus and Titkov 1984; Kityk 1991). This means that the high-field regime corresponds to the localized electrons. For localized electrons, the main relaxation mechanism under this condition is hyperfine coupling to the nuclei. The hyperfine interaction produces a fluctuation magnetic field, also known as the effective magnetic field, in which an electron precesses (Prinz 1998). As an external magnetic field is applied, the nuclear contribution to spin relaxation will be reduced when the external field exceeds the nuclear fluctuation field (Lampel and Weisbuch 1975; Ivchenko and Kiselev 1992).

However, the lifetime at zero fields should be equal to the inhomogeneous spin dephasing time, since there is no energy splitting between the two spin states. The spin relaxation decreases with the increase in strength of the magnetic field for the lower-field region. For the intermediate region, the motional averaging occurs and in this regime, spin memory time will increase with the external magnetic field with the spin relaxation (Pikus and Titkov 1984). This field dependence arises from motional averaging of the hyperfine effects for interacting electrons. However, the application of a magnetic field in the transverse configuration tends to localize electrons due to cyclotron motion (Dzhioev et al. 1997). As the external magnetic field is further increased, the motional averaging is no longer a constant and increases as the electrons become localized due to the field (Salimullah et al. 2003). As a result, the precession frequency increases with the field and becomes comparable to the inverse of the correlation time corresponding to the motional averaging at the fields of some T (Miah 2011b).

4. Conclusions

Spin memories of electrons in optoelectronic semiconductor devices were investigated. The dynamics of spins generated by an optical technique was studied in the presence of an external magnetic field applied perpendicularly to the spins. A time-resolved pump-probe photoluminescence measurement was performed to study the spin polarization and spin memory times in the host. The measured spin polarization was found to be an exponential decay with the time delay. It was also found that the spin memory times enhance with the strength of the externally applied transverse magnetic field. However, because of the hyperfine coupling for interacting electrons with the local nuclear field (an interaction between the electron spin and the nuclear spin), the spin memory times were found to get saturated at higher fields. The findings resulting from this investigation might have potential applications in semiconductor-based opto-spintronics devices.

Source: Miah M I. Optoelectronic spin memories of electrons in semiconductors [J]. Applied Nanoscience, 2015, 6:1–4.

References

[1] Awschalom DD, Loss D, Samarth N (eds) (2002) Semiconductor spintronics and quantum computation. Springer, Berlin.

[2] Cortez S, Krebs O, Laurent S, Senes M, Marie X, Voisin P, Ferreira R, Bastard G, Gerard J-M, Amand T (2002) Optically driven spin memory in n-doped InAs-GaAs quantum dots. Phys Rev Lett 89:207401–207404.

[3] Dyakonov MI, Khaetskii AV (2008) Spin Hall Effect. In: Dyakonov MI (ed) Spin physics in semiconductors. Springer, Berlin).

[4] Dzhioev RI, Zakharchenya BP, Korenev VL, Stepanova MN (1997) Spin diffusion of optically oriented electrons and photon entrainment in n-gallium Arsenide. Phys Solid State 39:1765–1768.

[5] Endo T, Sueoka K, Mukasa K (2000) High spin polarization in semiconductors. Jpn J Appl Phys 39:397–400.

[6] Gotoh H, Ando H, Kamada H, Chavez-Pirson A, Temmyo J (1998) Spin relaxation of excitons in zero-dimensional InGaAs quantum disks. Appl Phys Lett 72:1341–1343.

[7] Ivchenko EL, Kiselev AA (1992) Cyclotron energies of electrons in quantum wells. Sov Phys Semicond 26:827–831.

[8] Kityk IV (1991) Band energy structure calculations in semiconductors. Phys Solid State 33:1026–1030.

[9] Lampel G, Weisbuch C (1975) Proposal for an efficient source of polarized photoelectrons from semiconductors. Solid State Commun 16:877–880.

[10] Miah MI (2011a) Depolarization of the photoluminescence and spin relaxation in n-doped GaAs. Opt Commun 284:1254–1257.

[11] Miah MI (2011b) Bias induced reduction of the electron-hole coupling. Solid State Sci 13:1709–1714.

[12] Pikus GE, Titkov AN (1984) In optical orientation. In: Modern problems in condensed matter science, vol 8, North-Holland, Amsterdam Prinz GA (1998) Magnetoelectronics. Science 282:1660–1663.

[13] Salimullah M, Shukla PK, Ghosh SK, Nitta H, Hayashi Y (2003) Electron-phonon coupling effect on wakefields in piezoelectric semiconductors. J Phys D 36:958–960.

[14] Seymour RJ, Alfano RR (1980) Time-resolved measurement of the electron-spin relaxation kinetics in GaAs. Appl Phys Lett 37:231–233.

[15] Wagner J, Schneider H, Richards D, Fischer A, Ploog K (1993) Observation of extremely long electron-spin-relaxation times in p-type d-doped GaAs/AlxGa1-xAs double heterostructures. Phys Rev B 47:4786–4790.

[16] Wang T, Li A, Tan Z (2007) Band-filling in quantum wells. Proc SPIE 6838:683814–683817.

[17] Ziese M, Thornton MJ (eds) (2001) Spin electronics, vol. 569, Springer, Heidelberg.

Chapter 18

Optoelectronic System for the Determination of Blood Volume in Pneumatic Heart Assist Devices

Grzegorz Konieczny[1], Tadeusz Pustelny[1], Maciej Setkiewicz[1], Maciej Gawlikowski[2]

[1]Department of Optoelectronics, Silesian University of Technology, 2A Akademicka Str., 44–100 Gliwice, Poland
[2]Foundation of Cardiac Surgery Development Zabrze, 345a Wolności Str., 41–800 Zabrze, Poland

Abstract: Background: The following article describes the concept of optical measurement of blood volume in ventricular assist devices (VAD's) of the pulsatile type. The paper presents the current state of art in blood volume measurements of such devices and introduces a newly developed solution in the optic domain. The objective of the research is to overcome the disadvantages of the previously developed acoustic method—the requirement of additional sensor chamber. **Me- thods:** The idea of a compact measurement system has been introduced, followed by laboratory measurements. Static tests of the system have been presented, followed by dynamic measurements on a physical model of the human ventricular system. The results involving the measurements of blood chamber volume acquired by means of an optical system have been compared with the results acquired by means of the Transonic T410 ultrasound flow rate sensor (11PLX transducer, uncertainty ±5%). **Results:** Preliminary dynamic measurements conducted on the physical model of the human cardiovascular system show that the proposed optical measurement system may be used to measure the transient blood chamber volumes of pulsatile VAD's with the uncertainties (standard mean deviation) lower than 10%. **Conclusions:** The results show that the noninvasive measurements of the temporary blood chamber volume in the POLVAD prosthesis with the use of the developed optical system allows us to carry out accurate static and dynamic measurements.

Keywords: Ventricular Assist Device, Reflectance Measurements, Transient Blood Volume, Cardiac Output, Optoelectronic Sensor

1. Introduction

Heart problems are nowadays the major cause of death. Most heart and cardiovascular diseases are caused by hypertension, which is usually conditioned by the use of drugs. At the same time, we can observe a growing number of patients with III and IV stage heart failures (NYHA scale)[1], in which pressor amine drugs alone can be ineffective and a surgery procedure is required. In such cases, the ventricular assist device (VAD) can be used to support the heart muscle in the pumping process of blood. It releases the heart muscle from much effort, thus the life of the patient waiting for a heart transplant can be prolonged, or in some cases can bring about a partial or full recovery of the heart muscle[2]. There are two main types of VAD used in medical applications: pulsatile and non-pulsatile ones[3]. Recent research studies seem to prove that both types of devices can be successfully used in supporting the heart muscle. Non-pulsatile solutions can be smaller in size and easier to implant in a patient's body as compared with pulsatile solutions. In spite of that, there are still cases in which pulsatile solutions are preferred[4], especially in the cases when the patient is subjected to conditioning for heart transplantation. The pulsatile solutions are usually paracorporeal, and usually connected with the human ventricular system in parallel with the human heart, thus they can be used in some cases for the recovery of the heart muscle.

1.1. Objective of the Research

The main objective of our research studies is the elaboration of an optical system for measuring blood chamber volume in the paracorporeal pneumatic, pulsatile ventricular assist devices. The experiments involved the modified model of the ReligaHeart EXT (manufactured by the Foundation of Cardiac Surgery Development, Poland). It is a paracorporeal, pulsatile heart support device used in patients for over a decade now. The construction of the POLVAD and ReligaHeart EXT is similar to most solutions of the pulsatile ventricular assist devices utilized in the world (Berlin Heart EXCOR, Medos HIA-VAD, Abiomed BVS 5000 etc.)[3]. The prosthesis consists of two chambers: the blood and pneumatic chambers (separated by a flexible membrane), blood inflow and blood outflow connectors and two mechanical valves. The improved model of the POL-VAD, called Religa Heart EXT, can be seen in **Figure 1**. POLVAD and ReligaEeart EXT pumps are clinically utilized in Poland.

The changes of air pressure inside the pneumatic chamber cause the movement of the membrane, thus pumping the blood. Air pressure changes are induced using the pneumatic driving unit [**Figure 1(b)**]. The direction of the blood flow is determined by a relative configuration of the valves in the connectors.

1.2. State of Art

Cardiac recovery by mechanical heart supporting is a complex process. It occurs in about 20% of cases involving the mechanical supporting of the heart by means of blood pumps[5]. During the recovery, the cardiac function gradually improves and the total blood flow in the aorta increases[6]. It necessitates a gradual reduction of the mechanical supporting of the heart.

In the current state of art, the POLVAD and ReligaHeart EXT ventricular assist devices, as well as other solutions of such pneumatic assist devices, are not equipped with systems monitoring the actual (transient) blood volume inside the prosthesis. Its semitransparent casing allows the medical staff to evaluate the heart support process visually and by monitoring the patients vitals. This approach requires a continuous medical assistance. Patients need to be hospitalized all the time, which causes a major discomfort and significantly reduces their mobility. The solutions of ultrasound measurements are expensive and large in size. Therefore, their application is mostly limited to clinical conditions.

Polish experience involving the application of the cardiac assist system

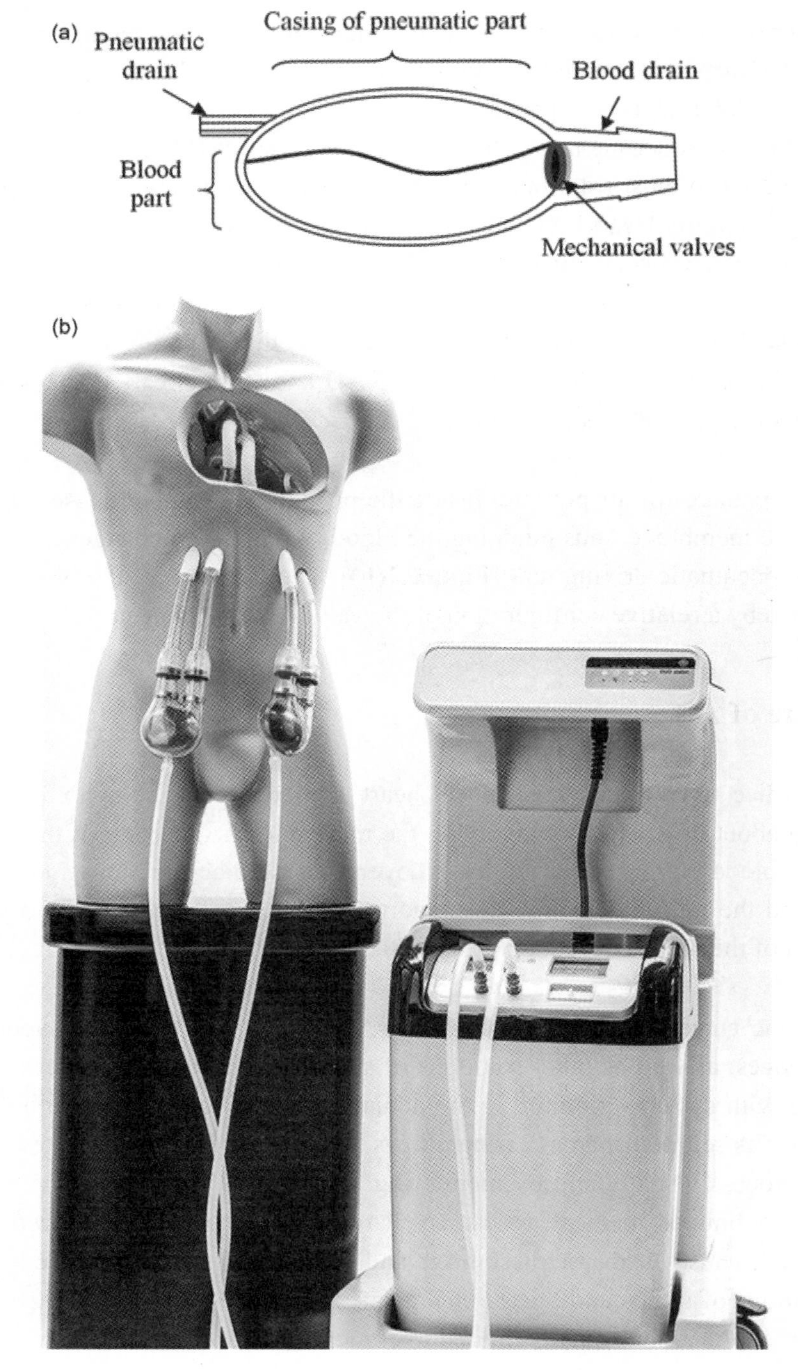

Figure 1. The newly developed RELIGA-EXT prosthesis (construction diagram and the complete heart support system).

POLCAS in more than 300 cases proved that the best clinical results were obtained by synchronous supporting[7][8], commonly with the manual reduction of heart supporting during the recovery. The measurement of the actual blood chamber volume might significantly improve the recovery process by providing feedback information about the cardiac output of the prosthesis. This information about the current state of VAD device would be a milestone in automating the heart support process completely.

Currently, all over the world, the most popular method for estimating the cardiac output of pneumatic VAD is the measurement of flow rate, using the ultrasound flow rate meters[9]. These devices are expensive and fairly large in their size. Although the measurements are accurate, the mobility of the patients is still considerably restricted.

The most important issue in supporting the heart with VAD is to empty completely the device in each cycle, so that the blood might circulate properly in the device. In the case of a partial ejection the volume of blood that remains in the blood chamber has by about 60% longer contact with the biomaterial than the portion of blood ejected directly from the VAD, which can lead to the formation of thrombus. Moreover, the washing of internal surface of the blood chamber is better in the case of a total ejection than in the case of partial ejection[10][11]. That is why the determination of the minimum-filling state is of crucial importance.

The existing solutions incorporated in the VAD device and used in patients allow us to determine only the full- and minimum-filling states of the prosthesis, using a simple optical or magnetic switch[12]. This is unfortunately only a partial solution of the problem. In that configuration, the sensor does not allow us to monitor the actual output of the assist device. At some point the air flow measurements in the pneumatic duct were used to measure the pneumatic part volume. The method provided the measurements with the 10% ~ 20% uncertainties[13]. The newer approach to the subject matter was based only on the measurements of pressure[14]. Both methods, however, required frequent calibration cycles, using the reference method. There were many other approaches to solve that problem in the pneumatic type of VAD[15]-[18]. One of the promising solutions developed in our Department was based on the concept of Helmholtz's acoustic resonator[17]-[19]. In that solution, the pneumatic part acts as an acoustic resonator. The elaborated method allows us to determine the actual volume of the prosthesis, but requires an additional sensor chamber over the pneumatic part of the prosthesis. It increases the total volume of the VAD. Although the method was successfully tested in laboratory conditions, it has not been incorporated in the new model of the VAD. To the Authors' knowledge, there is no solution incorporated in the pneumatic VAD used in patients, allowing to estimate online the blood volume in the VAD device. Another solution developed by the Authors of the manuscript is focused on the application of the ultrasound Doppler method[20]. It was proved that apart from the information about the velocity of the flow, the energy of ultrasound echo contains the information about micro-objects drifting in the blood[21][22]. This effect offers an opportunity to detect micro-emboli and larger clots, which can be then used in controlling the anticoagulation in order to increase the patient's safety.

The mentioned acoustic methods had some drawbacks that hindered their application in the prosthesis[17]-[19]. It motivated the Authors to look for an improved measurement solution.

1.3. Optical Measurement System

The experience gained with the acoustic method was used in the development of a new solution—an optical one. The proposed measurement system must be noninvasive to blood environment. That is why, most of transient blood volume measurement methods are taking advantage of the properties of the membrane, which separates the blood chamber and the pneumatic chamber of the VAD, at the pneumatic part. The proposed measurement system has no contact with blood environment. The blood volume in the blood part of the prosthesis is estimated by the measurement of the volume of the pneumatic part.

In the new, optical method, several light sources and light detectors are situated in the casing of the pneumatic chamber. The spatial configuration of the light emitting diodes and photodiodes was chosen basing on the previously conducted research studies at the Department of Optoelectronics. The light emitted by light emitting diodes (E), after direct reflection on the membrane at the pneumatic side, is detected by photodiodes (PD). The principle of the optical measurement system can be seen in **Figure 2**. For each volume of the pneumatic part of the VAD,

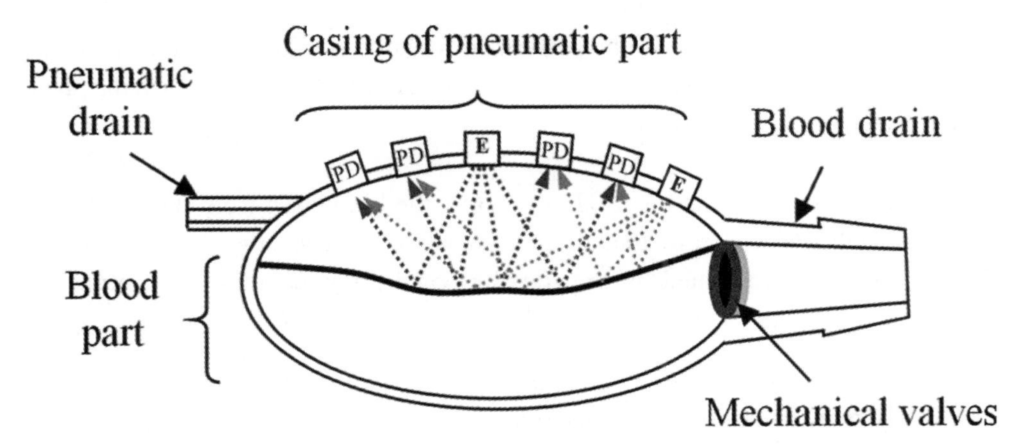

Figure 2. The idea of an optical measurement system with photodiodes (PD) and light emitting diode (E) in the pneumatic part casing.

all the related amplitude signals on the photodiodes, for all configurations of the light emitting diode-photodiode, are measured. The signal is then analyzed. Since the position of the membrane changes, the amplitudes of the detected signal for some of light emitting diode-photodiode pairs also change, and thus the actual blood chamber volume can be estimated. During the process of blood pumping the shape of the membranes surface changes in a random manner. That is why reflectance measurements based on a single sensor element alone[23][24] cannot be used. In preliminary research studies of the optical measurement system the membrane reflectance properties were extensively examined, using the spectral ellipsometry (Sentech SE850 ellipsometer) within the range 400–1000 nm, in order to find the optimal range of the wavelengths of light that could be utilized in the system[25]. It turned out that the maximum reflectance occurred at the UV range, but the difference of energy reflectance was less than 10% in the examined spectrum range, with the minimum value at 770nm. Due to the wide selection of light emitting diode and photodiode elements, as well as the availability of photodiodes with high-pass filters, allowing to immunize the optic system to visible light, it was decided that the IR range should be used. In order to minimize the effect of the shape of the membrane on the measurement results, wide-angle light emitters and light detectors were used. The complimentary pair of light emitting diode (SFH203FA) and photodiode (SFH203PFA) manufactured by Osram, operating at peak wavelength of 880nm, was used in the final solution[26]. The matrix of light emitters and light detectors used in the research studies was incorporated in the pneumatic part of the casing (**Figure 2**).

The VAD model used in the performed research studies was produced using rapid prototyping techniques. The internal dimensions of the model are the same as in the medically used VADs.

The membrane which is used to separate the blood part from the pneumatic part is exactly the same as that used in actual prosthesis: two layers made of *Chronoflex ARLT*, separated by the graphite powder for reducing the friction between the two layers.

1.4. Preliminary Research Studies

The preliminary research studies (performed at the Department of Optoelectronics, Silesian University of Technology, Poland) comprised static and low dynamicity testing (<5 ml/s), aiming to determine the behavior of the membrane and the development of the electronics and software for the measurement system[27][28]. Since the membrane in this construction changes its shape in a random manner during the process of blood pumping, it was decided that in the first stage of the investigations, the optical method would use 12 light emitting diodes and 32 photodiodes. In each measurement cycle, each IR light emitting diode was powered (one at a time), and the signal from all 32 photodiodes was detected. In order to achieve the sufficient speed of acquisition, fast multiple input A/D converters had to be used, along with precise amplifiers connected to each of the photodiodes. Four 8-channel A/D converters were used, interrogated by 2MHz SPI bus. The proper time delay between the consecutive interrogations introduced by the software, was based on research studies concerning the photodiode circuit response times, after the light emitting diodes had been lit. The time of one cycle for all single light emitting diode-single photodiode configurations (equal to 384) amounted to 100ms.

The relations between the volume and amplitude on some photodiode combinations for diode 1 are shown in **Figure 3(a)**. The arrangement of the photodiode and the light emitting diode in the pneumatic casing can be seen in **Figure 3(b)**. Some of the pair signals depended on the changing volume of the pneumatic part in the whole range of the volumes (e.g. E1-PD1), and some combinations could provide information about the volume in the limited range (e.g. E1-PD31, E1-PD4). Some of the E-PD combinations could not provide any meaningful information

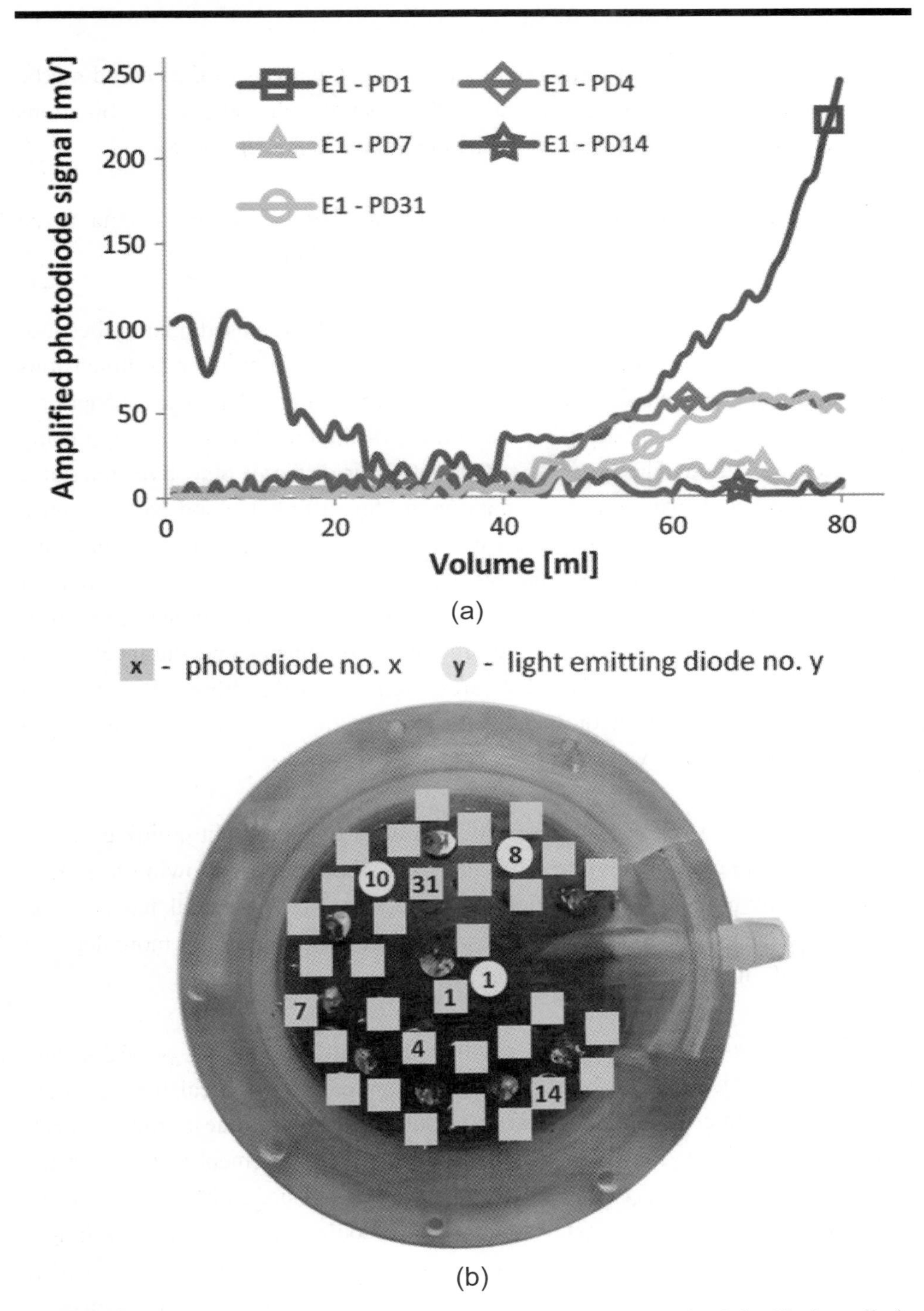

Figure 3. Relation between the actual volume and signals from each of the 32 photodiodes (a), when only light emitting diode E1 is used (b).

(e.g. E1-PD7, E1-PD14). It turned out that the direct analysis of data could not be used—none of the single light emitting diode-single photodiode combinations could provide direct information about the pneumatic chamber volume.

Due to the above and because of the significant quantity of data to be processed, it was decided that data-mining techniques should be used.

The PCA (principal components analysis) method was used, mainly because of its nonparametric operation, the possibility to reduce the number of dimensions of the input data, as well as to identify of redundant data[26]. The measurement results were then classified basing on the volume of the blood part, measured using the reference method. The process was repeated for different objective functions, aiming at the best separation of the data points for different blood part volumes. This process resulted in a 3D space, formed by three first principal components, where selected blood part volumes were represented (each 5ml in a range of 0–80ml). The process of finding the actual volume required the transformation of the input data into the 3D space of the principal components, and the localization of the nearest volume using k-NN (k Nearest Neighbors) algorithm. The algorithm allowed us to determine unequivocally the volume of the pneumatic part, when three adjacent neighbors were localized.

The analysis resulted in the selection of a set of three light emitting diodes (E1, E8, E10) [marked in **Figure 3(b)**] and some photodiodes, allowing to reduce the required number of configurations (features) to 34. The calculated function allowed us to determine the volume in low dynamic conditions[26]. A more detailed description of this problem can be found in[27].

The successful preliminary results obtained by means of PCA and k-NN[26][27] provided a solid basis, allowing to prepare the optical measurement system for dynamic measurements at the Foundation of Cardiac Surgery Development (FRK). The system was modified to acquire measurement results with the required rate. The measuring time of a single cycle was reduced from 100 to 35ms for all combinations (384 E-PD pairs). Additionally, the configurations of the chosen light emitting diodes and photodiodes were programmed on the basis of the data acquired in the course of the static measurements. It enabled a further increase of the acquisition speed, resulting in 10 ms measurement cycles for the selected combinations.

2. Methods

Preliminary dynamic measurements of the optical measurement system were performed at the FRK. The physical model of the human cardiovascular system was used[28]. The independent volume measurement method consisting of two ultrasound flow rate meters (at the inflow and outflow of the prosthesis) was used as the reference. Transonic T410 (11PLX transducer—5% uncertainty) flow rate meters were applied in the experiment. The ultrasound sensors used in the measurements were previously calibrated using the Fluxus ultrasound flowmeter (2% uncertainty).

The blood-like liquid used during the measurements was a 60% water solution of glycerin mixed with benzoic aldehyde (capillary viscometer measurement result: 8.5cP at 25°C).

The synchronization of the time of volume measurements obtained by the application of the developed optical system and volume measurements derived from the flow rate was obtained using the pressure signal in the pneumatic drain. The pressure signal was measured simultaneously in both measurement stands. In the case of dynamic tests, the prosthesis was controlled by the pneumatic driving unit PDU-502 (produced by the FRK), utilized in clinical trials. The concept of the measurement stand used in the testing can be seen in **Figure 4**.

During the tests, several heart support speeds (average heart rate—AHR) and blood duct impedances were used. This allowed us to analyze various working modes of the prosthesis: full filling/full ejecting, full filling/partial ejecting, partial filling/partial ejecting and partial filling/full ejecting.

3. Results and Discussion

The results were the same as in the case of previous research studies with low dynamics[25][26]. The photodiode signal amplitudes were repeatable with respect to the blood chamber volume characteristics over the whole range of working

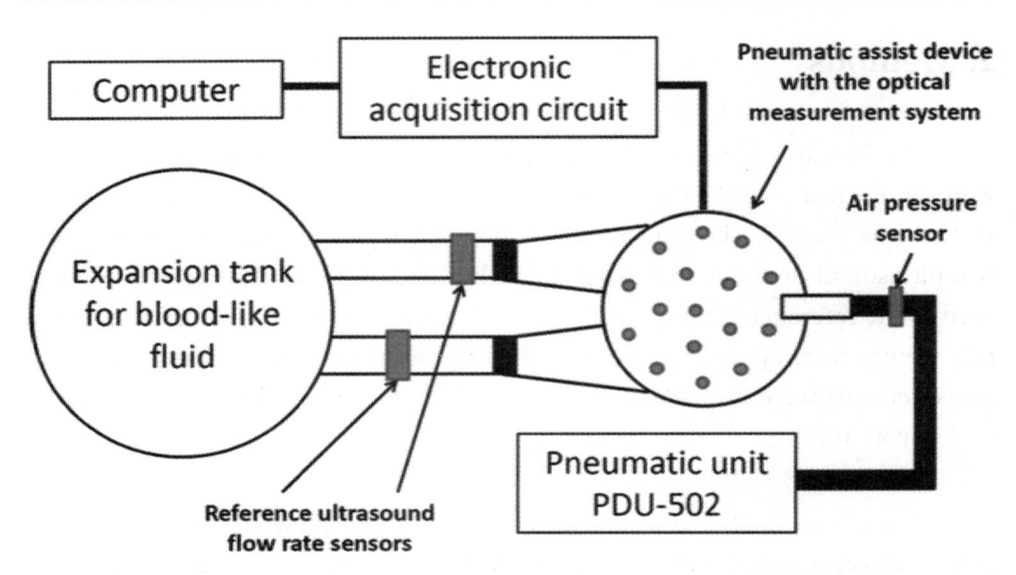

Figure 4. The idea of the measurement stand at the Foundation of Cardiac Surgery Development, Zabrze, Poland.

conditions of the prosthesis. **Figure 5** shows the exemplary characteristic of a chosen photodiode signal in the time domain, accompanied by the volume derived from the flow rate.

The photodiode signal shown in **Figure 5** was achieved by using the centrally localized light emitting diode and photodiode shown in **Figure 6(a)** (light emitting diode 1-photodiode 1).

Figure 5 shows the amplitude of the detected signal for all of the volumes of the blood part of the VAD. It can be seen that the characteristic is equivocal. When the membrane reaches the casing of the pneumatic part, the light emitting diode and the photodiode are partially covered by the membrane—this results in two maximum peaks in this area. At very low volumes the membrane starts acting as a concave mirror structure, resulting in a slight increase of the detected amplitude for this specific configurations.

Dynamic measurements were realized under various driving pressures and various blood duct impedances at the inflow and outflow cannulas, which allowed us to modify the filling and ejecting process of the prosthesis under similar driving pressures.

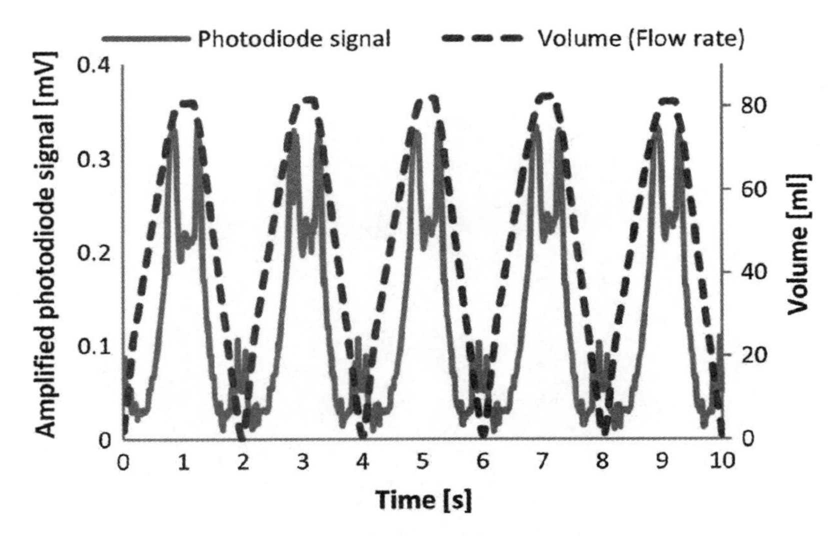

Figure 5. Volume resulting from the flow rate and an exemplary optical signal amplitude (light emitting diode1-photodiode 1) characteristics in time.

The complete set of optical measurements and the ultrasound measurements of the derived flow rate data was analyzed by means of the numerical methods (PCA and k-NN).

The PCA was used to select the light emitting diode-photodiode configurations that could provide an unequivocal relation between the amplitude on the detectors and the volume of the blood part.

During the dynamic tests, realized on the model of human cardiovascular system, the elaborated system acquired data from all 384 configurations. Additionally, the reduced number of configurations (34 features preselected by the PCA) was used in a faster acquisition program.

Due to the requirement of limited changes in the construction of the prosthesis, inflicted by the manufactures of the cardiac support system, in order to maintain the integrity of the construction of the prosthesis casing, it was decided that the number of sensor elements should be limited. Basing on the manufacturers' suggestions, in order not to hinder the mechanical properties of the casing of the prosthesis, it was agreed that the maximum number of light emitting diode and photodiode elements incorporated into the casing of the prosthesis should not exceed eight.

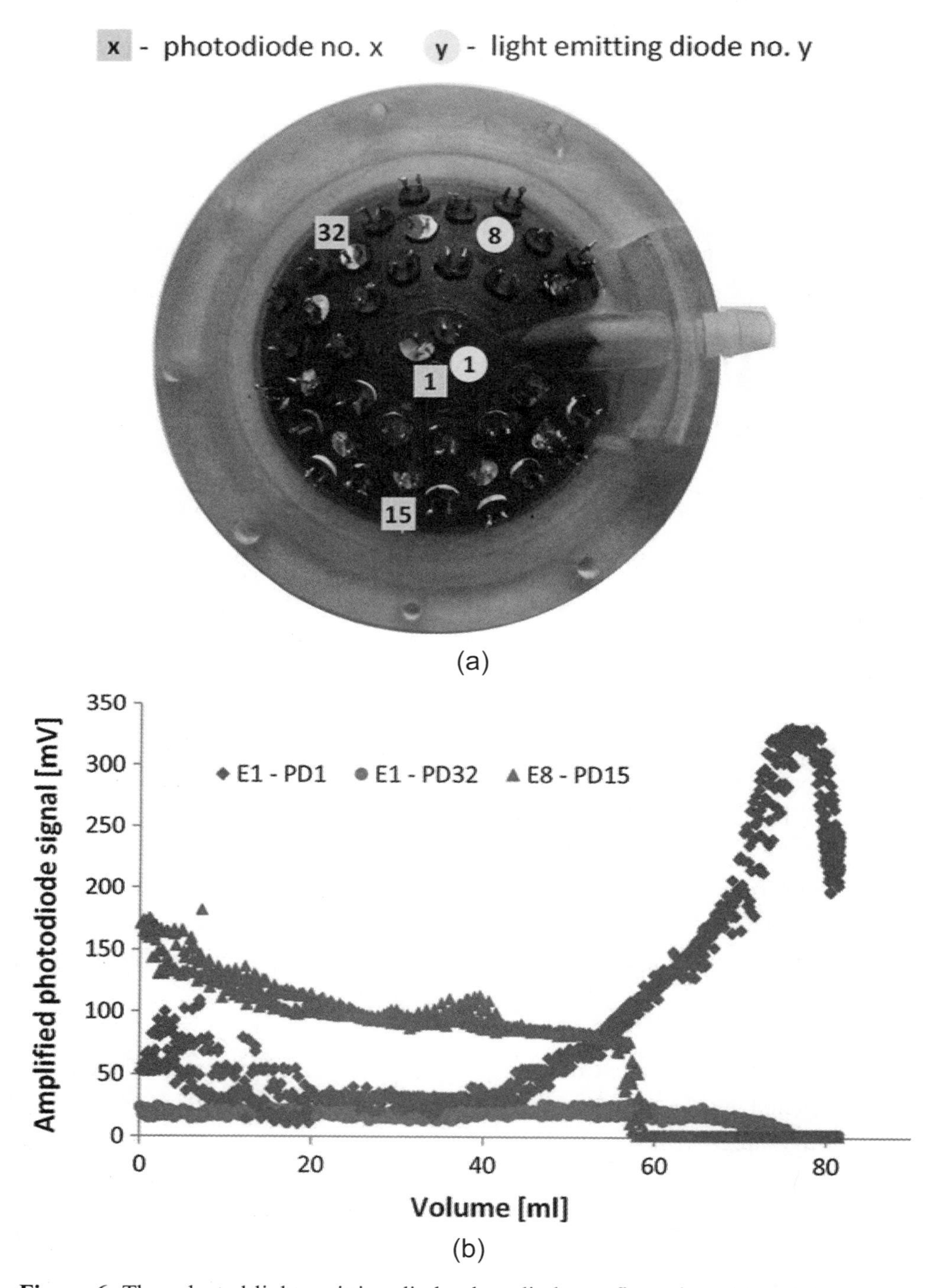

Figure 6. The selected light emitting diode-photodiode configurations (a) and the relating photodiode signals as a function of the volume of the blood part (b).

Measurement Results

The Authors were able to select three configurations out of 34 preselected previously using the PCA, that provided the possibility to determine an unequivocal transfer function.

The combinations that were used in the final solution are: E1-PD1, E1-PD32, E8-PD15 [**Figure 6(a)**].

Basing on the amplitude of the signal on the photodiode for the particular configurations, it is possible to determine the volume of the blood part. The configuration E8-PD15 is used to determine the lower volumes (<35ml); E1-PD1 is used for the range 35–75ml. The E1-PD32 configuration is used to detect the equivocal part of the E1-PD1 configuration at higher volumes (>75ml) [**Figure 6(b)**].

All of the configurations turned out to behave differently in the filling and emptying cycle, otherwise being repeatable in all cycles. This dependence is visible in mid-range volumes (30–40ml) [**Figure 7(b)**]. These results in the artefacts marked in **Figure 8(b)**.

Knowing the required number of configurations, the size of the measurement system was reduced; the electronic circuit was redesigned to a more compact one. In order to reduce the size of the electronic circuit, a 4-layer transducer board was used. In **Figure 7(a)** the reduced preliminary system is shown. Finally, the optical system was reduced to two light emitting diodes (E1, E8) and three photodiodes (PD1, PD15, PD32) marked in **Figure 6(a)**.

Figure 8 presents the volume calculated with the use of flow rate measurements and the volume results from the optical system.

We can observe in **Figures 8(a)-8(c)** that the blood volume measurements obtained by means of the new optic method coincide with the results calculated with the use of flow rate measurements. The exemplary results show three different test series: 8a—operating with AHR = 30, at approximately half of the maximum stroke volume of the VAD; 8b—operating with the same AHR and pneumatic

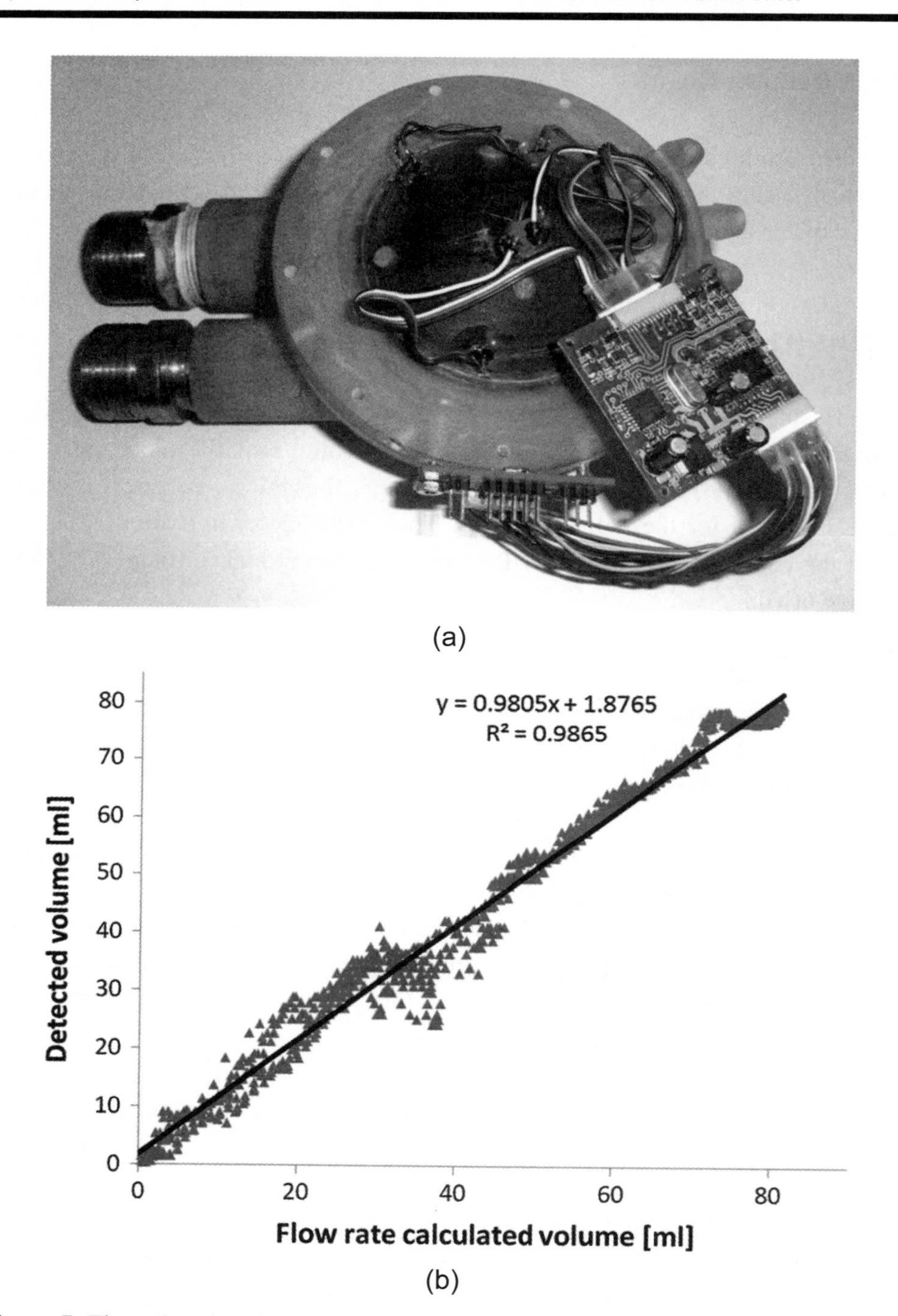

Figure 7. The reduced optical measurement system (a), and the relation between the volume detected by the optical system and derived using the flow rate based measurement system (b).

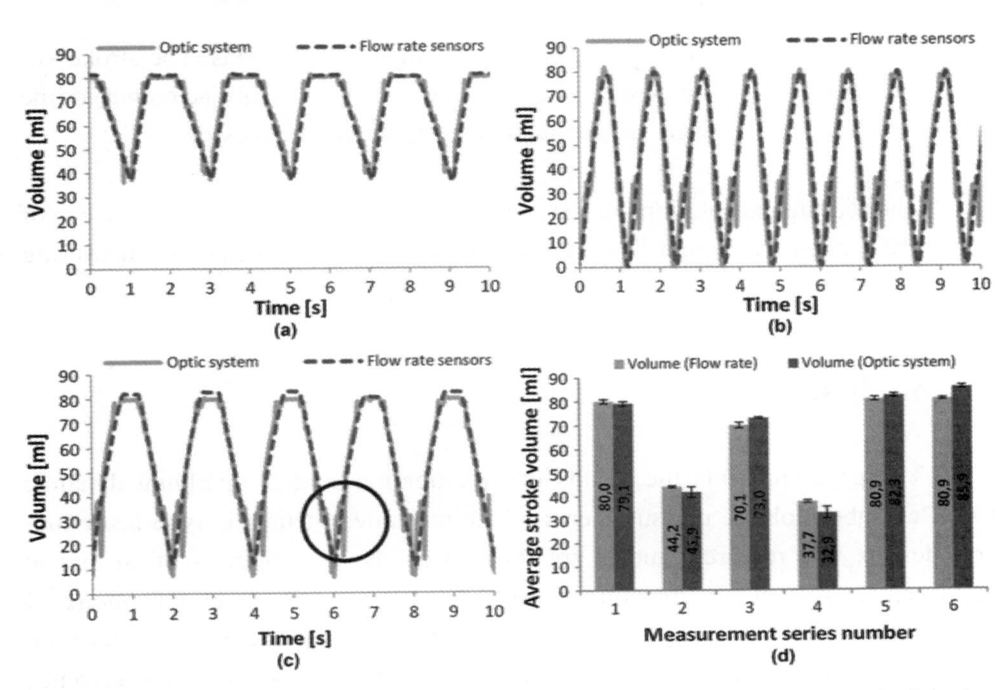

Figure 8. Exemplary results of dynamic tests carried out on the physical model of the human circulatory system in various conditions of heart support in the time domain (a, b AHR = 30, c AHR = 50) and comparison of the results of measurements acquired using the optical system and the system based on flow rate sensors (d).

driving pressures, but with slightly reduced blood duct impedances resulting in larger stroke volume of the VAD; 8c—VAD operating at AHR = 50bpm with full stroke volume). The measurements of mid-range (30 - 40 ml) volumes showed the artefacts marked in **Figure 8(b)**. They can be related to the rapid change in the shape of the membrane (from convex to concave character in relation to the pneumatic part casing) and ought to be improved. When the impedances of the blood duct change, which takes place in the case of the human cardiovascular system, the same driving pressures of the VAD result in a different stroke volume [**Figure 8(a)**, **Figure 8(b)**]. By comparing the driving pressures with the volumes of the blood part acquired using the proposed method, it would be possible to detect the changes in the impedance of the blood duct, providing important information concerning the state of the supported cardiovascular system.

In **Figure 8(d)**, we compared the average SV (stroke volume) results obtained in the optic system with average SV results obtained using the flow rate measurements at five experiments under different heart support conditions. The stroke volume measurements can be used for the estimation of the CO (cardiac output) of the VAD, which is a well-recognized parameter in heart support processes.

The redesigned solution will be a subject of extended tests in the presence of clinically applied ultrasound flow rate sensors as a reference method of measuring the volume.

4. Conclusions

The paper presents the results of investigation studies involving dynamic blood chamber volume measurements of the pulsatile, pneumatic type heart support device. The research studies are focused on the development of an optical measurement system. The dynamic measurements were performed on the model of the human cardiovascular system at the Foundation of Cardiac Surgery Development in Zabrze, Poland[28]. Ultrasound flow rate sensors were used as a reference volume measurement method. The method was tested in changing operating conditions of the VAD device with various driving pressures of the heart assist device, various impedances in the blood duct, varying heart rates.

The results of the performed measurements, making use of the optical system and the reference method, show that the proposed optical method can be applied for accurate blood chamber volume measurements. A direct comparison of the average blood output volume during consecutive cycles using both methods confirms the above declaration.

Future research studies will comprise the examination of a miniaturized measurement system as well as an improvement of the measurement algorithm. The optical measurement system meets all requirements for a noninvasive transient blood volume measurement system.

The proposed method is designed for pulsatile type VADs, which are usually extracorporeal solutions. In that case, the device will be directly powered. The elaborated system at the current state of development requires approximately 100mW of electric power. Further investigations aiming at miniaturizing the measurement system should provide an additional essential reduction of power consumption, which would be important if used in the future in intracorporeal pulsatile VADs, or battery powered solutions.

The proposed optical solution of transient blood volume measurements has overcome the issues existing in measurement systems based on the acoustic method[17][18].

According to an extended search in literature and the Authors' knowledge, the proposed measurement method (and the elaborated system) is an original and novel solution of blood chamber volume measurements in pneumatic heart assist devices. The optical solution has been submitted to be patented[29].

The optical method of blood volume measurement presented in the paper may be used as a method to assess the location of the membrane (e.g. to detect the full filling and full ejection of blood from the blood chamber). In conjunction with an ultrasound Doppler assessment of micro-emboli it may constitute an automated extracorporeal cardiac assist system.

Authors' Contributions

GK participated in all of researches and development of the measurement method, carried out the data analysis and drafted the manuscript. TP participated in the design and coordination of the study and revised the manuscript. MS participated in all of researches and development of the miniaturized measurement circuit and carried out data analysis. MG participated in the dynamic measurements on the measurement stand at the Foundation of Cardiac Surgery Development and participated in data analysis. All authors read and approved the final manuscript.

Acknowledgements

The work is partially financed by the Silesian University of Technology, Faculty of Electrical Engineering within the BKM/514/RE4/2014 and BKM/534/RE4/2015 and The National Centre of Research and Development (Warsaw, Poland), within the grant No. PBS1/A3/11/2012.

Competing Interests

The authors declare that they have no competing interests.

Source: Konieczny G, Pustelny T, Setkiewicz M, et al. Optoelectronic system for the determination of blood volume in pneumatic heart assist devices [J]. Biomedical Engineering Online, 2015, 14(1):1–14.

References

[1] Schartel SA, Pai SS. Mechanical hemodynamic support. In: Criner GJ, Barnette RE, D'Alonzo GE, editors. Critical care study guide, vol XXII. 2nd ed. New York: Springer; 2010. p. 918–949.

[2] Young JB. Healing the heart with ventricular assist device therapy: mechanisms of cardiac recovery. Ann Thorac Surg. 2001; 71:S210–219.

[3] Ravishankar C, Dominguez TE, Rosenthal TM, Gaynor JW. Mechanical circulatory support in the patient with congenital heart disease. In: Shaddy RE, editor. Heart failure in congenital heart disease: from fetus to adult, 1st ed. London: Springer; 2010. p. 123–154.

[4] Pacholewicz J, Zakliczyński M, Barańsk-Kosakowska A, Religa G, Siondalski P, Kucewicz E, Nadziakiewicz P, Zembala M. Efficacy of treatment with pulsatile pump as a bridge to transplantation in patients with a congestive heart failure—polish experiences with POLVAD. J Heart Lung Transpl. 2009; 28:S98.

[5] Maybaum S, Mancini D, Xydas S, et al. Cardiac improvement during mechanical circulatory support a prospective multicenter study of the Lvad Working Group. Circulation. 2007; 115:2497–2505.

[6] Gawlikowski M, Pustelny T, Kustosz R. The physical parameters estimation of physiologically worked heart prosthesis. J PHYS-PARIS IV. 2006; 137:73–78.

[7] Pacholewicz J, Zakliczyński M, Kowalik V, Kalarus Z, Zembala M. Bridge to recovery in two cases of dilated cardiomyopathy after long-term mechanical circulatory support. Kardiochir I Torakochir Polska. 2014; 11(2):69–172.

[8] Pacholewicz J, Hrapkowicz T, Chodor B, Zembala M, Religa Z. Good long-term result of biventricular assisted circulation using Polcas-Religa system in 16-year old boy with a irreversible heart failure—a case report. Kardiochir I Torakochir Polska.

2009; 6(4):377–383.

[9] Slaughter MS, Bartoli CR, Sobieski MA, Pantalos GM. Intraoperative evaluation of the HeartMate II Flow estimator. J Heart Lung Transpl. 2009; 28(1):39–43.

[10] Deutsch S, Tarbell J, Manning K, Rosenberg G, Fontaine A. Experimental fluid mechanics of pulsatile artificial blood pumps. Annu Rev Fluid Mech. 2006; 8:65–86.

[11] Wootton D, Ku D. Fluid mechanics of vascular systems, diseases, and thrombosis. Annu Rev Biomed Eng. 1999; 1: 299–329.

[12] Reichenbach SH, Farrar DJ, Hill JD. A Versatile intracorporeal ventricular assist device based on the thoratec VAD system. Ann Thorac Surg. 2001; 71:171–5.

[13] Shaw WJ, Pantalos GM, Everett S, Olsen DB. Factors influencing the accuracy of the cardiac output monitoring and diagnostic unit for pneumatic artificial hearts. ASAIO Trans. 1990; 36(3):M264–M268.

[14] Nishinaka T, Taenaka Y, Tatsumi E, Ohnishi H, et al. Development of a compact portable driver for a pneumatic ventricular assist device. J Artif Organs. 2007;10(4):236–239.

[15] Komorowski D, Gawlikowski M. Preliminary investigations regarding the blood volume estimation in pneumatically controlled ventricular assist device by pattern recognition. Comput Recogn Syst 2 Adv Soft Comput. 2007; 45:558–565.

[16] Sobotnicki A, Pałko T, Mocha J, Czerw M. Evaluation of volumetric parameters of the ventricular assist device using bioimpedance method. J Med Inform Technol. 2012; 19:117–123.

[17] Opilski Z, Konieczny G, Pustelny T, Gacek A, Kustosz R. Noninvasive acoustic blood volume measurement system for the POLVAD prosthesis. Bull Pol Acad Sci Tech Sci. 2011; 59(4):429–433.

[18] Konieczny G, Opilski Z, Pustelny T, Gawlikowski M. Acoustic system for the estimation of the temporary blood chamber volume of the POLVAD heart supporting prosthesis. Biomed Eng Online. 2012; 11:72. doi:10.1186/1475-925X-11-72

[19] Konieczny G, Pustelny T, Opiski Z, Gawlikowski M. Acoustic system of determining the instantaneous volume of the blood part of the ventricular assist device POL-VAD-EXT. Arch Acoust. 2014; 39(4):541-8.

[20] Gawlikowski M, Lewandowski M, Nowicki A, et al. The application of ultrasonic methods to flow measurement and detection of microembolus in heart prostheses. Acta Phys Polonica A. 2013; 124(3):417–420.

[21] Palanchon P, Boukaz A, Klein J, De Jong N. Subharmonic and ultraharmonic emissions for emboli detection and characterization. Ultrasound Med Biol. 2003;29(3):417–425.

[22] Cowe J, Gittins J, Naylor AR, Evans DH. RF signals provide additional information on embolic events recorded during TCD monitoring. Ultrasound Med Biol.

2005;31(5):613–623.

[23] Missoffe A, Chassagne L, Topçu S, et al. New simple optical sensor: from nanometer resolution to centimeter displacement range. Sensor Actuator A Phys. 2012; 176: 46–52.

[24] Fleming AJ. A review of nanometer resolution position sensors: operation and performance. Sensor Actuator A-Phys. 2013; 190:106–126.

[25] Konieczny G, Pustelny T, Marczyński P. Optical sensor for measurements of the blood chamber volume in the POL-VAD prosthesis—static measurements. Acta Phys Pol A. 2013; 124(3):479–482.

[26] Konieczny G, Pustelny T, Marczyński P. Quasi-dynamic testing of an optical sensor for measurements of the blood chamber volume in the POLVAD prosthesis. Acta Phys Pol, A. 2013; 124(3):483–485.

[27] Marczynski P, Konieczny G, Pustelny T. Research on pattern recognition applied for volume estimation of blood chamber with matrix of optical sensors. Acta Phys Pol A. 2013; 124(3):498–501.

[28] Ferrari G, Kozarski M, De Lazzari C, et al. Development of a hybrid (numerical-hydraulic) circulatory model: prototype testing and its response to IABP assistance. Int J Artif Organs. 2005; 28(7):750–759.

[29] Pustelny T, Konieczny G. Method for measuring instantenous blood volume in cardiac support chamber. Pol Pat Off Off Gaz. 1055; 13(2014):9.

Chapter 19

Sol-gel Synthesized Zinc Oxide Nanorods and Their Structural and Optical Investigation for Optoelectronic Application

Kai Loong Foo, Uda Hashim, Kashif Muhammad, Chun Hong Voon

Nano Biochip Research Group, Institute of Nano Electronic Engineering (INEE), Universiti Malaysia Perlis (UniMAP), Kangar, Perlis 01000, Malaysia

Abstract: Nanostructured zinc oxide (ZnO) nanorods (NRs) with hexagonal wurtzite structures were synthesized using an easy and low-cost bottom-up hydrothermal growth technique. ZnO thin films were prepared with the use of four different solvents, namely, methanol, ethanol, isopropanol, and 2-methoxyethanol, and then used as seed layer templates for the subsequent growth of the ZnO NRs. The influences of the different solvents on the structural and optical properties were investigated through scanning electron microscopy, X-ray diffraction, Fourier transform infrared spectroscopy, ultraviolet-visible spectroscopy, and photoluminescence. The obtained X-ray diffraction patterns showed that the synthesized ZnO NRs were single crystals and exhibited a preferred orientation along the (002) plane. In addition, the calculated results from the specific models of the refractive index are consistent with the experimental data. The ZnO NRs that grew from the 2-methoxyethanol seeded layer exhibited the smallest grain size (39.18nm), largest diffracted intensities on the (002) plane, and highest bandgap (3.21eV).

Keywords: Zinc Oxide Nanorods, Hydrothermal Growth, Solvent, Refractive Index, Bandgap

1. Introduction

Top-down and bottom-up methods are two types of approaches used in nanotechnology and nanofabrication[1]. The bottom-up approach is more advantageous than the top-down approach because the former has a better chance of producing nanostructures with less defects, more homogenous chemical composition, and better short- and long-range ordering[2]. Semiconductor nanorods (NRs) and nanowires possess convenient and useful physical, electrical, and optoelectronic properties, and thus, they are highly suitable for diverse applications[3][4].

ZnO, one of the II-VI semiconductor materials, has attracted considerable interest because of its wide bandgap (approximately 3.37eV), high exciton binding energy (approximately 60meV), and long-term stability[5][6]. ZnO has been applied in various applications, such as in light-emitting diode[7], gas and chemical sensors[8]-[10], ultraviolet (UV) detector[11][12], solar cell[13][14], and bio- molecular sensors[15][16]. To create high-quality ZnO NRs, various techniques have been proposed, such as the aqueous hydrothermal growth[10], metal-organic chemical vapor deposition[17], vapor phase epitaxy[18], vapor phase transport [19], and vapor-liquid-solid method[20].

Among these methods, the aqueous hydrothermal technique is an easy and convenient method for the cultivation of ZnO NRs. In addition, this technique had some promising advantages, like its capability for large-scale production at low temperature and the production of epitaxial, anisotropic ZnO NRs[21][22]. By using this method and varying the chemical use, reaction temperature, molarity, and pH of the solution, a variety of ZnO nanostructures can be formed, such as nanowires (NWs)[16][23], nanoflakes[24], nanorods[25], nanobelts[26], and nanotubes[27].

In this study, we demonstrated a low-cost hydrothermal growth method to synthesize ZnO NRs on a Si substrate, with the use of different types of solvents. Moreover, the effects of the solvents on the structural and optical properties were investigated. Studying the solvents is important because this factor remarkably affects the structural and optical properties of the ZnO NRs. To the best of our knowledge, no published literature is available that analyzed the effects of different seeded layers on the structural and optical properties of ZnO NRs. Moreover, a comparison of such NRs with the specific models of the refractive index has not been published.

2. Methods

2.1. ZnO Seed Solution Preparation

Homogenous and uniform ZnO nanoparticles were deposited using the sol-gel spin coating method[28]. Before seed layer deposition, the ZnO solution was prepared using zinc acetate dihydrate [Zn $(CH_3COO)_2 \cdot 2H_2O$] as a precursor and monoethanolamine (MEA) as a stabilizer. In this study, methanol (MeOH), ethanol (EtOH), isopropanol (IPA), and 2-methoxyethanol (2-ME) were used as solvents. All of the chemicals were used without further purification. ZnO sol (0.2M) was obtained by mixing 4.4g of zinc acetate dihydrate with 100ml of solvent. To ensure that the zinc powder was completely dissolved in the solvent, the mixed solution was stirred on a hot plate at 60°C for 20min. Then, 1.2216 g of MEA was gradually added to the ZnO solution, while stirring constantly at 60°C for 2h. The milky solution was then changed into a homogenous and transparent ZnO solution. The solution was stored for 24h to age at room temperature (RT) before deposition.

2.2. ZnO Seed Layer Preparation

In this experiment, a p-type Si (100) wafer was used as the substrate. Prior to the ZnO seed layer deposition process, the substrate underwent standard cleaning processes, in which it was ultrasonically cleaned with hydrochloric acid, acetone, and isopropanol. The native oxide on the substrate was removed using a buffered oxide etch solution, and then, the substrate was rinsed with deionized water (DIW). Subsequently, a conventional photoresist spin coater was used to deposit the aged ZnO solution on the cleaned substrates at 3,000rpm for 20s. A drying process was then performed on a hot plate at 150°C for 10min. The same coating process was repeated thrice to obtain thicker and more homogenous ZnO films. The coated films were annealed at 500°C for 2h to remove the organic component and solvent from the films. The annealing process was conducted in the conventional furnace. The preparation of the ZnO thin films is shown in **Figure 1**.

2.3. ZnO NRs Formation

After the uniform coating of the ZnO nanoparticles on the substrate, the ZnO NRs were obtained through hydrothermal growth. The growth solution consisted of an aqueous solution of zinc nitrate hexahydrate, which acted as the Zn^{2+} source, and hexamethylenetetramine (HMT). The concentration of the $Zn(NO_3)_2$ was maintained at 35mM, and the molar ratio of the $Zn(NO_3)_2$ to HMT was 1:1. For the complete dissolution of the $Zn(NO_3)_2$ and HMT powder in DIW, the resultant solution was stirred using a magnetic stirrer for 20min at RT. The ZnO NRs were grown by immersing the substrate with the seeded layer that was placed upside down in the prepared aqueous solution. During the growth process, the aqueous solution was heated at 93°C for 6h in a regular laboratory oven. After the growth process, the samples were thoroughly rinsed with DIW to eliminate the residual salts from the surface of the samples and then dried with a blower. Finally, the ZnO NRs on the Si substrate were heat-treated at 500°C for 2h. The growth process of the ZnO NRs is presented in **Figure 2**.

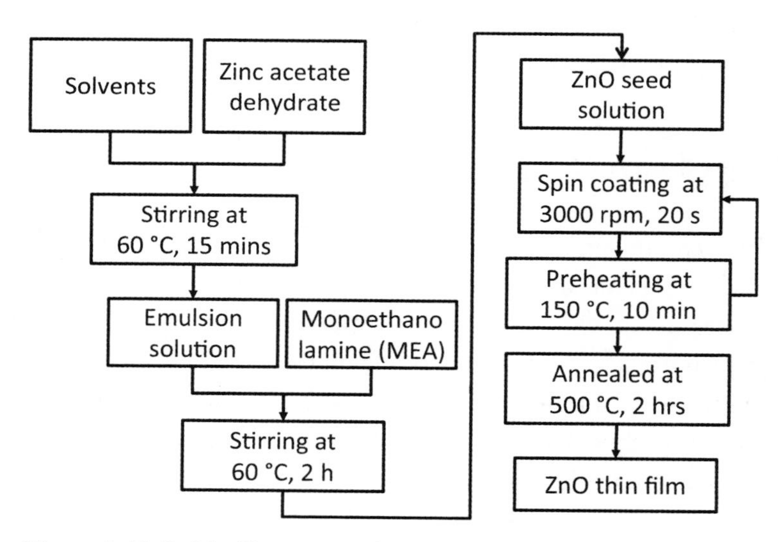

Figure 1. ZnO thin film preparation process flow.

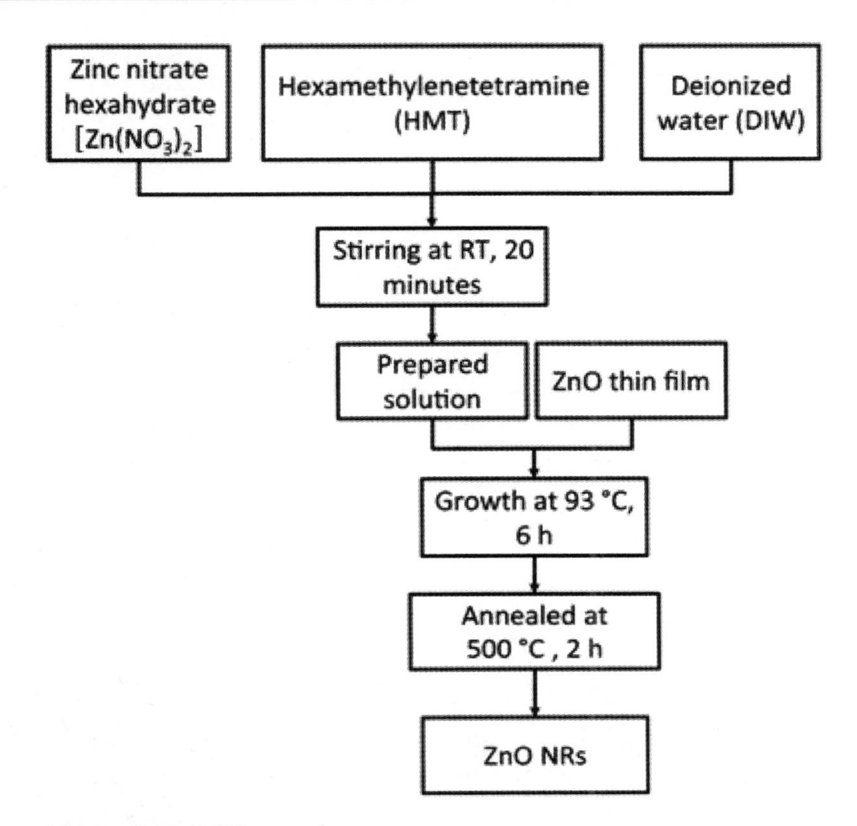

Figure 2. ZnO NR growth process.

2.4. Material Characterization

The surface morphology of the ZnO NRs was analyzed using scanning electron microscopy (SEM, Hitachi SU-70, Hitachi, Ltd, Minato-ku, Japan). X-ray diffraction (XRD, Bruker D8, Bruker AXS, Inc., Madison, WI, USA) with a Cu Kα radiation (λ = 1.54Å) was used to study the crystallization and structural properties of the NRs. The absorbed chemical compounds that exited on the surface of the ZnO NRs and SiO_2/Si substrate were identified using the Fourier transform infrared spectroscopy (FTIR, PerkinElmer Spectrum 400 spectrometer, PerkinElmer, Waltham, MA, USA). A UV-visible-near-infrared spectrophotometer from PerkinElmer was used to study the optical properties of the ZnO NRs at RT. In addition, the optical and luminescence properties of the ZnO NRs were studied through photoluminescence (PL, Horiba Fluorolog-3 for PL spectroscopy, HORIBA Jobin Yvon Inc., USA).

3. Results and Discussion

3.1. SEM Characterization

The top-view SEM images of the ZnO NRs that were synthesized with the use of different solvents are shown in **Figure 3**. All of the synthesized ZnO NRs showed a hexagonal-faceted morphology. The diameter of the obtained ZnO NRs was approximately 20 to 50nm. The NRs covered the entire surface of the substrate, and most of these NRs grew into an unchain-like and branched structure. On the basis of the SEM images, the utilization of different solvents evidently resulted in different diameters of the synthesized ZnO NRs. The ZnO NRs that were synthesized using 2-ME provided the smallest diameter, whereas those synthesized with EtOH displayed the largest diameters. The size of the ZnO NRs in diameter is strongly dependent on the grain size of the ZnO seed layer[29]. As the grain size of the seed layer increases, larger sizes of ZnO NRs in diameter are produced.

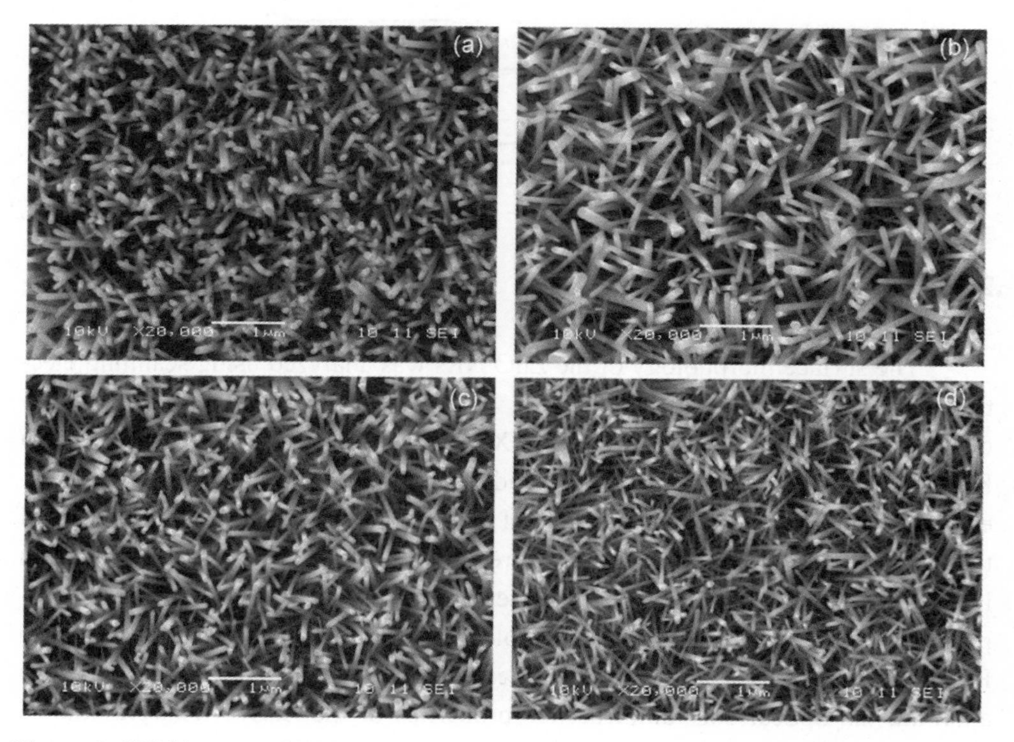

Figure 3. SEM images of ZnO NRs prepared with different solvents: (a) MeOH, (b) EtOH, (c) IPA, and (d) 2-ME.

3.2. XRD Characterization

The crystal structure and microstructure of the assynthesized ZnO NRs were studied through XRD. **Figure 4** shows the XRD patterns of the ZnO NRs that were synthesized on the silicon substrate with the aqueous solutions and different seeded layers. All of the diffraction peaks are consistent with the standard card Joint Committee on Powder Diffraction Standards (JCPDS) 36-1451. The peak intensities were measured in the range of 30° to 70° at 2θ. The result showed that the ZnO NRs that were prepared through the hydrothermal growth method presented a remarkably strong diffraction peak at the (002) plane, which is located between 34.5° and 34.6°[30][31]. This finding indicated that all of the ZnO samples possessed pure hexagonal wurtzite structures with high c-axis orientations.

Among the peaks, the ZnO NRs that were prepared with EtOH resulted in

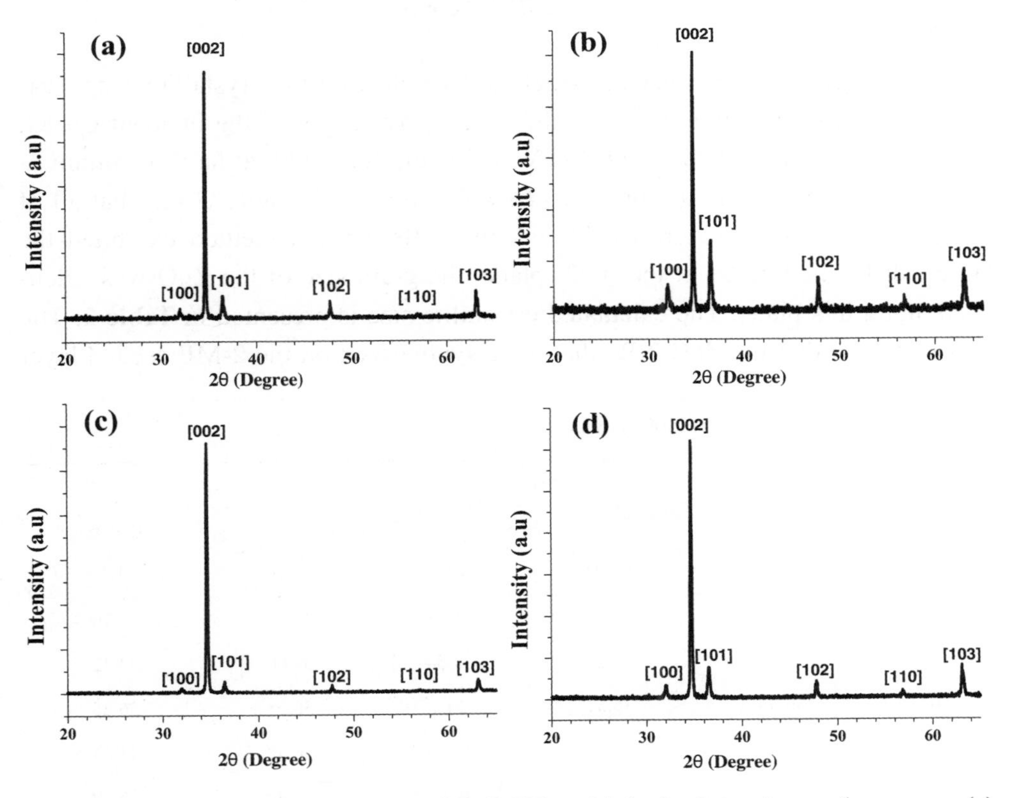

Figure 4. X-ray diffraction patterns of ZnO NRs with hydrothermal growth process: (a) MeOH, (b) EtOH, (c) IPA, and (d) 2-ME.

the narrowest peak of full width at half maximum (FWHM). By contrast, the ZnO NRs that were prepared with 2-ME showed the largest peak of FWHM. Simultaneously, the 2-ME solvent also showed the highest peak intensities on the (002) plane. Compared with the standard diffraction peaks of ZnO, the clear and sharp peaks indicated that the ZnO NRs possessed an excellent crystal quality, with no other diffraction peaks and characteristic peaks of impurities in the ZnO NRs. Therefore, all of the diffraction peaks were similar to those of the bulk ZnO. **Table 1** shows the ZnO XRD data from the JCPDS card compared with the measured ZnO XRD results.

The average grain size of the ZnO NRs was estimated using Scherrer's formula[32]:

$$D = \frac{k\lambda}{FWHM \cos\theta}$$

where κ is the Scherrer constant, which is dependent on the crystallite shape and can be considered as 0.9[33][34]; λ is the X-ray wavelength of the incident Cu Kα radiation, which is 0.154056 nm[35]; FWHM is the full width at half maximum of the respective peak; and θ represents the diffraction peak angle. Given that all of the ZnO NRs that were grown through the hydrothermal method exhibited the largest diffraction peaks at the (002) plane, the grain size of the ZnO was calculated along this plane. The calculated crystallite size is presented in **Table 2**. The result showed that the ZnO NRs that were synthesized on the 2-ME seeded layer

Table 1. XRD parameters of ZnO NRs.

hkl	2θ (°)				
	Observed				JCPDS
	MeOH	EtOH	IPA	2-ME	
100	32.02	31.98	31.98	32.10	31.76
002	34.52	34.62	34.64	34.68	34.42
101	36.46	36.52	36.5	36.58	36.25
102	47.76	47.8	47.74	47.8	47.53
110	56.94	56.78	56.96	56.86	56.60
103	63.08	63.06	63.08	63.06	62.86

Table 2. Measured structural properties of ZnO NRs using XRD for different solvents.

Solvent	XRD (100) peak position	XRD (002) peak position	*a*(Å) (100)	*c*(Å) (002)	Grain size (nm)
MeOH	32.02	34.52	3.225	5.192	54.84
EtOH	31.98	34.62	3.229	5.178	58.75
IPA	31.98	34.64	3.229	5.175	45.70
2-ME	32.10	34.68	3.217	5.169	39.18

produced the smallest crystallite size of 39.18nm. This result is consistent with the SEM images. However, the largest crystallite size of 58.75nm was observed when the ZnO NRs were synthesized on the seeded EtOH layer. This finding may be due to the higher viscosity of the EtOH solvent than those of the other solvents.

The lattice constants a and c of the ZnO wurtzite structure can be calculated using Bragg's law[36]:

$$a = \sqrt{\frac{1}{3}}\frac{\lambda}{\sin\theta}$$

$$c = \frac{\lambda}{\sin\theta}$$

where λ is the X-ray wavelength of the incident Cu Kα radiation (0.154056nm). For the bulk ZnO from the JCPDS data with card number 36–1451, the pure lattice constants a and c are 3.2498 and 5.2066Å, respectively. Based on the results shown in **Table 2**, all of the ZnO NRs had lower lattice constant values compared with the bulk ZnO. The ZnO NRs prepared with MeOH (a = 3.23877Å and c = 5.20987Å) were closest to the bulk ZnO. This phenomenon can be attributed to the high-temperature annealing condition. Similar results were observed by Lupan *et al.*[37], in which the increase in temperature decreases the lattice constant of ZnO.

3.3. FTIR Characterization

Figure 5 illustrates the FTIR spectra of the as-deposited four representative

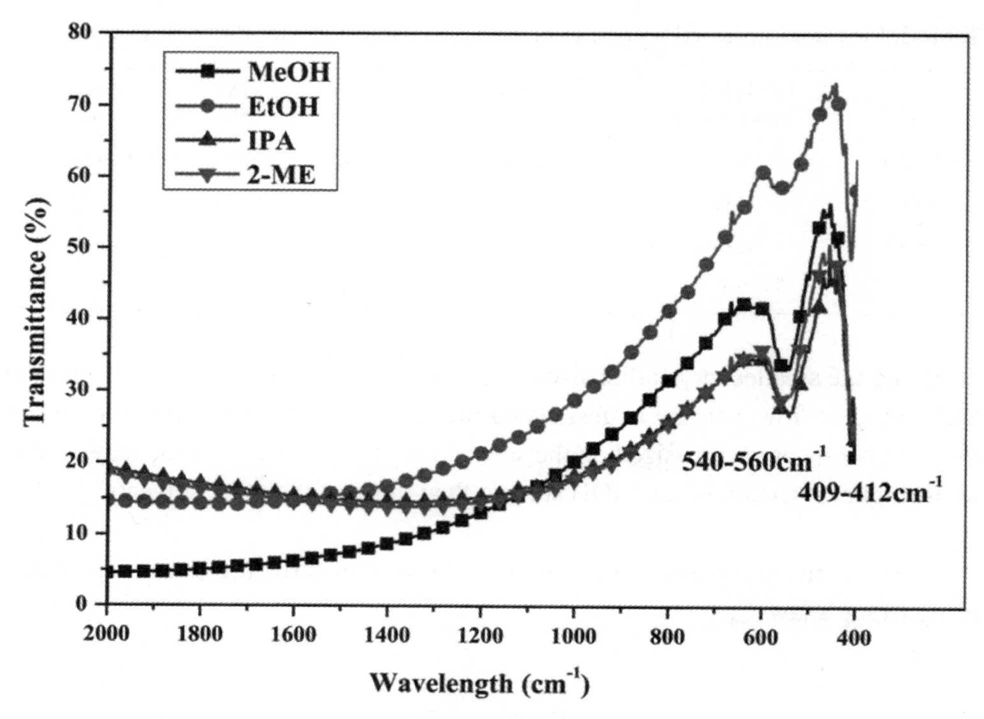

Figure 5. FTIR absorption spectrum of ZnO NRs using various solvents.

ZnO NRs prepared using four different solvents. Given that the wavelength of the finger-print of the material ranged from 400 to 2,000cm^{-1}[38], the absorption region was fixed in this region. Overall, the spectrum showed two significant peaks and all of the ZnO NRs that were prepared using different solvents exhibited the same peaks. The ZnO NR morphologies that are grown via wet chemical synthesis prefer the c-axis growth[39]. Thus, the ZnO NRs usually had a reference spectrum at around 406cm^{-1}[40]. However, this absorption spectra is found at 410, 412, 409, and 410cm^{-1} for the ZnO NRs prepared with the use of MeOH, EtOH, IPA, and 2-ME solvents, respectively, because these solvents caused a blueshift in the spectra of as-prepared ZnO NRs. The band from 540 to 560cm^{-1} is also a stretching mode that is correlated with the ZnO[41][42].

3.4. UV-vis Characterization

The transmittance spectra and optical properties of the ZnO NRs in the wavelength range of 300 to 800nm were investigated through UV-visible spectroscopy at RT. The UV-visible transmittance spectra of the ZnO NRs are shown in **Figure 6**. The inset of **Figure 6** shows the magnified view of transmittance spectrum in the wavelength range of 350 to 450nm. The results showed that all of the ZnO NRs that were prepared using different solvents exhibited strong excitonic absorption peaks at 378nm. These peaks indicated that the grown ZnO NRs possessed good optical quality and large exciton binding energy.

The absorption coefficient (α) for the direct transition of the ZnO NRs was studied using Equation (4)[43]:

$$\alpha = \frac{\ln(1/T)}{d} \tag{4}$$

where T is the transmittance of the ZnO films, and d is the film thickness. The

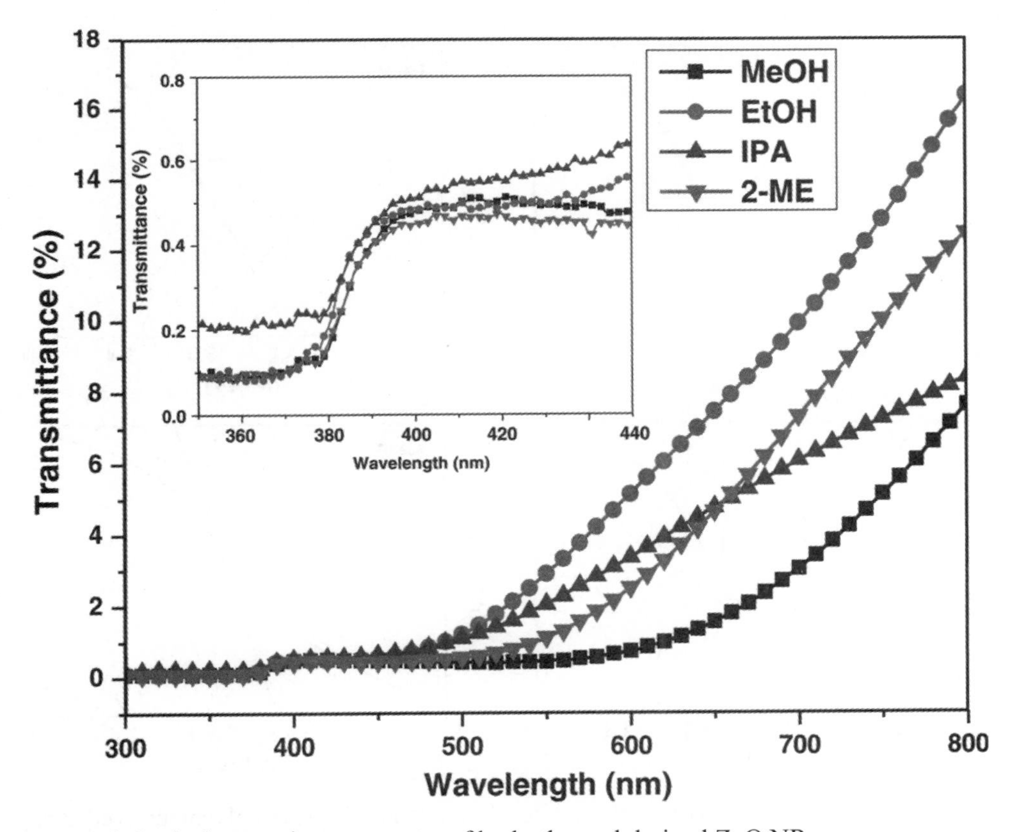

Figure 6. Optical transmittance spectra of hydrothermal derived ZnO NRs.

optical bandgap (αhv) dependence on the absorption coefficient (α) over the energy range of 3 to 3.5eV at RT was calculated using the following relation[44]:

$$\alpha hv = B\left(hv - E_g\right)^n \tag{5}$$

where hv is the photon energy, B is the constant, E_g is the bandgap energy, and n is the allowed direct band with the value of 1/2. The direct bandgap energies for the different solvents used were determined by plotting the corresponding Tauc graphs, that is, $(\alpha hv)^2$ versus hv curves. This method was used to measure the energy difference between the valence and conduction bands. The direct bandgap of the ZnO films was the interception between the tangent to the linear portion of the curve and the hv-axis (**Figure 7**). The optical bandgaps determined from the curves are summarized in **Table 3**. The results indicated that the ZnO NRs that were grown with 2-ME for the seed layer preparation showed the highest bandgap (3.21eV),

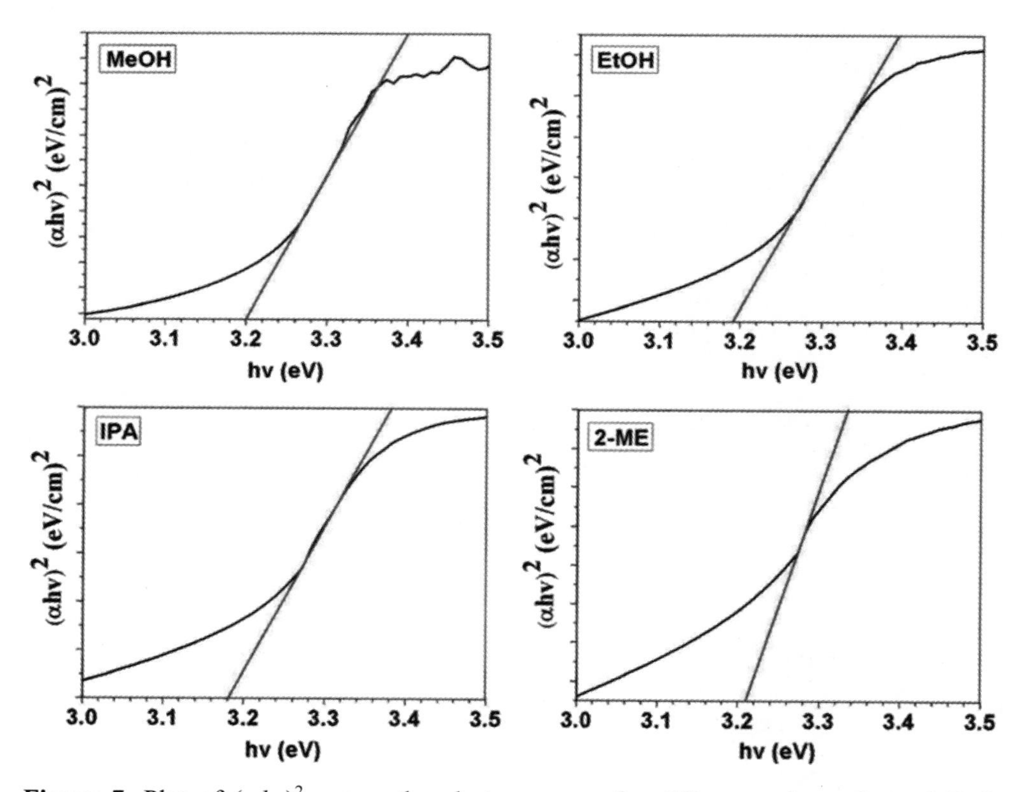

Figure 7. Plot of $(\alpha hv)^2$ versus the photon energy for different solvent derived ZnO thin films.

Table 3. Direct bandgap, calculated refractive indices of ZnO NRs corresponding to optical dielectric constant.

Solvent	Bandgap (eV)			Refractive index (n)			Optical constant (ε_∞)		
MeOH	3.20	3.28[a]	3.25[b]	2.064[i]	2.290[j]	2.329[k]	4.260[i]	5.246[j]	5.426[k]
EtOH	3.19	3.31[c]	3.10[d]	2.070[i]	2.293[j]	2.331[k]	4.286[i]	5.259[j]	5.436[k]
IPA	3.18	3.29[e]	3.27[f]	2.076[i]	2.296[j]	2.334[k]	4.311[i]	5.272[j]	5.445[k]
2-ME	3.21	3.28[g]	3.39[h]	2.058[i]	2.288[j]	2.327[k]	4.235[i]	5.233[j]	5.417[k]

[a]Yi *et al.*[64]. [b]Cao *et al.*[58]. [c]Karami *et al.*[59]. [d]Gowthaman *et al.*[60]. [e]Shakti *et al.*[61]. [f]Mejía- García *et al.*[62]. [g]Kashif *et al.*[23]. [h]Abdullah *et al.*[63]. [i]Ravindra *et al.*[51]. [j]Herve and Vandamme[52]. [k]Ghosh *et al.*[53].

whereas those grown with the IPA exhibited the lowest bandgap (3.18eV), which is believed to possess a better conductivity. According to the corresponding bandgap energy (E_g) and absorption band edge (λ) of the bulk ZnO, that is, 367nm and 3.36eV, respectively[45], the as-grown ZnO NRs possessed a significantly lower bandgap or exhibited a redshift of E_g from 0.15 to 0.18eV. This shift can be attributed to the optical confinement effect of the formation of ZnO NRs[46] and the size of the ZnO NRs[47].

Many attempts have been made to relate the refractive index (n) and E_g through simple relationships[48]-[51]. However, these relationships of n are independent of the temperature and incident photon energy. Herein, the various relationships between n and E_g were reviewed. Ravindra *et al.*[51] presented a linear form of n as a function of E_g:

$$n=\alpha+\beta E_g \tag{6}$$

where $\alpha = 4.048\text{eV}^{-1}$ and $\beta = -0.62\text{eV}^{-1}$. Moreover, light refraction and dispersion were inspired. Herve and Vandamme[52] proposed an empirical relation as follows:

$$n=\sqrt{1+\left(\frac{A}{E_g+B}\right)^2} \tag{7}$$

where A = 13.6eV and B = 3.4eV. For group IV semiconductors, Ghosh *et al.*[53] published an empirical relationship based on the band structure and quantum dielectric considerations of Penn[54] and Van Vechten[55]:

$$n^2 - 1 = \frac{A}{\left(E_g + B\right)^2} \tag{8}$$

where $A = 25\ E_g + 212$, $B = 0.21\ E_g + 4.25$, and $(E_g + B)$ refer to an appropriate average E_g of the material. The calculated refractive indices of the end-point compounds and Eg are listed in **Table 3**. In addition, the relation $\varepsilon_\infty = n^2$[56] was used to calculate the optical dielectric constant ε_∞. Our calculated refractive index values are consistent with the experimental values[23][57]–[63], as shown in **Table 3**. Therefore, Herve and Vandamme model is an appropriate model for solar cell applications.

3.5. PL Characterization

The effects of solvents on the luminescence properties of ZnO NRs were studied via PL spectroscopy, with excitation of a xenon lamp at 325nm. **Figure 8** shows the typical spectra for the photoluminescence of ZnO NRs that were grown on different seeded substrates. All the samples demonstrated two dominant peaks, which had UV emissions of 300 to 400nm and visible emissions at 400 to 800nm. The first emission band that was located in that UV range was caused by the recombination of free excitons through an exciton-exciton collision process[24][64][65]. In addition, the second emission band, which was a broad intense of green emission, originated from the deep-level emission. This band revealed the radiative recombination of the photogenerated hole with the electrons that belonged to the singly ionized oxygen vacancies[66]–[68].

UV luminescence can be used to evaluate the crystal quality of a material, whereas visible luminescence can be used to determine structural defects[69]. A study by Abdulgafour[70]. indicates that a higher ratio of UV/visible is an indicative index of a better crystal quality. In the current study, the UV/visible ratios for the ZnO NRs prepared with the use of IPA, MeOH, 2-ME, and EtOH were 13.34, 12.15, 8.32, and 5.14, respectively. Therefore, the UV/visible ratio trend confirms the improvements in crystal quality of the ZnO NRs that were prepared using different solvents.

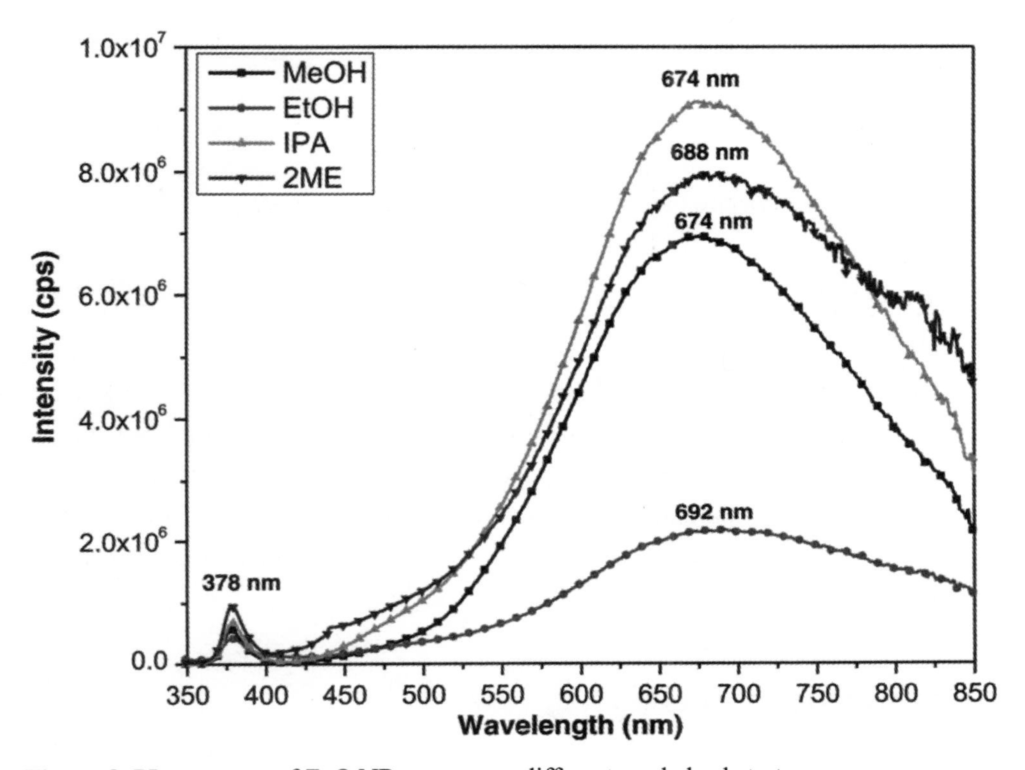

Figure 8. PL spectrum of ZnO NRs grown on different seeded substrate.

4. Conclusions

In this study, ZnO NRs with a highly crystalline structure were synthesized via a low-cost and convenient hydrothermal technique. The SEM images of the samples demonstrated that the diameters of the hydrothermally synthesized ZnO NRs range from 20 to 50nm. The XRD patterns exhibited that all of the ZnO NRs had remarkably excellent crystal qualities and high c-axis orientations. The calculated bandgap values of the synthesized ZnO NRs were lower than that of the bulk ZnO. The crystal qualities, grain size, diameter, and optical bandgap of the ZnO NRs were affected by the type of solvent used in the ZnO seed layer preparation. The ZnO NRs that were synthesized with the use of 2-ME, a solvent, exhibited the most improved results, in terms of structural and optical properties; these ZnO

NRs showed the smallest grain size, smallest crystallite size, and highest bandgap values. The method developed in this study provides a simple and low-cost approach to fabricate ZnO NRs with the desired properties.

Competing Interests

The authors declare that they have no competing interests.

Authors' Contributions

KLF conducted the sample fabrication and took part in the ZnO NR preparation and characterization and manuscript preparation. UH initialized the research work and coordinated and supervised this team's work. MK carried out the ZnO NR preparation and characterization. CHV conducted the ZnO NR characterization and manuscript preparation. All authors read and approved the final manuscript.

Acknowledgements

The authors wish to acknowledge the financial support of the Malaysian Ministry of Higher Education (MOHE) through the FRGS grant no. 9003-00276 to Prof. Dr. Uda Hashim. The author would also like to thank the technical staff of the Institute of Nano Electronic Engineering and School of Bioprocess Engineering, University Malaysia Perlis for their kind support to smoothly perform the research.

Source: Kai L F, Hashim U, Muhammad K, *et al.* Sol-gel synthesized zinc oxide nanorods and their structural and optical investigation for optoelectronic application. [J]. Nanoscale Research Letters, 2014, 9(1):1–10.

References

[1] Wang ZM: One-Dimensional Nanostructures. Springer Science + Business Media, LLC, 233 Spring Street, New York, NY 10013, USA: Springer; 2008.

[2] Cao GZ, Wang Y: Nanostructures and Nano-materials: Synthesis, Properties, and Applications. 2nd edition. Singapore 596224: World Scientific Publishing Co. Pte. Ltd; 2010.

[3] Ghosh R, Fujihara S, Basak D: Studies of the optoelectronic properties of ZnO thin films. J Electron Mater 2006, 35:1728–1733.

[4] Fan J, Freer R: The electrical properties and d.c. degradation characteristics of silver doped ZnO varistors. J Mater Sci 1993, 28:1391–1395.

[5] Jie J, Wang G, Wang Q, Chen Y, Han X, Wang X, Hou JG: Synthesis and characterization of aligned ZnO nanorods on porous aluminum oxide template. J Phys Chem B 2004, 108:11976–11980.

[6] Johnson JC, Knutsen KP, Yan H, Law M, Zhang Y, Yang P, Saykally RJ: Ultrafast carrier dynamics in single ZnO nanowire and nanoribbon lasers. Nano Lett 2003, 4:197–204.

[7] Kim K, Moon T, Lee M, Kang J, Jeon Y, Kim S: Light-emitting diodes composed of n-ZnO and p-Si nanowires constructed on plastic substrates by dielectrophoresis. Solid State Sci 2011, 13:1735–1739.

[8] Foo KL, Kashif M, Hashim U, Ali M: Fabrication and characterization of ZnO thin films by sol-gel spin coating method for the determination of phosphate buffer saline concentration. Curr Nanosci 2013, 9:288–292.

[9] Foo KL, Hashim U, Kashif M: Study of zinc oxide films on SiO2/Si substrate by sol-gel spin coating method for pH measurement. Appl Mech Mater 2013, 284:347–351.

[10] Kashif M, Ali M, Ali SMU, Foo KL, Hashim U, Willander M: Sol-gel synthesis of ZnO nanorods for ultrasensitive detection of acetone. Adv Sci Lett 2013, 19:3560–3563.

[11] Chai G, Lupan O, Chow L, Heinrich H: Crossed zinc oxide nanorods for ultraviolet radiation detection. Sensor Actuat A Phys 2009, 150:184–187.

[12] Foo KL, Kashif M, Hashim U, Ali M: Sol-gel derived ZnO nanoparticulate films for ultraviolet photodetector (UV) applications. Optik-Int J Light Electron Optics 2013, 124:5373–5376.

[13] Guillen E, Azaceta E, Peter LM, Zukal A, Tena-Zaera R, Anta JA: ZnO solar cells with an indoline sensitizer: a comparison between nanoparticulate films and electrodeposited nanowire arrays. Energy Environ Sci 2011, 4:3400–3407.

[14] Matsubara K, Fons P, Iwata K, Yamada A, Sakurai K, Tampo H, Niki S: ZnO transparent conducting films deposited by pulsed laser deposition for solar cell applications. Thin Solid Films 2003, 431–432:369–372.

[15] Fulati A, Ali SMU, Asif MH, Alvi NH, Willander M, Brännmark C, Strålfors P, Börjesson SI, Elinder F, Danielsson B: An intracellular glucose biosensor based on nanoflake ZnO. Sensor Actuat B Chem 2010, 150:673–680.

[16] Ali SMU, Nur O, Willander M, Danielsson B: A fast and sensitive potentiometric glucose microsensor based on glucose oxidase coated ZnO nanowires grown on a thin silver wire. Sensor Actuat B Chem 2010, 145:869–874.

[17] Lee W, Sohn H, Myoung JM: Prediction of the structural performances of ZnO nanowires grown on GaAs (001) substrates by metalorganic chemical vapour deposition (MOCVD). Mater Sci Forum 2004, 449–452:1245–1248.

[18] Park WI, Kim DH, Jung S-W, Yi G-C: Metalorganic vapor-phase epitaxial growth of vertically well-aligned ZnO nanorods. Appl Phys Lett 2002, 80:4232–4234.

[19] Bakin A, Che Mofor A, El-Shaer A, Waag A: Vapour phase transport growth of ZnO layers and nanostructures. Superlattice Microst 2007, 42:33–39.

[20] Suh D-I, Byeon CC, Lee C-L: Synthesis and optical characterization of vertically grown ZnO nanowires in high crystallinity through vapor- liquid-solid growth mechanism. Appl Surf Sci 2010, 257:1454–1456.

[21] Xia Y, Yang P, Sun Y, Wu Y, Mayers B, Gates B, Yin Y, Kim F, Yan H: One- dimensional nanostructures: synthesis, characterization, and applications. Adv Mater 2003, 15:353–389.

[22] Hossain M, Ghosh S, Boontongkong Y, Thanachayanont C, Dutta J: Growth of zinc oxide nanowires and nanobelts for gas sensing applications. J Metastable Nanocrystalline Mater 2005, 23:27–30.

[23] Kashif M, Hashim U, Ali ME, Foo KL, Ali SMU: Morphological, structural, and electrical characterization of sol-gel-synthesized ZnO nanorods. J Nano Mat 2013, 2013:7.

[24] Kashif M, Ali SMU, Ali ME, Abdulgafour HI, Hashim U, Willander M, Hassan Z: Morphological, optical, and Raman characteristics of ZnO nanoflakes prepared via a sol-gel method. Phys Status Solid A 2012, 209:143–147.

[25] Kashif M, Hashim U, Ali SMU, Ala'eddin AS, Willander M, Ali ME: Structural and impedance spectroscopy study of Al-doped ZnO nanorods grown by sol-gel method. Microelectron Int 2012, 29:1–1.

[26] Li YB, Bando Y, Sato T, Kurashima K: ZnO nanobelts grown on Si substrate. Appl Phys Lett 2002, 81:144–146.

[27] Ali SMU, Kashif M, Ibupoto ZH, Fakhar-e-Alam M, Hashim U, Willander M: Functionalised zinc oxide nanotube arrays as electrochemical sensors for the selective determination of glucose. Micro & Nano Lett 2011, 6:609–613.

[28] Foo KL, Kashif M, Hashim U, Liu W-W: Effect of different solvents on the structural and optical properties of zinc oxide thin films for optoelectronic applications. Ceram Int 2014, 40:753–761.

[29] Kenanakis G, Vernardou D, Koudoumas E, Katsarakis N: Growth of c-axis oriented ZnO nanowires from aqueous solution: the decisive role of a seed layer for controlling the wires' diameter. J Cryst Growth 2009, 311:4799–4804.

[30] Jing-Shun H, Ching-Fuh L: Controlled growth of zinc oxide nanorod array in aqueous solution by zinc oxide sol-gel thin film in relation to growth rate and optical property. In Nanotechnology, 2008 NANO '08 8th IEEE Conference on; 18-21 Aug. 2008; Arlington, Texas USA. The Institute of Electrical and Electronics Engineer; 2008:135–138.

[31] Li Z, Huang X, Liu J, Li Y, Li G: Morphology control and transition of ZnO nanorod arrays by a simple hydrothermal method. Mater Lett 2008, 62:1503–1506.

[32] Jenkins R, Snyder R: Introduction to X-Ray Powder Diffractometry. Canada: John Wiley & Sons, Inc; 2012.

[33] Metin H, Esen R: Annealing effects on optical and crystallographic properties of CBD grown CdS films. Semicond Sci Technol 2003, 18:647.

[34] Pearton SJ, Norton DP, Ip K, Heo YW, Steiner T: Recent advances in processing of ZnO. J Vac Sci Technol B 2004, 22:932–948.

[35] Kaneva N, Dushkin C: Preparation of nanocrystalline thin films of ZnO by sol-gel dip coating. Bulg Chem Commun 2011, 43:259–263.

[36] Suryanarayana C, Norton G: X-Ray Diffraction: A Practical Approach. Springer Science + Business Media, LLC, 233 Spring Street, New York, NY 10013, USA: Plenum Press; 1998.

[37] Lupan O, Pauporté T, Chow L, Viana B, Pellé F, Ono L, Roldan Cuenya B, Heinrich H: Effects of annealing on properties of ZnO thin films prepared by electrochemical deposition in chloride medium. Appl Surf Sci 2010, 256:1895–1907.

[38] Feng L, Liu A, Ma Y, Liu M, Man B: Fabrication, structural characterization and optical properties of the flower-like ZnO nanowires. Acta Physiol Pol 2010, 117:512–517.

[39] Verges MA, Mifsud A, Serna CJ: Formation of rod-like zinc oxide microcrystals in homogeneous solutions. J Chem Soc 1990, 86:959–963.

[40] Kleinwechter H, Janzen C, Knipping J, Wiggers H, Roth P: Formation and properties of ZnO nano-particles from gas phase synthesis processes. J Mater Sci 2002, 37:4349–4360.

[41] Khun K, Ibupoto ZH, Nur O, Willander M: Development of galactose biosensor based on functionalized ZnO nanorods with galactose oxidase. J Sensors 2012, 2012:7.

[42] Wang J, He S, Zhang S, Li Z, Yang P, Jing X, Zhang M, Jiang Z: Controllable synthesis of ZnO nanostructures by a simple solution route. Mater Sci Poland 2009, 27:477–484.

[43] W-n M, X-f L, Zhang Q, Huang L, Zhang Z-J, Zhang L, Yan X-J: Transparent conductive In2O3: Mo thin films prepared by reactive direct current magnetron sputtering at room temperature. Thin Solid Films 2006, 500:70–73.

[44] Singh S, Kaur H, Pathak D, Bedi R: Zinc oxide nanostructures as transparent window layer for photovoltaic application. Dig J Nanomater Bios 2011, 6:689–698.

[45] Klingshirn C: The luminescence of ZnO under high one- and two- quantum excitation. Phys Status Solidi B 1975, 71:547–556.

[46] Lee GJ, Lee Y, Lim HH, Cha M, Kim SS, Cheong H, Min SK, Han SH: Photoluminescence and lasing properties of ZnO nanorods. J Korean Phys Soc 2010, 57:1624–1629.

[47] Samanta P, Patra S, Chaudhuri P: Visible emission from ZnO nanorods synthesized by a simple wet chemical method. Int J Nanosci Nanotech 2009, 1:81–90.

[48] Moss T: A relationship between the refractive index and the infra-red threshold of sensitivity for photoconductors. Proc Phys Soc Sect B 2002, 63:167.

[49] Gupta VP, Ravindra NM: Comments on the moss formula. Phys Status Solid B 1980, 100:715–719.

[50] Hervé P, Vandamme LKJ: General relation between refractive index and energy gap in semiconductors. Infrared Phys Technol 1994, 35:609–615.

[51] Ravindra NM, Auluck S, Srivastava VK: On the Penn Gap in semiconductors. Phys Status Solid B 1979, 93:K155–K160.

[52] Herve PJL, Vandamme LKJ: Empirical temperature dependence of the refractive index of semiconductors. J Appl Phys 1995, 77:5476–5477.

[53] Ghosh D, Samanta L, Bhar G: A simple model for evaluation of refractive indices of some binary and ternary mixed crystals. Infrared Physics 1984, 24:43–47.

[54] Penn DR: Wave-number-dependent dielectric function of semiconductors. Phys Rev 1962, 128:2093–2097.

[55] Van Vechten JA: Quantum dielectric theory of electronegativity in covalent systems. I Electron Dielectric Constant Phys Rev 1969, 182:891.

[56] Samara GA: Temperature and pressure dependences of the dielectric constants of semiconductors. Phys Rev B 1983, 27: 3494–3505.

[57] Yang Z, Liu QH: The structural and optical properties of ZnO nanorods via citric acid-assisted annealing route. J Mater Sci 2008, 43:6527–6530.

[58] Cao HL, Qian XF, Gong Q, Du WM, Ma XD, Zhu ZK: Shape- and size- controlled synthesis of nanometre ZnO from a simple solution route at room temperature. Nanotechnology 2006, 17: 3632.

[59] Karami H, Fakoori E: Synthesis and characterization of ZnO nanorods based on a new gel pyrolysis method. J Nanomater 2011, 2011:11.

[60] Gowthaman P, Saroja M, Venkatachalam M, Deenathayalan J, Senthil TS: Structural and optical properties of ZnO nanorods prepared by chemical bath deposition method. Aust J Basic Appl Sci 2011, 5:1379–1382.

[61] Shakti N, Kumari S, Gupta PS: Structural, optical and electrical properties of ZnO nanorod array prepared by hydrothermal process. J Ovonic Res 2011, 7:51–59.

[62] Mejía-García C, Díaz-Valdés E, Ortega-Cervantes G, Basurto-Cazares E: Synthesis of hydrothermally grown zinc oxide nanowires. J Chem Chem Eng 2012, 6:63–66.

[63] Abdullah H, Selmani S, Norazia MN, Menon PS, Shaari S, Dee CF: ZnO:Sn deposition by sol-gel method: effect of annealing on the structural, morphology and optical properties. Sains Malays 2011, 40:245–250

[64] Yi S-H, Choi S-K, Jang J-M, Kim J-A, Jung W-G: Low-temperature growth of ZnO nanorods by chemical bath deposition. J Colloid Interface Sci 2007, 313:705–710.

[65] Kashif M, Hashim U, Ali ME, Ali SMU, Rusop M, Ibupoto ZH, Willander M: Effect of different seed solutions on the morphology and electrooptical properties of ZnO nanorods. J Nanomater 2012, 2012:6.

[66] Heo YW, Norton DP, Pearton SJ: Origin of green luminescence in ZnO thin film grown by molecular-beam epitaxy. J Appl Phys 2005, 98:073502.

[67] Lin B, Fu Z, Jia Y: Green luminescent center in undoped zinc oxide films deposited on silicon substrates. Appl Phys Lett 2001, 79:943–945.

[68] Zeng H, Duan G, Li Y, Yang S, Xu X, Cai W: Blue luminescence of ZnO nanoparticles based on non-equilibrium processes: defect origins and emission controls. Adv Funct Mater 2010, 20:561–572.

[69] Mridha S, Basak D: Effect of concentration of hexamethylene tetramine on the structural morphology and optical properties of ZnO microrods grown by low emperature olution approach. Phys Status Solid A 2009, 206:1515–1519.

[70] Abdulgafour HI, Hassan Z, Al-Hardan N, Yam FK: Growth of zinc oxide nanoflowers by thermal evaporation method. Phys B-Condensed Matter 2010, 405:2570– 2572.

Chapter 20

Effect of 2-Mercaptoethanol as Capping Agent on ZnS Nanoparticles: Structural and Optical Characterization

Abbas Rahdar

Department of Physics, Faculty of Science, University of Zabol, Zabol 98615538, Iran

Abstract: In this work, we report the effect of a capping agent on the structural and optical properties of nanocrystalline ZnS particles, which have been synthesized by co-precipitation method. The structural properties of ZnS nanoparticles have been characterized by X-ray diffraction (XRD) analysis. The XRD patterns show a hexagonal structure in the nanoparticles. The mean crystallite size calculated from the XRD patterns has been found to be in the range of 1.80 to 2.45nm with the increase in molar concentration of the capping agent. Absorption spectra have been obtained using a UV-vis spectrophotometer to find the optical direct band gap. We also found that optical band gap (Eg) increases with the increase in molar concentration of the capping agent. This behavior is related to size quantization effect due to the small size of the particles.

Keywords: ZnS Nanoparticles, Chemical Co-Precipitation Method, Capping Agent, Optical Band Gap

1. Introduction

Semiconductor nanocrystals represent a class of materials that have hybrid molecular and bulk properties. They have attracted much attention over the past few years because of their novel properties which originate from quantum confinement effect[1]-[4]. Quantum confinement effect modifies the electronic structure of the nanocrystals when the sizes of the nanoparticles are comparable to that of the Bohr excitonic radius of those materials. When the particle radius falls below the excitonic Bohr radius, the band gap energy is widened, leading to a blueshift in the band gap, emission spectra, etc. On the other hand, the surface states will play a more important role in the nanoparticles, due to their large surface-to- volume ratio with a decrease in particle size (surface effects). In the case of semiconductor nanoparticles, radiative or nonradiative recombination of an exciton at the surface states becomes dominant in its optical properties with a decrease of particle size. Therefore, the decay of an exciton at the surface states will influence the qualities of the material for an optoelectronic device. These size-dependent optical properties have many potential applications in the areas of solar energy conversion, light-emitting devices, chemical/biological sensors, and photocatalysis[5]-[9].

Wide band gap II-VI semiconductors are expected to be the novel materials for the optoelectronic devices. ZnS, which is an important member of this family, has been extensively investigated as it has numerous applications to its credit. ZnS has been used widely as an important phosphor for photoluminescence, electroluminescence, and cathodoluminescence devices due to its better chemical stability compared to other chalcogenides such as ZnSe. In optoelectronics, it finds use as light-emitting diode, reflector, dielectric filter, and window material. The synthesis of ZnS remains a topic of interest for researchers, as new synthetic routes are being explored to get a single phase material via an economically and technically viable method. In the present paper, chemical method has been used to synthesize the ZnS nanoparticles. In the present paper, ZnS nanoparticles are synthesized by using mercaptoethanol as capping agent. This paper is organized as follows: The structural and optical properties of synthesized nanoparticles are presented in the 'Results and discussion' section in detail, followed by a 'Conclusion' section. In the 'Methods' section, we give a description of materials and synthesis method of ZnS nanoparticles with different molar concentrations of the capping agent.

2. Results and Discussion

2.1. Structural Characterization

The X-ray diffraction (XRD) patterns of the prepared ZnS nanoparticles with different molar concentrations of the capping agent (HOCH2CH2SH) are shown in **Figure 1**. All of the crystalline Bragg peaks in the XRD pattern [(111),

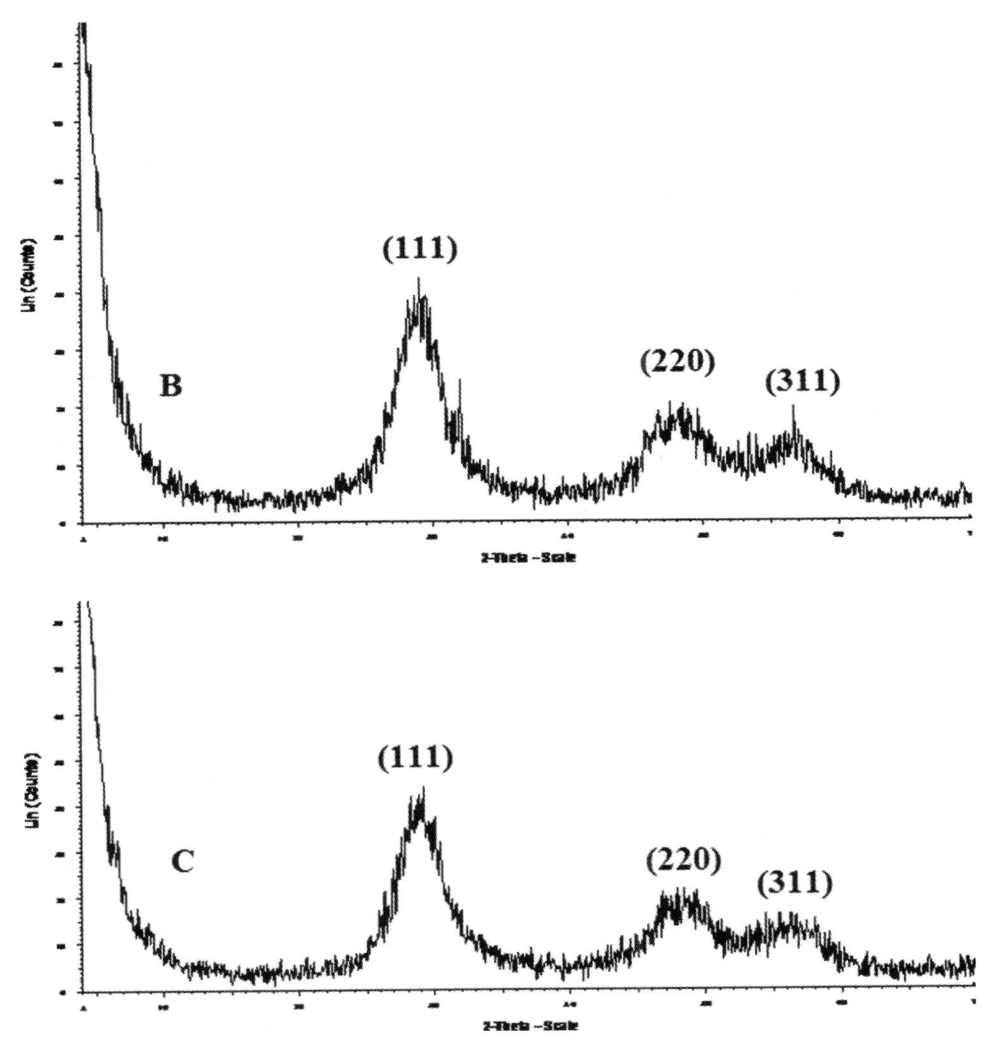

Figure 1. XRD patterns of ZnS nanoparticles. With different molar (M) concentrations of capping agent: 0.1 M (B) and 0.3 M (C).

(220), and (311) planes] are in good agreement with the diffraction data of a hexagonal structure with cell parameters a = 3.800Å and c = 6.230Å from the JCPDS card. Furthermore, the peak broadening in the XRD patterns clearly indicates the formation of ZnS nanocrystals with a very small size.

The peak broadening at a lower angle is more meaningful for the calculation of particle size. The mean crystallite size (D) of nanoparticles was also estimated using the Scherrer formula (Scherrer 1918) using (111) reflection from the XRD pattern as follows:

$$D = \frac{0.9\lambda}{B\cos\theta} \tag{1}$$

where λ, B, and θ are the X-ray wavelength of the radiation used [Kα(Cu) = 0.154056nm], the full width at half maximum of the diffraction peak, and the Bragg diffraction angle, respectively.

The values of mean crystallite size obtained from XRD for different molar concentrations of the capping agent are listed in **Table 1**.

2.2. Optical Characterization

The absorption spectra of the different samples are shown in **Figure 2**. The absorption edge is observed in the range of 324 to 270nm, which is blueshifted compared to bulk ZnS. As the molar concentration of the capping agent increases, the absorption edge shifts to a lower wavelength compared to ZnS (A). This blueshift of the absorption edges for different-sized nanocrystals is related to the size

Table 1. Mean crystallite size and optical band gap variation of ZnS nanoparticles.

Sample	Molar concentration of capping agent	Mean crystallite size (nm)	Optical band gap E_g(eV)
ZnS (A)	0.0	2.45	3.83
ZnS (B)	0.1	2.38	4.27
ZnS (C)	0.3	2.22	4.49
ZnS (D)	0.5	1.80	4.59

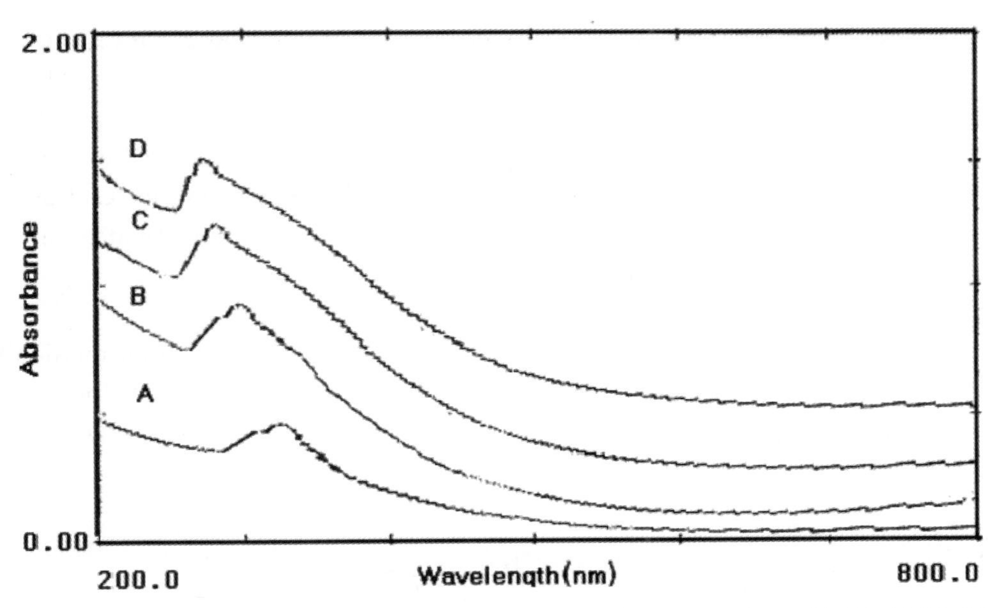

Figure 2. Absorption spectra of different samples of ZnS nanoparticles. 0 M (A), 0.1M (B), 0.3 M(C), and 0.5M (D).

decrease of particles and is attributed to the quantum confinement limit reaching of nanoparticles. The quantum confinement effect is expected for semiconducting nanoparticles, and the absorption edge will be shifted to a higher energy when the particle size decreases[10][11].

The obtained direct optical band gap values for different samples are shown in **Table 1**. It is necessary to mention that the optical direct band gap values of the ZnS nanoparticles were determined using Tauc's relation[12][13]:

$$\alpha hv = \alpha_0 \left(hv - E_g \right)^{1/2},$$

where hv, α_o, and E_g are the photon energy, a constant, and optical band gap of the nanoparticles, respectively. Absorption coefficient (α) of the powders at different wavelengths can be calculated from the absorption spectra. Finally, the values of E_g were determined using extrapolations of the linear regions of the plot of $(\alpha hv)^2$ versus hv.

As seen in **Table 1**, the values of optical band gap 'E_g' increase with the increase in molar concentration of the capping agent, and therefore, the decrease in particle size as mentioned earlier is due to the quantum confinement effect. Also, the variation of optical band gap with particle size is shown in **Figure 3**. As illustrated in **Figure 3**, the optical band gap values of nanoparticles have changed from 3.83 to 4.59eV by decreasing the particle sizes. The increase in band gap with increase in molar concentration of the capping agent is attributed to size quantization effect due to the small size of the particles[14].

These semiconductor nanocrystals were then subjected to transmission electron microscopy (TEM; JEOL JEM3010, JEOL Ltd., Tokyo, Japan) for characterization. **Figure 4** shows the TEM image of ME-capped ZnS semiconductor nano-crystals.

2.3. Relation Particle size with Chemical Reaction rate

Chemical reaction rate directly affects the time evolution of the number of nuclei, which determines both nucleation and growth process. First, the influence on nucleation is obvious: nucleation is faster when the chemical reaction is faster.

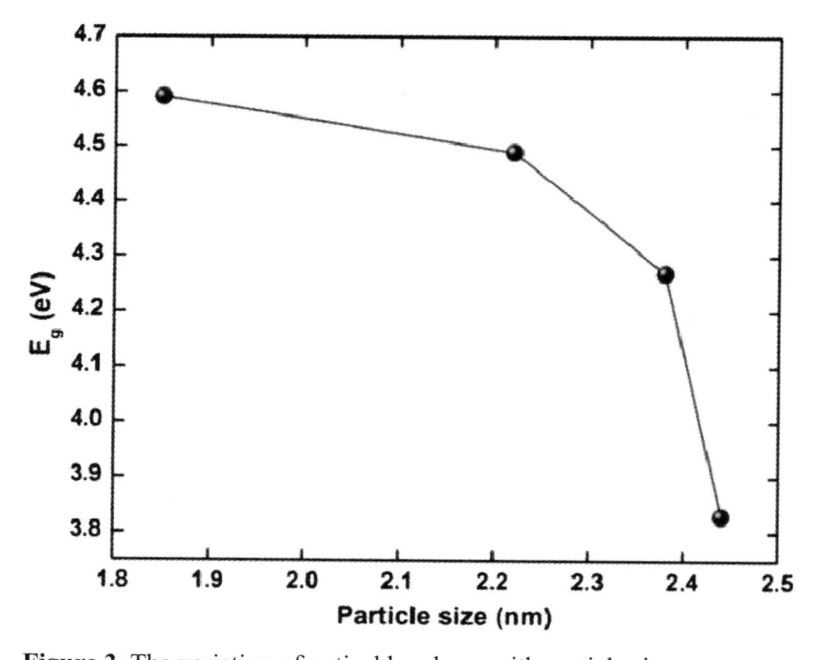

Figure 3. The variation of optical band gap with particle size.

Figure 4. TEM image of ME-capped ZnS semiconductor nanocrystals [0.1 M (B)].

Second, growth will be strongly influenced by the nuclei number which already formed at a given time. A great number of nucleation favor a fast autocatalytic growth, giving rise to a large number of small particles. Chemical reaction controls this kind of growth, in which the autocatalytic growth is faster when the chemical reaction is faster. However, in nanoparticle formation, there is another contribution to the growth molecules on the surface of a small particle which will tend to diffuse through the solution and add to the surface of a larger particle (growth by ripening). A slow chemical reaction favors continuous nuclei, keeping always a certain number of nuclei in the system. As a result, growth by ripening can take place during the whole process. This fact explains the bigger particle size obtained from a slow reaction. One can conclude that a slow chemical reaction rate is associated with a more important ripening contribution to the growth. A high number of nuclei are still forming at this stage when the reaction is slow at the same time; some particles have already grown to the final value of size. This means that in this case (slow reaction rate), nucleation and growth take place simultaneously. This overlapping of nucleation and growth processes, which is more pronounced as the chemical reaction is slower, leads to larger nanoparticle sizes[15].

Rate of reaction depends on the molar concentration of the reactant solution and increases with the increase in molar concentration of the reactant solution. In the present study, the molar concentration of the capping agent varies from 0 to 0.5M; the reaction rate is highest for the 0.5M solution, and hence, the particle size obtained is smallest for the 0.5M solution as compared to other materials in the series, which is in consistent with the above made argument.

3. Conclusions

It is possible to produce different-sized ZnS nanoparticles using a simple chemical method using different molar concentrations of the capping agent. XRD and optical band gap data have been obtained to confirm the nanosize of these materials. It is also observed that the particle size depends on the molar concentration of the capping agent. A decrease in formation rate of nanoparticles gives rise to a larger final particle size for all the studied synthesis conditions. As the particle size depends on the molar concentration of the doping agent, a decrease in the size of particle is observed with the increase of molar concentration of the capping agent. The mean crystallite size range of particles was between 1.85 and 2.44nm, depending on molar concentration of the capping agent. The optical band gap values of ZnS nanoparticles have changed from 3.83 to 4.59eV by increasing the molar concentration of the capping agent. These values exhibit a blueshift in E_g which is related to the size decrease of the particles and to the quantum confinement limit reaching of nanoparticles.

Considering these results, the chemical co-precipitation method using 2-mercaptoethanol as a capping agent is very efficient for the preparation of ZnS nanoparticles in order to control the particle size and also for modern optoelectronic technology and other industrial applications.

4. Methods

4.1. Materials

Zinc chloride ($ZnCl_2$) and sodium sulfide (Na_2S) as starting materials, 2-mercaptoethanol ($HOCH_2CH_2SH$) as a capping agent for the control particle size, and double-distilled water as dispersing solvent were used to prepare ZnS nanoparticles.

4.2. Synthesis of ZnS nanoparticles by co-precipitation method

The ZnS nanoparticles were synthesized using the chemical co-precipitation method as follows. First, $ZnCl_2$ was dissolved in double-distilled water (0.1M), and then, the obtained molar solution was stirred for 20min at room temperature to achieve complete dissolution. Sodium sulfide (Na_2S) was also dissolved in double-distilled water separately as per molar concentration. Afterwards, first, sodium sulfide solution was added drop by drop to the zinc chloride solution, and then, an appropriate amount of 2-mercaptoethanol as a capping agent was added to the reaction medium to control the particle size of ZnS. As a result of this, the white precipitate of the ZnS nanoparticles is formed slowly in the solution. In the final step, the obtained precipitate was filtered and dried at room temperature to remove both water and organic capping and other byproducts which formed during the reaction process. After sufficient drying, the precipitate was crushed to fine powder with the use of mortar and pestle. It is mentioned that the synthesis was done by passing nitrogen gas.

For the study of the effect of the capping agent on structural and optical properties of ZnS nanoparticles, different samples of nanoparticles have been obtained by changing the molar concentration of the capping agent, namely ZnS (A), ZnS (B), ZnS (C) and ZnS (D), as the amount of molar concentration of the capping agent used in the preparation is 0, 0.1, 0.3 and 0.5M, respectively. The synthesis process of ZnS nanoparticles by the co-precipitation method is summarized in a flow chart shown in **Figure 5**.

4.3. Powder Characterization

The XRD patterns of ZnS nanoparticles were recorded using a Bruker system (Bruker Optik GmbH, Ettlingen, Germany) using CuKα radiation ($\lambda = 0.154056$nm) with 2θ ranging from 5° to 70°. The optical absorption spectra of nanoparticles were measured using a USB-2000 UV-vis spectrophotometer (Ocean Optics, Inc., Dunedin, FL, USA). Therefore, the obtained nanopowders have been suspended in

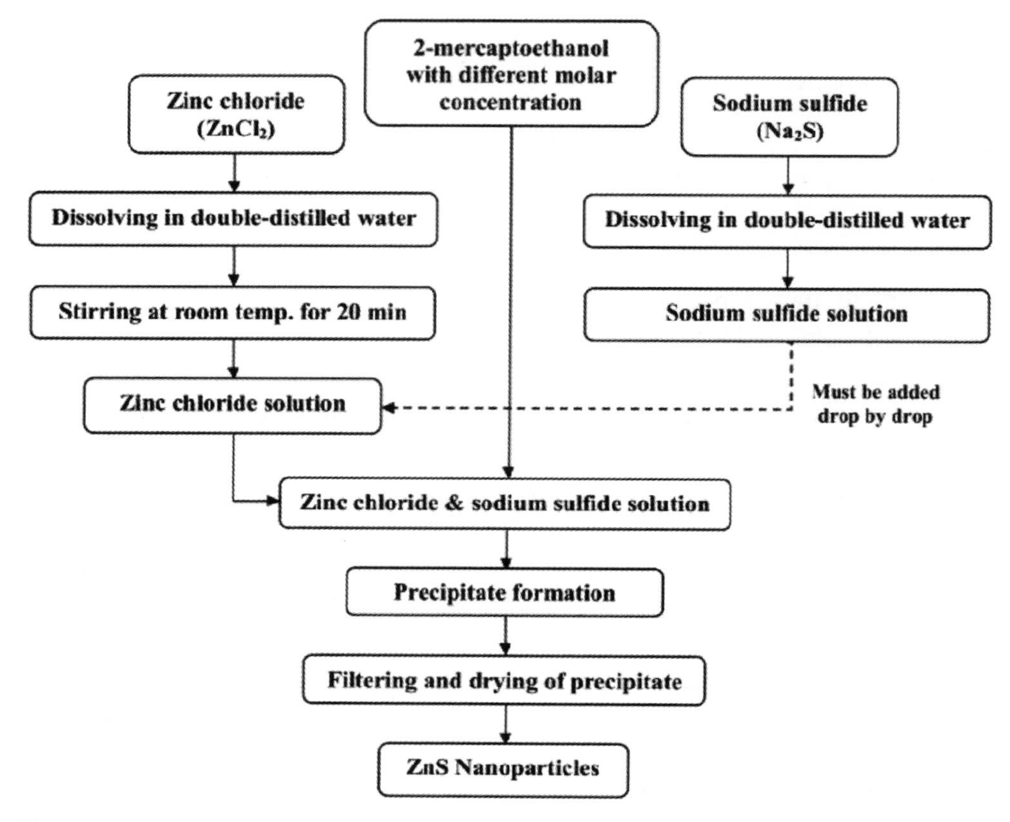

Figure 5. Flow chart of the preparation ZnS nanoparticles by chemical co-precipitation method.

glycerol using a magnetic stirrer, and their optical absorption spectra have been recorded at room temperature over the range of 200 to 800nm to determine the optical band gap values.

Competing Interests

The author has no competing interests.

Acknowledgement

The author would like to thank Mr. A. Rahdar, Mr. R. Hakimi-Pooya, Mrs.

M. Asudeh, Mr. H. Rahdar, and Mrs. Heidari Mokarrar for their support and assistance with this project.

Source: Rahdar A. Effect of 2-mercaptoethanol as capping agent on ZnS nanoparticles: structural and optical characterization[J]. Journal of Nanostructure in Chemistry, 2013, 3(1):1–5.

References

[1] Rossetti, R, Ellison, JL, Gibson, JM, Brus, LE: Size effects in the excited electronic states of small colloidal CdS crystallites. J. Chem Phys. 80, 4464 (1984).

[2] Sang, W, Qian, Y, Min, J, Li, D, Wang, L, Shi, W, Liu, Y: Microstructural and optical properties of ZnS:Cu nanocrystals prepared by an ion complex transformation method. Solid State Commun. 121, 475–478 (2002).

[3] Prabhu, RR, Abdul Khadar, M: Characterization of chemically synthesized CdS nanoparticles. Pramana J. Phys. 5, 801 (2005).

[4] Wageh, S, Ling, ZS, Xu-Rong, X: Growth and optical properties of colloidal ZnS nanoparticles. J. Cryst. Growth 255, 332–337 (2003).

[5] Rajeswar, K, Tacconi, NR, Chenthamarakshan, CR: Semiconductor-based composite materials: preparation, properties, and performance. Chem. Mater. 13, 2765 (2001).

[6] Alivisatos, AP: Semiconductor clusters, nanocrystals, and quantum dots. Science 271, 933–937 (1996).

[7] Anderson, MA, Gorer, S, Penner, RM: A hybrid electrochemical/chemical synthesis of supported, luminescent cadmium sulfide nanocrystals. J. Phys. Chem. B 101, 5895 (1997).

[8] Henshaw, G, Parkin, IP, Shaw, G: Convenient low-energy synthesis of metal sulfides and selenides; PbE, Ag2E, ZnE, CdE (E=S, Se). Chem. Commun. 10, 1095–1096 (1996).

[9] Hirai, T, Bando, Y, Komasawa, I: Immobilization of CdS nanoparticles formed in reverse micelles onto alumina particles and their photocatalytic properties. J. Phys. Chem. B 106, 8967–8970 (2002).

[10] Zhu, H, Yang, D, Yu, G, Zhang, H, Yao, K: A simple hydrothermal route for synthesizing SnO2 quantum dots. Nanotechnology 17, 2386 (2006).

[11] Sarhaddi, R, Shahtahmasebi, N, Rezaee Rokn-Abadi, M, Bagheri-Mohagheghi, MM: Effect of post-annealing temperature on nano-structure and energy band gap of indium tin oxide (ITO) nano-particles synthesized by polymerizing-complexing sol-gel

method. Physica E 43, 452–457 (2010).

[12] Tauc, J: Optical Properties of Solids. Academic Press, New York (1966).

[13] Liu, CM, Zu, XT, Wei, QM, Wang, LM: Fabrication and characterization of wire-like SnO2. J. Phys. D: Appl. Phys. 39, 2494 (2006).

[14] Amaranatha Reddy, D, Divya, A, Murali, G, Vijayalakshmi, RP, Reddy, BK: Synthesis and optical properties of Cr doped ZnS nanoparticles capped by 2-mercaptoethanol. Physica B 406, 1944-1949 (2011).

[15] de Dios, M, Barroso, F, Tojo, C, Blanco, MC, Lopez-Quintela, MA: Effects of the reaction rate on the size control of nanoparticles synthesized in microemulsions. Colloids and SurfacesA:P hysicochem.E ng.Aspects 83, 270 (2005).

Chapter 21

Structural and Optical Properties of ZnO Nanorods by Electrochemical Growth Using Multi-Walled Carbon Nanotube-Composed Seed Layers

Yeong Hwan Ko, Myung Sub Kim, Jae Su Yu

Department of Electronics and Radio Engineering, Kyung Hee University, 1 Seocheon-dong, Giheung-gu, Yongin-si, Gyeonggi-do, 446–701, Republic of Korea

Abstract: We reported the enhancement of the structural and optical properties of electrochemically synthesized zinc oxide [ZnO] nanorod arrays [NRAs] using the multi-walled carbon nanotube [MWCNT]-composed seed layers, which were formed by spin-coating the aqueous seed solution containing MWCNTs on the indium tin oxide-coated glass substrate. The MWCNT-composed seed layer served as the efficient nucleation surface as well as the film with better electrical conductivity, thus leading to a more uniform high-density ZnO NRAs with an improved crystal quality during the electrochemical deposition process. For ZnO NRAs grown on the seed layer containing MWCNTs (2wt.%), the photoluminescence peak intensity of the near-band-edge emission at a wavelength of approximately 375nm was enhanced by 2.8 times compared with that of the ZnO nanorods grown without the seed layer due to the high crystallinity of ZnO NRAs and the surface plasmon-meditated emission enhancement by MWCNTs. The effect of the MWCNT-composed seed layer on the surface wettability was also investigated.

Keywords: ZnO Nanorod Arrays, Multi-Walled Carbon Nanotubes, Electrochemical Growth, Crystallinity, Photoluminescence

1. Introduction

In the past decade, various ZnO nanostructures including nanowires, nanorods, nanosheets, nanoflowers, and nanotubes have received intensive attention because of their excellent physical properties for a wide range of practical device applications such as ultraviolet photodetectors, field-effect transistors, light-emitting diodes, and biological and chemical sensors[1]-[3]. Among many fabrication approaches, a hydrothermal or electrochemical deposition method has been considered as an efficient way to grow ZnO nanostructures since it is a simple, low- temperature, large-scale, and cost-effective process[4][5]. In order to grow high- quality ZnO nanostructures in such chemical synthesis methods, the seed layer is very important because the orientation and crystallinity of ZnO nanostructures depend on the conditions of the underlying seed layer[6]. For this reason, the radio frequency [rf] sputtering or atomic layer deposition and subsequent thermal annealing treatment have been employed to form a good seed layer with excellent step coverage, thickness controllability, and reproducibility. However, it requires a somewhat complicated procedure and high-vacuum environment.

Meanwhile, carbon nanotubes [CNTs] have been one of the most advanced functional materials because of their superior electronic property, good thermal/ chemical stability, high mechanical strength, and large surface area[7]-[9]. Recently, the ZnO/CNT composites and hybrid nanostructures have been considered as a promising candidate for improving the device efficiency in the electronic and optoelectronic devices because these structures can provide the enhanced electrical and optical properties by the cooperative physical interaction between ZnO nano- structures and CNTs[10]. However, achieving good control over the size and mor- phology of the ZnO/CNT hybrid structures is still difficult. Thus, the investigation of the seed layer containing CNTs for growing the ZnO nanostructures is very interesting. In this work, the electrochemically synthesized ZnO nanorod arrays [NRAs] after spin-coating an aqueous seed solution containing MWCNTs, which can be expected to be a facile and efficient process for the fabrication of high- quality ZnO NRAs, were studied. Their structural and optical properties were also evaluated.

2. Experimental Details

All chemicals were purchased from Sigma-Aldrich Corporation (St. Louis, MO, USA) and Kojundo Chemical Laboratory Co., Ltd. (Saitama, Japan), which were of analytical grade and used without further purification. The MWCNTs and indium tin oxide [ITO]-coated soda-lime glasses were also purchased from Hanwha-Nanotech (Incheon, Republic of Korea) and Samsung Corning (Seoul, Republic of Korea), respectively. The ITO coated on the soda-lime glass (*i.e.*, ITO/glass) substrate was fabricated by rf magnetron sputtering. The samples were cleaned by acetone, methanol, and deionized [DI] water under sonication. To prepare an aqueous seed solution, the 0.1M zinc acetate dihydrate [$Zn(CH_3COO)_2 \cdot 2H_2O$] was dissolved in DI water. A sonication process was performed while slowly adding the MWCNT paste which was grown by a thermal chemical vapor deposition method. Then, the ITO/glass substrate was spin-coated with this seed solution at 3,000rpm for 90s and dried at a hot plate of 80°C for 10min. After the spin-coating-and-drying pro- cedure was repeated successively five times for a uniform seed layer coating, the sample was heated at 200°C for 1h to increase the adhesion between the seed layer containing MWCNTs and the substrate. In order to electrochemically grow the ZnO NRAs, the seed layer-coated ITO/glass and platinum [Pt] electrode were immersed into the electrolyte solution containing 2mM zinc nitrate hexahydrate [$Zn(NO_3)_2 \cdot 6H_2O$], 2mM hexamethylenetetramine ($C_6H_{12}N_4$), and DI water. During the electrochemical growth, the temperature of the electrolyte solution and the applied cathodic voltage were kept at 80°C and −2V, respectively. After the synthesis of ZnO NRAs, the sample was rinsed with flowing DI water and dried by flowing nitrogen gas.

The morphology and structural properties of the fabricated samples were analyzed using a field-emission scanning electron microscope [FE-SEM] (LEO SUPRA 55, Carl Zeiss, Oberkochen, Baden-Württemberg, Germany) with an operating voltage of 15kV. To prevent or reduce the electric charge accumulation, the samples were coated by Pt sputtering. The orientation and crystallinity of the samples were characterized using an X-ray diffractometer (M18XHF-SRA, Mac Science, Yokohama, Japan) with a monochromated Cu Ka line source (l = 0.154178nm). The photoluminescence [PL] measurements were performed using a PL mapping system (RPM 2000, Accent Optics, Denver, CO, USA) with a laser source emitting at the wavelength of 266nm at room temperature. The macroscopic surface property on wettability was characterized from the measurement of the contact angles with the water droplet on the surface of the samples using a contact angle measurement system (Phoenix-300, SEO Co., Ltd., Gyeonggi-do, Republic of Korea) with a measurement accuracy of ±0.1°.

3. Results and Discussion

Figure 1 shows the cross-sectional SEM images of the electrochemically synthesized (a) ZnO nanorods on ITO/glass without the seed layer, (b) ZnO NRAs on the zinc acetate seed layer coated on ITO/glass (*i.e.*, ZnO NRAs/seed layer on ITO/glass), (c) ZnO NRAs/MWCNT (2wt.%)-composed seed layer on ITO/glass, and (d) ZnO NRAs/MWCNT (5wt.%)-composed seed layer on ITO/glass. The insets of **Figure 1** show the top-view SEM images of the corresponding samples.

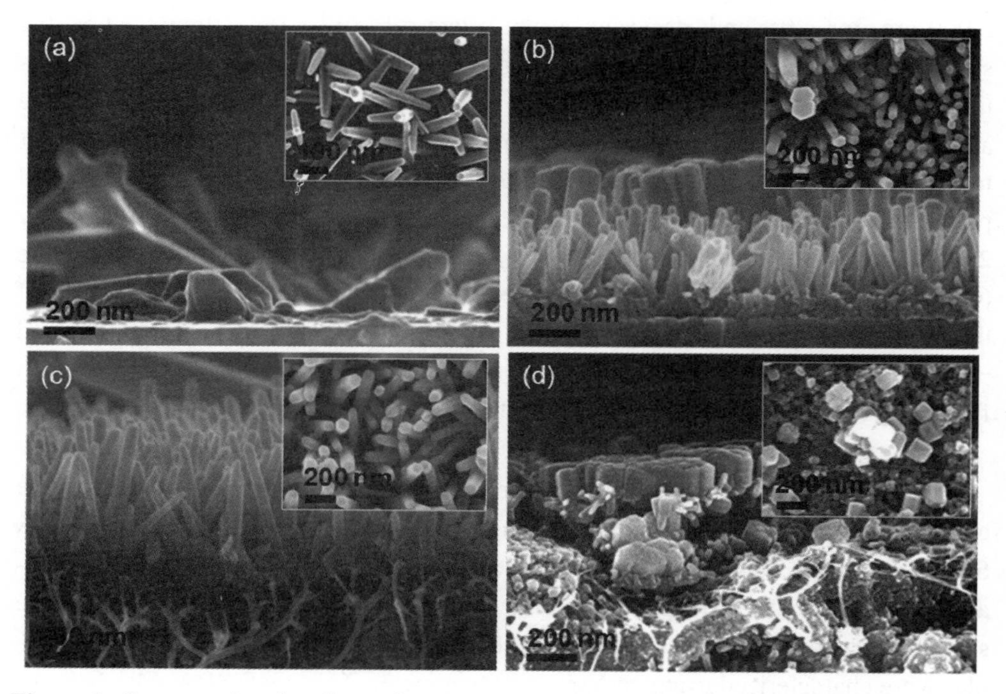

Figure 1. Cross-sectional and top-view SEM images of the samples. Cross-sectional SEM images of the electrochemically synthesized (a) ZnO nanorods on ITO/glass without the seed layer, (b) ZnO NRAs/zinc acetate seed layer on ITO/glass, (c) ZnO NRAs/MWCNT (2wt.%)-composed seed layer on ITO/glass, and (d) ZnO NRAs/MWCNT (5wt.%)- composed seed layer on ITO/glass. The insets show the top-view SEM images of the corresponding samples.

When the ZnO nanorods were synthesized on ITO/glass without the seed layer, they were sparsely populated and randomly orientated on the bare ITO surface. As shown in **Figure 1(a)**, ZnO nanorods with no preferred orientation were observed. In contrast, the use of the seed layer gives rise to the ZnO NRAs aligned along a dominant c-axis orientation of the wurzite structure as can be seen in **Figures 1(b)-1(d)**, which indicates that the spin-coated seed layer can efficiently provide the nucleation sites on the ITO surface[11]. For ZnO NRAs/zinc acetate seed layer on ITO/glass, the ZnO nanorods were densely assembled, but their height and size were not uniformly distributed and at approximately 100 to 800 nm and 20 to 150nm, respectively. Whereas, the ZnO NRAs/MWCNT (2wt.%)-composed seed layer on ITO/glass exhibited a more uniform distribution in the height and size. The height and size were about 600 to 1,000nm and 20 to 50nm, respectively. It is noted that the ZnO nanorods were more densely and uniformly grown using the MWCNT (2wt.%)-composed seed layer. This is probably attributed to a uniform electric field between the sample surface and electrolyte solution by the enhanced electrical conductivity of the seed layer. Herein, the electric field plays a key role in the chemical synthesis. It can be expected to improve the electrical and optical properties of ZnO NRAs by forming the ZnO/ MWCNT hybrid nanostructure. However, the MWCNT (5wt.%)-composed seed layer made the ZnO nanostructures shorter and thicker, which creates somewhat aggregated ZnO/MWCNT composites as shown in Figure 1d. This means that such excess MWCNTs in the composed seed layer may prevent the successful formation of ZnO nanorods.

Figure 2 shows the 2θ scan X-ray diffraction [XRD] patterns of the (a) ZnO nanorods with no seed layer, (b) ZnO NRAs zinc acetate seed layer, (c) ZnO NRAs/MWCNT (2wt.%)-composed seed layer, and (d) ZnO NRAs/MWCNT (5wt.%)-composed seed layer on ITO/glass. From the XRD patterns, the (100), (222), (400), (440), and (622) XRD peaks of ITO were clearly observed, and they exhib- ited almost similar intensities for all the samples. For ZnO nanorods grown directly on ITO/glass, the weak peak intensity of ZnO was observed at the (002) and (101) planes due to the poor orientation and low density in **Figure 1(d)**. For the ZnO NRAs in **Figures 2(b)-2(d)**, the XRD peak intensity at the (002) plane of ZnO was increased. This confirms that the spin-coated seed layer enables the crystallization of ZnO nanorods with a hexagonal wurtzite structure and a preferred orientation along the c-axis as mentioned in **Figure 1**. For ZnO NRAs/MWCNT-composed seed layer as shown in **Figure 2(c)**, **Figure 2(d)**, the (100) and (004)

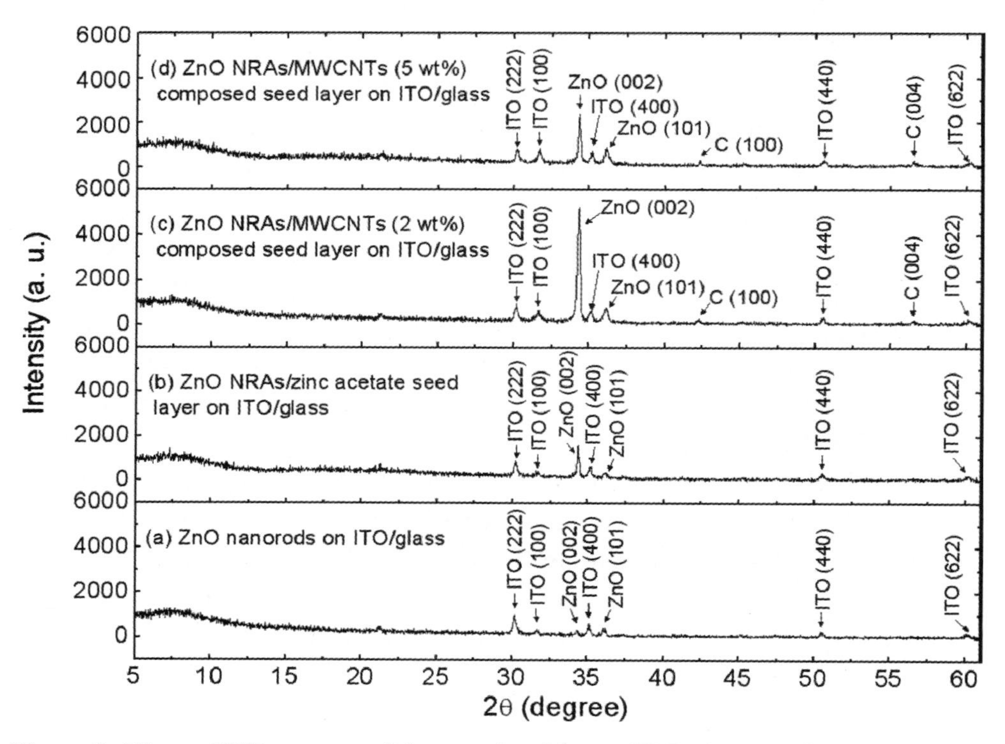

Figure 2. 2θ scan XRD patterns of the samples. 2θ scan XRD patterns of the (a) ZnO nanorods with no seed layer, (b) ZnO NRAs/zinc acetate seed layer, (c) ZnO NRAs/MWCNT (2wt.%)-composed seed layer, and (d) ZnO NRAs/MWCNT (5wt.%)-composed seed layer on ITO/glass.

XRD peaks of carbon were also observed. By incorporating the MWCNTs into the zinc acetate seed layer, the (002) XRD peak intensity of ZnO was largely enhanced for 2wt.% MWCNTs, but it was decreased for 5wt.% MWCNTs. It is clear that the MWCNT-composed seed layer enhances the crystallinity of ZnO nanorods, but the aggregated ZnO/MWCNT composites synthesized on the excess MWCNT-composed seed layer have a degraded crystallinity.

Figure 3 shows the room-temperature PL spectra of the (a) ZnO nanorods with no seed layer, (b) ZnO NRAs/zinc acetate seed layer, (c) ZnO NRAs/MWCNT (2wt.%)-composed seed layer, and (d) ZnO NRAs/MWCNT (5wt.%)-composed seed layer on ITO/glass. The inset shows the PL intensity and the full width at half maximum [FWHM] value of the corresponding samples. The peak positions of the near-band-edge [NBE] UV emission for the synthesized ZnO nanorods were observed at wavelengths of 373 to 375nm. The PL peak intensity of

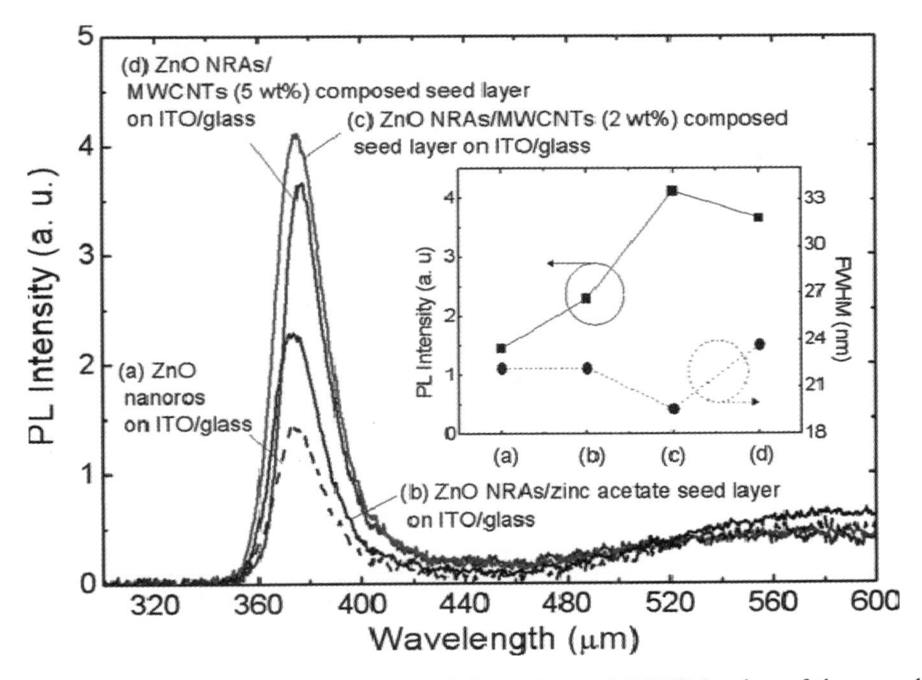

Figure 3. Room-temperature PL spectra, PL intensity, and FWHM value of the samples. Room-temperature PL spectra of the (a) ZnO nanorods with no seed layer, (b) ZnO NRAs/ zinc acetate seed layer, (c) ZnO NRAs/MWCNT (2wt.%)-composed seed layer, and (d) ZnO NRAs/MWCNT (5wt.%)-composed seed layer on ITO/glass. The inset shows the PL intensity and the FWHM value of the corresponding samples.

the NBE UV emission was relatively low. The weak and broad visible emissions are related to the structural defects in ZnO nanostructures. For ZnO NRAs with the seed layer, the PL peak intensity of the NBE UV emission was increased, which exhibits a similar tendency with the XRD data. This indicates that the optical property of ZnO nanostructures is also strongly dependent on the seed layer. For the ZnO NRAs/ MWCNT (2wt.%)-composed seed layer, the PL peak intensity was significantly increased, and the NBE UV emission became somewhat more sharp. The PL peak intensity was enhanced by 2.8 times, and a narrow FWHM value of 19.6nm was obtained compared to the ZnO nanorods grown without the seed layer as shown in the inset of **Figure 3**. However, when the concentration of MWCNTs in the composed seed layer was increased to 5wt.%, the structure exhibited an increased FWHM value as well as a reduced PL peak intensity. The enhanced PL peak intensity of the NBE UV emission would be caused by the high crystallinity of ZnO NRAs and surface plasmon-meditated emission enhancement by MWCNTs. Since the estimated surface plasmon energy is about 3.1 to 3.3eV for the dielectric constant of ZnO, the NBE UV emission of ZnO NRAs may be coupled with the surface plasmon resonance of the CNT[10][12].

Figure 4 shows the photographic images of the water droplet on the surface of (a) ZnO nanorods with no seed layer, (b) ZnO NRAs/zinc acetate seed layer, (c) ZnO NRAs/MWCNT (2wt.%)-composed seed layer, and (d) ZnO NRAs/MWCNT (5wt.%)-composed seed layer on ITO/glass. The measured contact angles of the corresponding samples are also shown. The ZnO nanorods grown directly on ITO/glass exhibited a surface wettability with a contact angle of 82.07° because the ZnO nanorods were sparsely distributed and not well aligned. It is noticeable that the surface of ZnO is known to be hydrophilic[13]. For ZnO NRAs with the seed layer, however, the contact angle was gradually decreased as 58.73°, 35.99°, and 31.95°, as can be seen in **Figures 4(b)-4(d)**, respectively. The surface macroscopic

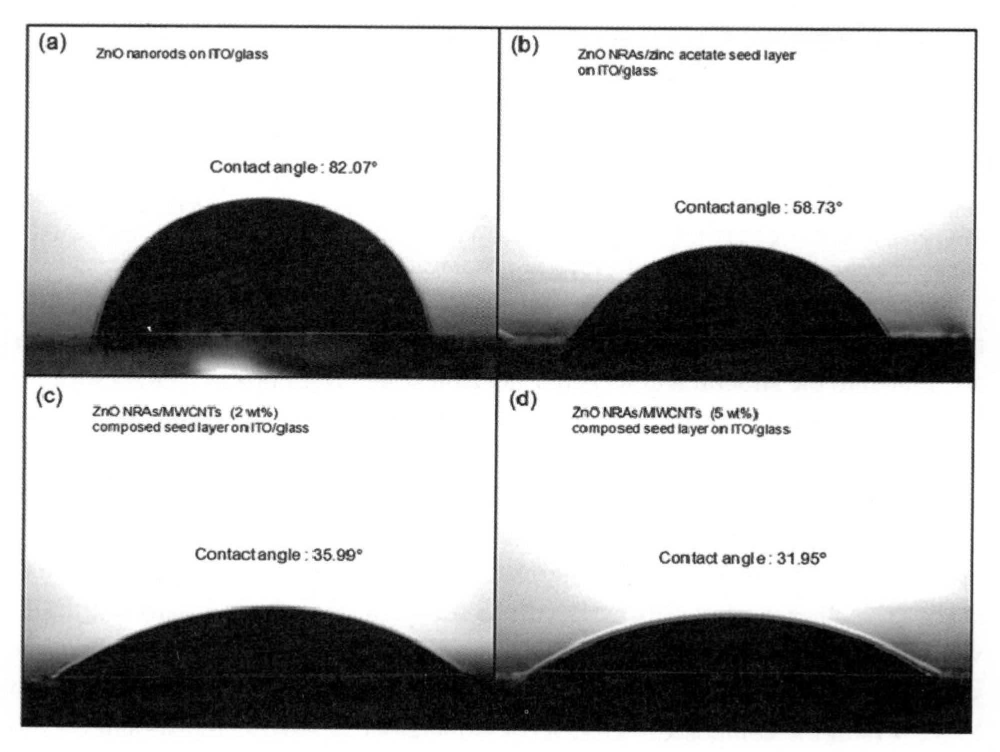

Figure 4. Photographic images of the water droplet on the surface and contact angles of the samples. Photographic images of the water droplet on the surface of (a) ZnO nanorods with no seed layer, (b) ZnO NRAs/zinc acetate seed layer, (c) ZnO NRAs/MWCNT (2wt.%)-composed seed layer, and (d) ZnO NRAs/MWCNT (5wt.%)-composed seed layer on ITO/glass and the measured contact angles of the corresponding samples.

property on the wettability became more hydrophilic for high-density, ordered ZnO NRAs. Therefore, the surface-modified structure with a better hydrophilic property can be expected to be used for microfluidic device applications.

4. Conclusion

The structural and optical properties of electrochemically synthesized ZnO NRAs on the ITO/glass substrate using MWCNT-composed seed layers formed by a simple method were investigated. It was found that the morphology of ZnO NRAs strongly depends on the kinds of seed layers. The optimized MWCNT-composed seed layer resulted in the high-density, well-aligned ZnO NRAs with a high crystallinity. The PL peak intensity of the NBE UV emission in ZnO/MWCNT hybrid nanostructures was significantly increased due to the surface plasmonme-ditated emission enhancement by MWCNTs as well as the improved crystalliza-tion property. Also, the surface macroscopic property on the wettability could be modified with more hydrophilic characteristics. This simple electrochemical fa-brication method using the seed layer containing CNTs is very useful to grow high-quality ZnO NRAs on an ITO/glass substrate for various optoelectronic device applications.

Acknowledgements

This research was supported by the Basic Science Research Program through the National Research Foundation of Korea (NRF) funded by the Ministry of Education, Science and Technology (no. 2010-0016930 and no. 2010-0025071).

Authors' Contributions

YHK designed and analyzed the composed nanostructures by performing the measurements (FE-SEM, XRD, PL, contact angles), and MSK assisted in synthesizing the composed seed layer, growing the nanostructure, and analyzing each sample. The experiment and the writing of the manuscript were carried out under the instruction of JSY. All authors read and approved the final manuscript.

Competing Interests

The authors declare that they have no competing interests.

Source: Ko Y H, Kim M S, Yu J S. Structural and optical properties of ZnO nanorods by electrochemical growth using multi-walled carbon nanotube-composed seed layers[J]. Nanoscale Research Letters, 2012, 7(1):1–6.

References

[1] Ahmad M, Zhu J: ZnO based advanced functional nanostructures: synthesis, properties and applications. J Mater Chem 2011, 21:599–614.

[2] Zhang Q, Dandeneau CS, Zhou X, Cao G: ZnO nanostructures for dye-sensitized solar cells. Adv Mater 2009, 21:4087–4108.

[3] Wang ZL: Zinc oxide nanostructures: growth, properties and applications. J Phys Condens Matter 2004, 16:R829–R858.

[4] Ko YH, Leem JW, Yu JS: Controllable synthesis of periodic flower-like ZnO nanostructures on Si subwavelength grating structures. Nanotechnology 2011, 22:205604.

[5] Chander R, Raychaudhuri AK: Electrodeposition of aligned arrays of ZnO nanorods in aqueous solution. Sol Stat Commun 2008, 145:81–85.

[6] Liu Z, Ya J, E L: Effects of substrates and seed layers on solution growing ZnO nanorods. J Solid State Electrochem 2010, 14:957–963.

[7] Andrews R, Jacques D, Qian D, Dickey EC: Purification and structural annealing of multiwalled carbon nanotubes at graphitization temperatures. Carbon 2001, 39:1681–1687.

[8] Guo G, Qin F, Yang D, Wang C, Xu H, Yang S: Synthesis of platinum nanoparticles supported on poly(acrylic acid) grafted MWNTs and their hydrogenation of citral. Chem Mater 2008, 20:2291–2297.

[9] Xu CX, Sun XW: Field emission from zinc oxide nanopins. Appl Phys Lett 2003, 83:3806–3808.

[10] Kim S, Shin DH, Kim CO, Hwang SW, Choi SH, Ji SM, Koo JY: Enhanced ultraviolet emission from hybrid structures of single-walled carbon nanotubes/ZnO films. Appl Phys Lett 2009, 94:213113.

[11] Tang X, Ma ZQ, Zhao WG, Wang DM: Synthesis of ordered ZnO nanorod film on ITO substrate using hydrothermal method. In Proceedings of the Sixth International Conference on Thin Film Physics and Applications: February 29, 2008. Edited by:

Wenzhong Shen. Junhao Chu: SPIE; 2008:698421–698424.

[12] Dutta M, Basak D: Multiwalled carbon nanotubes/ZnO nanowires composite structure with enhanced ultraviolet emission and faster ultraviolet response. Chem Phys Lett 2009, 480:253–257.

[13] Pesika NS, Hu Z, Stebe KJ, Searson PC: Quenching of growth of ZnO nanoparticles by adsorption of octanethiol. J Phys Chem B 2002, 106:6985–6990.

Chapter 22

Synthesis of ZnSe Quantum Dots with Stoichiometric Ratio Difference and Study of Its Optoelectronic Property

Uzma B. Memon[1], U. Chatterjee[2], M. N. Gandhi[3], S. Tiwari[1], Siddhartha P. Duttagupta[2]

[1]Pt. Ravishankarshukla University Raipur(C.G), India-492001.
[2]Department of Electrical Engineering Indian Institute of Technology Bombay, India-400076.
[3]Center for Research in Nanotecnology and Science, Indian Institute of Technology Bombay, India-400076

Abstract: ZnSe quantum dots are key research interest of many groups from past few decades because of their novel opto-electronic properties. It is a direct wide band gap semiconductor having a band gap of 2.70eV. ZnSe based quantum dots have many application in various fields such as solar cell, scintillators, schottky diodes etc. Considering its multifarious applications and asserts of ZnSe nano-structures, it is a worldwide research interest. In this study we are going to report a conventional wet chemical method to synthesize ZnSe quantum dots with stoichiometric ratio difference of 3:2, 3:1 and 1:1 between Zn and Se. Characterization method such as Photoluminescence, X-Ray Diffraction (X.R.D), Transmission Electron Microscopy (T.E.M) were carried out. TEM imaging was executed to find out the size distribution, which was in the range of 2 to 10nm (diameter). A wurtzite crystal structure of quantum dots has been confirmed by X.R.D also a band edge emission with a distinct blue shift in terms of wavelength is observed from Photoluminescence studies.

Keywords: ZnSe, Nano Dots, XRD, TEM, PL

1. Introduction

Quantum dots have intrigued voluminous deliberation from past few decades both in elementary research and as well as in application research as reported by Biao Niel *et al.* (2013) and Amit D *et al.* (2007) attributed to their novel optical and electronic properties. As a consequence of Quantum confinement effect properties such as size tunable emission, broad absorption spectra, narrow and symmetric emission spectra are exhibited. Taking leverage of these properties they are widely used in optoelectronic devices proposed by Biao Niel *et al.* (2013) and Hsueh shih chen (2005).

Many analysis had been put forth in synthesis of II-VI compound semiconductor quantum dots, among them were nanostructure based on CdS, CdSe, ZnO, ZnS and CdTe as reported by Zhai TY *et al.* (2010), N. A. Hamizi and M.R Johan (2012), B. Liu and H. C. Zeng (2003), S.K Mandal *et al.* (2007), S. K Das and F. D'Souza (2009). On the other hand ZnSe nanoparticles were beyond the bounds of researchers due to synthesis challenges in realizing high quality ZnSe nanoparticles. Some of the phenomenal properties of ZnSe are their wide band gap, abundantly available, highly stable and good doping effect as reported by M. Tahashi *et al.* (2009). Up till now many approaches have been reported for synthesizing ZnSe quantum dots using MBE, MOCVD/MOVPE, Chemical synthesis and other methods as reported by X.Fanga *et al.* (2011), P. Kumar and K Singh (2009). In this study we are going to report a parallel optical and electronic study of three batches of ZnSe quantum synthesized by different stoichiometric ratio difference implementing a conventional wet chemical method.

2. Experiment

ZnSe quantum dots were synthesized by modifying the method reported by

Kumar and K. Singh (2009). The raw materials were used anhydrous zinc acetate (99.9%), selenium (99.999%), ethylene glycol and hydrazine hydrate. All the chemicals were procured from Sigma-Aldrich and used without further purification. 100ml of solution in 6:3:1 ratio of deionized water, ethylene glycol and hydrazine hydrate were taken. This solution was then divided into 80ml and 20ml. Three batches of sample were synthesized having stoichiometric difference of ratio between zinc acetate and selenium. **Table 1** identifies samples with different stoichiometric ratio.

In the separated 20ml of solution selenium was added. Anhydrous zinc acetate was mixed in the remaining 80ml. Then both the solutions were mixed and refluxed at 95°C for 8hrs. After the reactions precipitates were collected through centrifuge and washed with ethanol and deionizer water for several times. Collected precipitates were kept for drying in a furnace at 50°C for 6hrs.

The structural studies of quantum dots were performed by X-Pert^3 powder by PAN analytical with CuKa as radiation source with λ = 1.5444 from 10° to 90° and a scanning step of 0.02°/s. The optical studies were carried out by Photoluminescence spectra recorded by Flurolog 3 by Horiba spectrometer at room temperature using xenon lamp was used as excitation source and PMT as detector. Crystallinity of the quantum dots were studied by HR TEM using JEM2100F by JOEL having a resolution of 0.19nm with typical acceleration voltage of 200Kv.

3. Result and discussions

3.1. Morphological Studies

HR TEM-HR TEM were carried out to find the size distribution of ZnSe

Table 1. Sample preparation.

Different sample with stoichiometric ratio difference	Zinc Acitate (mg)	Selinide (mg)
Samplel 1 (3:2)	600	400
Samplel 2 (3:1)	600	200
Samplel 3 (1:1)	100	100

quantum dots it indicate formation of cluster of ZnSe quantum dots. **Figure 1** shows the T.E.M images of quantum dots.

3.2. Structural Characterization

X-Ray Diffraction (X.R.D)-**Figures 2–4** shows X-Ray diffraction pattem of ZnSe power Sample. The XRD peaks of sample 1 and 2 are broad indicating formation of low dimensional quantum dots and wurtzite crystal structure whereas X.R.D pattern of sample 3 does not show formation of low dimensional quantum

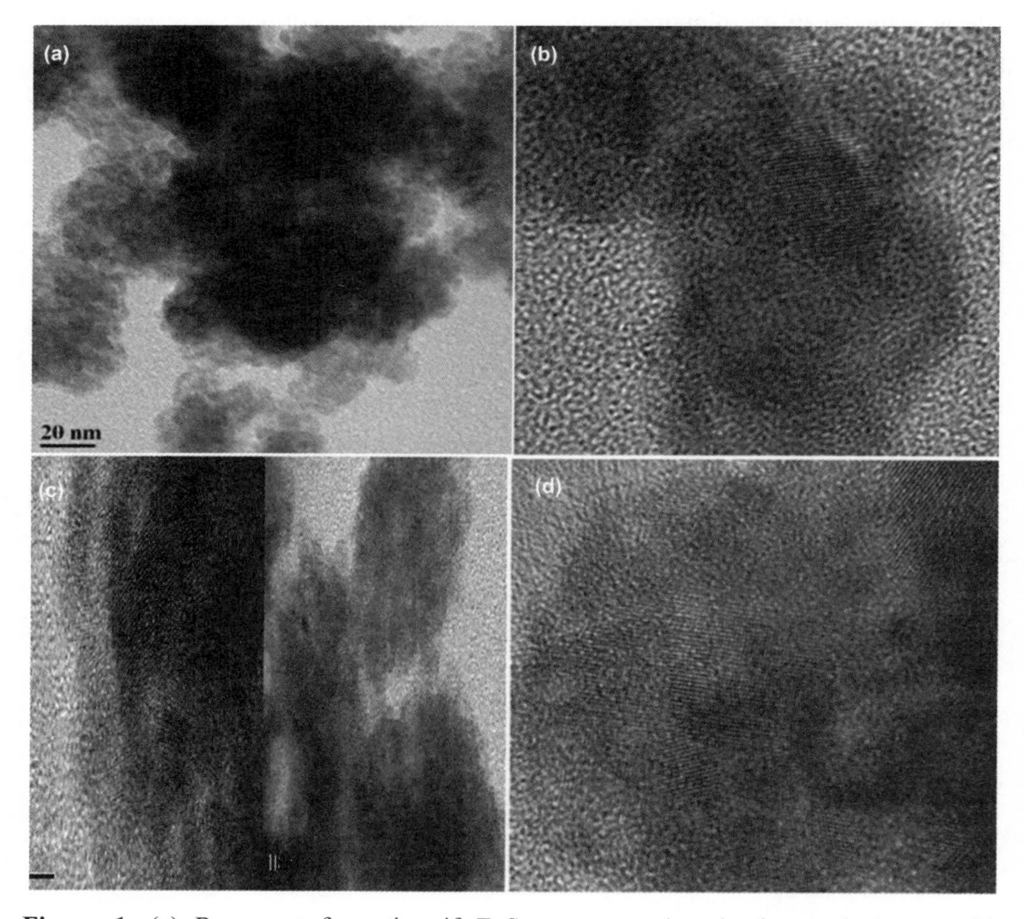

Figure 1. (a) Represent formation if ZnSe quantum dots in form of clusters. (b) Represents sample 1 with crystal plane arrangement. (c) Represent crystal plane arrangement and nearly mono dispersed ZnSe quantum dots. (d) Represents sample 3 with crystal plane arrangement.

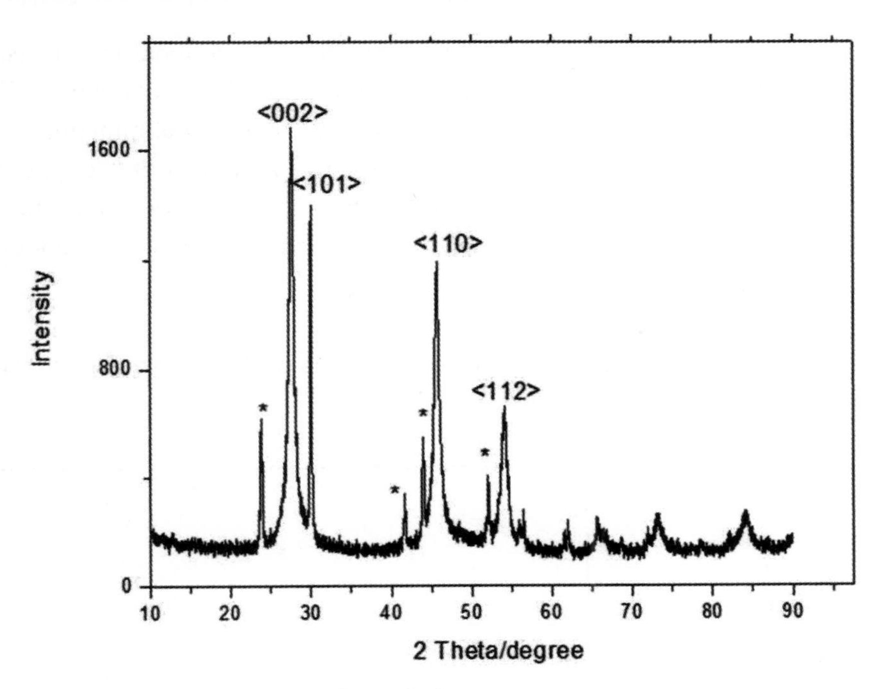

Figure 2. X. R. D analysis of sample 1.

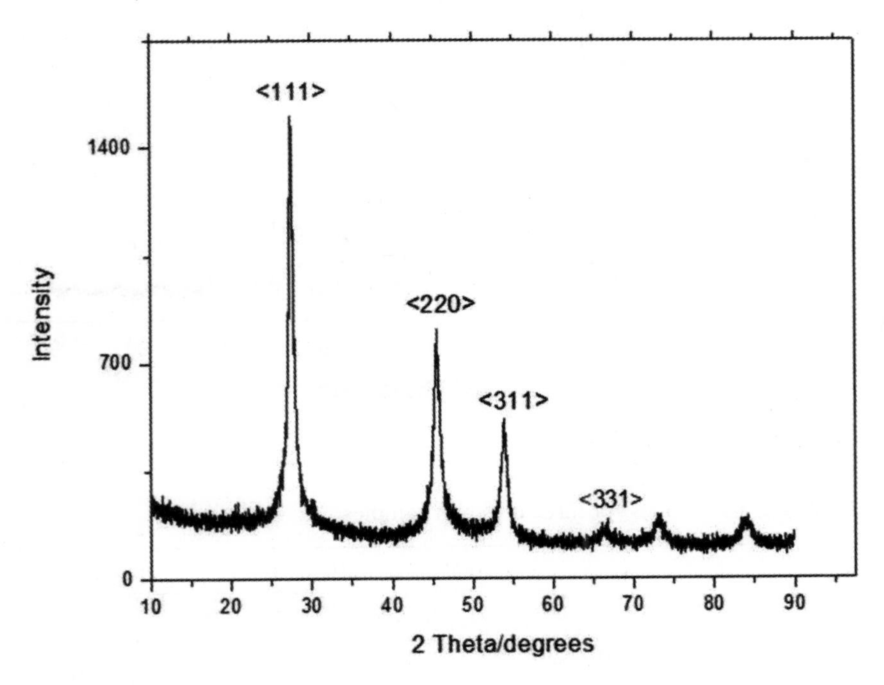

Figure 3. X. R. D analysis of sample 2.

dots. XRD analysis of Sample 1 suggests presence of additional unreduced ingredients at 20 values of 23.30, 30.07, 43.97, and 56.94. P. Kumar and K. Singh (2009) had reported same peaks. Sample 2 analysis suggests elimination of unreduced ingredients and formation of pure ZnSe quantum dots thus giving indication of more optimized method of synthesizing ZnSe quantum dot where as Sample 3 analysis suggest limited formation of ZnSe presence of large amount of unreduced ingredients.

The XRD pattern exhibits prominent and broad peaks at following 20 values for the three samples.

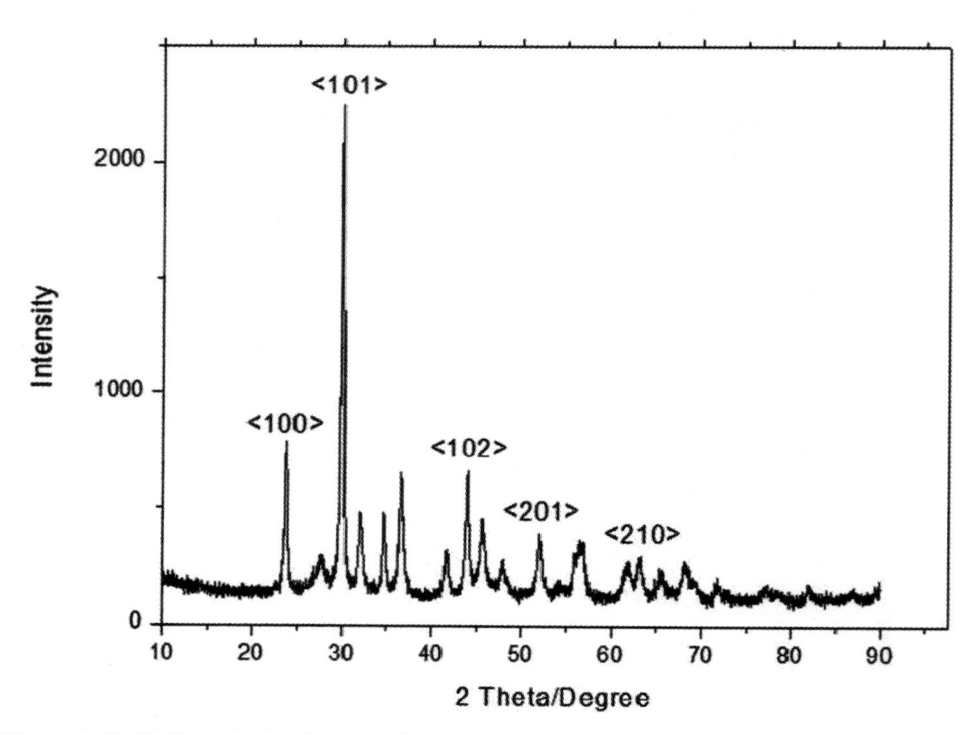

Figure 4. X. R. D analysis of sample 3.

Table 2. Different 2θ values for three different samples.

Sample	Angle of diffraction (θ)
Samplel 1	23.79, 27.68, 30.07, 43.93, 45.71, 54.17, 56.44, 65.59
Samplel 2	27.56, 45.58, 53.90, 66.30, 73.17, 83.98
Samplel 3	23.75, 27.65, 32.03, 34.65, 36.47, 43.94, 45.72, 47.82, 52.02, 54.19, 56.90, 63.17.

The size distribution of quantum dots were calculated by Scherrer's equation

$$D = k\lambda / \beta \cos\theta \qquad (1)$$

where *D*- Size of the particle;

K- Scherrer's constant, it varies with the shape of the crystallites;

λ- Wavelength of radiation;

β- FWHM in radians;

θ- Angle of diffraction.

The size of three samples were calculated as by using scherrer's equation

3.3. Optical Characterization

Photoluminescence Spectrum (PL) **Figures 4–6** shows the photoluminescence spectra of ZnSe quantum dots at 240nm of excitation wavelength. The photoluminescence peak is described as follow for the three samples-Sample 1 exhibited low intensity peak at 390nm and 470nm, this additional peak at 470nm results from the recombination of photon generated hole with a charge state of specific defect probably due to presence of unreduced ingredients. Sample 2 exhibits a high intensity peak at 340nm without any additional peak this implicate removal of unreduced ingredients. Sample 3 exhibits low intensity peaks at approx 430nm which is probably due to formation of small amount of ZnSe quantum dots and unreduced ingredients.

Table 3. Average size of quantum dots.

Sample	Size calculated (*nm*)
Samplel 1	5.66
Samplel 2	1.54
Samplel 3	9.54

Table 4. Band gap energy calculated and blue shift.

Sample	Band gap (*ev*)	Band Shift (*ev*)
Samplel 1	3.10	0.40
Samplel 2	3.64	0.94
Samplel 3	2.85	0.18

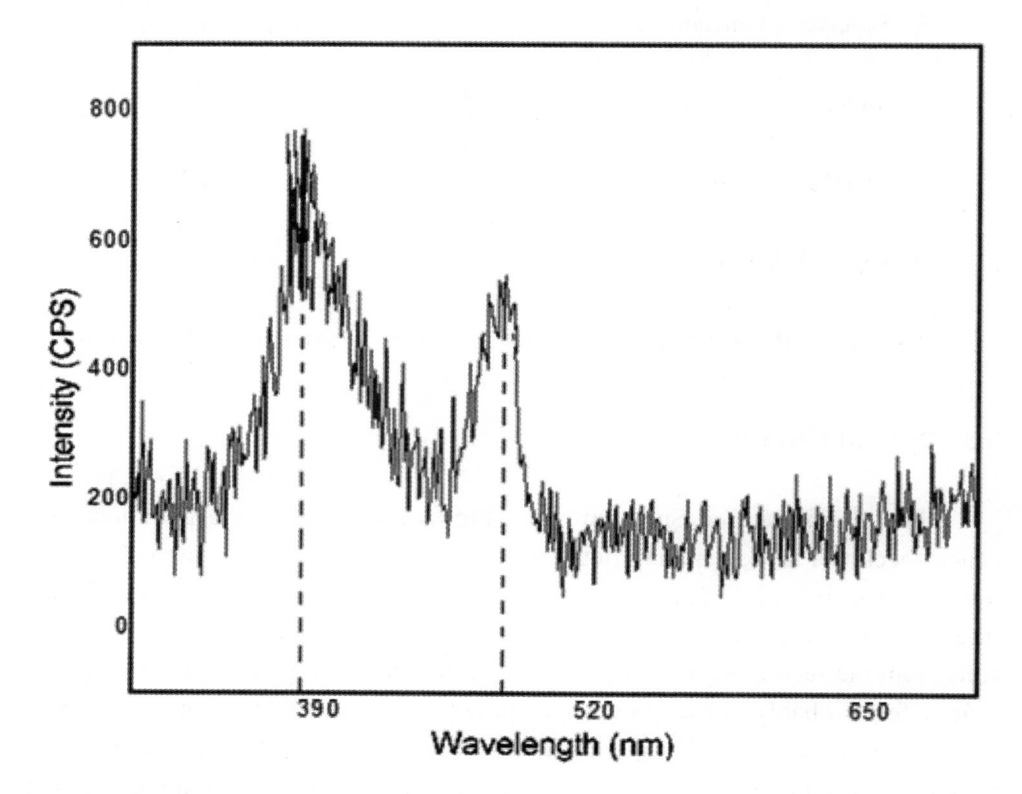

Figure 5. Photoluminescence spectrum of sample 1.

The band gap of ZnSe quantum dots were found to be as follows for three different samples.

4. Conclusion

Quantum dots of ZnSe were synthesized through wet chemical method. Structural characterization was studied through X.R.D and the size of nano particle

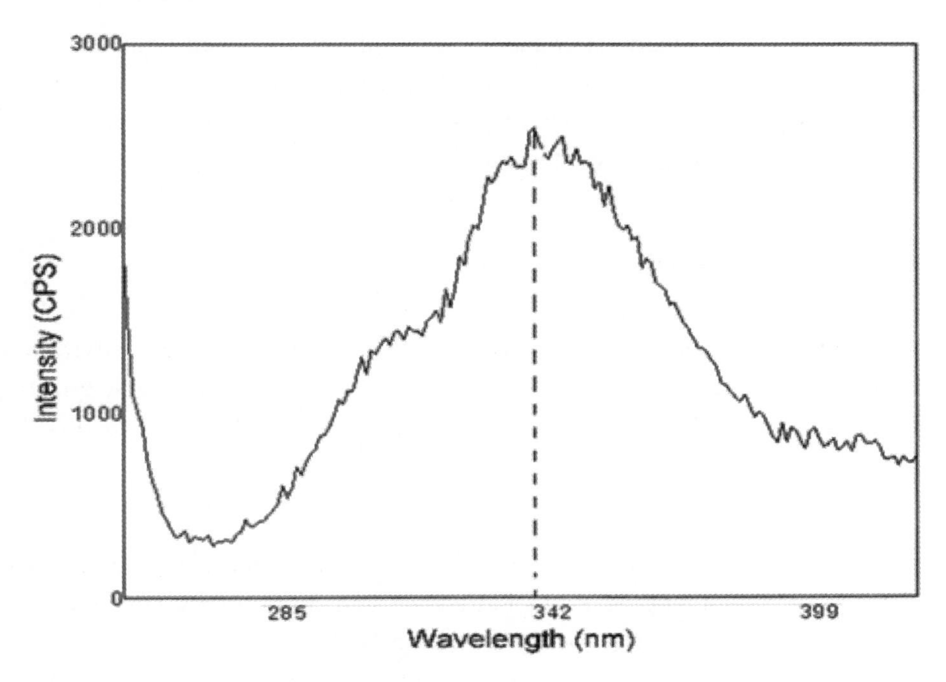

Figure 6. Photoluminescence spectrum of sample 2.

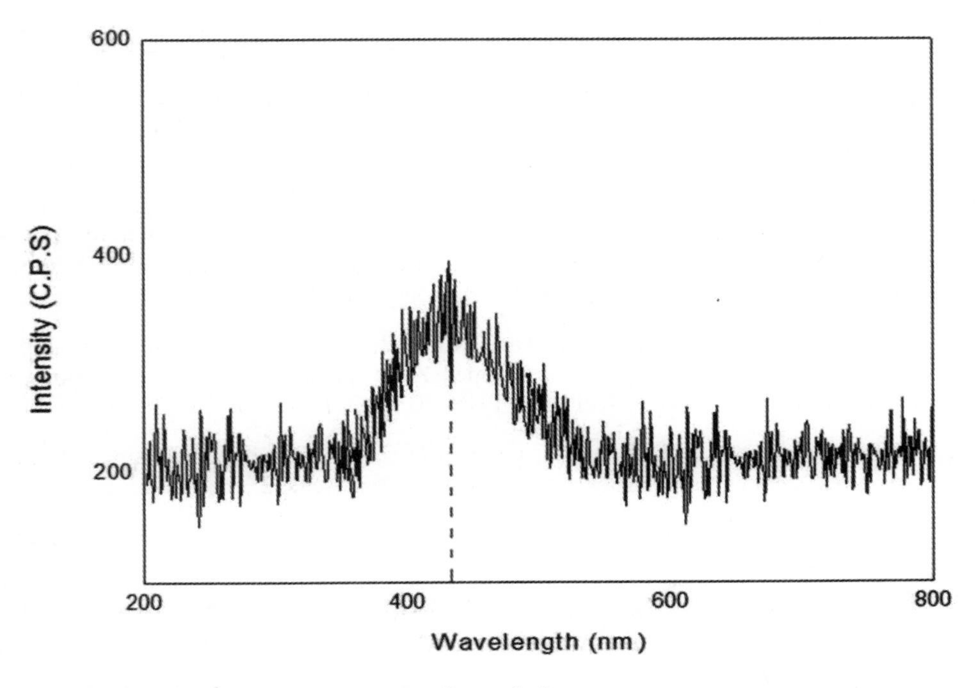

Figure 7. Photoluminescence spectrum of sample 3.

was calculated using Scherrer's equation which closely matched with the T.E.M results. Optical studies were done by Photoluminescence Spectroscopy, distinct blue shift in wavelength were observed which indicate formation of quantum dots as a result of quantum confinement effect.

Acknowledgements

We sincerely thank Sophisticated Analytical Instrument Facility(SAIF), IIT-Bombay, Powai, Mumbai INDIA for letting us use various characterization technique. This work was supported by IIT-Bombay Nano fabrication facility, Powai, Mumbai INDIA.

Source: Memon U B, Chatterjee U, Gandhi M N, *et al.* Synthesis of ZnSe Quantum Dots with Stoichiometric Ratio Difference and Study of its Optoelectronic Property [J]. Procedia Materials Science, 2014, 5:1027–1033.

References

[1] Amit D. Lad, Ch. Rajesh, Mahmud Khan, Naushad Ali, I. K. Gopalakrishnan, S. K. Kulshreshtha, and ShailajaMahamuni 2007, Magnetic behavior of manganese-doped ZnSe quantum dots, Journal of Applied Physics 101, 103906.

[2] Biao Niel, Lin-Bao Luol, Jing-Jing Chenl, Ji-Gang Hul, Chun-Yan Wul, Li Wangl, Yong-Qiang Yul, Zhi-Feng Zhul and Jian-Sheng Jie 2013.Fabrication of p-type ZnSe:Sb nanowires for high-performance ultraviolet light photodetector application, IOP Publishing Nanotechnology 24 (2013) 095603 (8pp).

[3] Hsueh Shih Chen, ShianJyJassy Wang, Chun Jeu Lo, and Jim Yong Chi 2005, White-light emission from organics-capped ZnSe quantum dots and application in white-light-emitting diodes, APPLIED PHYSICS LETTERS 86, 131905.

[4] Zhai TY, Fang XS, Li L, Bando Y, Golberg D (2010). One-dimensional CdS nanostructures: synthesis, properties, and applications.Nanoscale;2: 168–187.

[5] Nor Aliya Hamizi, Mohd Rafie Johan (2012), Optical Properties of CdSe Quantum Dots via Non-TOP based Route. Int. J. Electrochem. Sci., Vol.

[6] Bin Liu and Hua Chun Zeng (2003), Hydrothermal Synthesis of *ZnO* Nanorods in the Diameter Regime of 50 nm. J. AM. CHEM. SOC. 9 VOL. 125, NO. 15.

[7] S.K. Mandal, A.R. Mandal, S. Das, B. Bhattacharjee J. Appl. Phys., 101 (2007), p.

114315.

[8] Sushanta K Das and Francis D'Souza, Synthesis of CdTe Quantum Dots of Different Sizes and their Interactions with Water Soluble Porphyrins, Proceedings of the 5th Annual GRASP Symposium, Wichita State University, 2009.

[9] Masahiro Tahashi, Zunyi Wu, Hideo Goto, Youji Hayashi and Toshiyuki Ido (2009),Effect of Vanadium Doping on Structure and Properties of ZnSe Films Prepared by Metal-Organic Vapor Phase Epitaxy. Materials Transactions, Vol. 50, No. 4 pp. 719 to 722.

[10] X. Fanga, T. Zhai, U. K. Gautamb, L.Li ,L.Wua, Y. Bando, D. Golberg (2011), ZnS nanostructures: From synthesis to applications, Progress in Materials Science 56 175–287.

[11] P. Kumar and K. Singh 2009. Wurtzite ZnSe quantum dots: Synthesis, characterization and PL properties, Journal of optoelectronic and Biomedical Materials, Volume 1, Issue 1, p. 59 –69.

Chapter 23

Growth and Optical Properties of ZnO Nanorod Arrays on Al-doped ZnO Transparent Conductive Film

Suanzhi Lin[1], Hailong Hu[2], Weifeng Zheng[1], Yan Qu[1], Fachun Lai[1]

[1]College of Physics and Energy, Fujian Normal University, Fuzhou 350108, Fujian China
[2]Analytical and Testing Center, Southwest University of Science and Technology, Mianyang 621010, China.

Abstract: ZnO nanorod arrays (NRAs) on transparent conductive oxide (TCO) films have been grown by a solution-free, catalyst-free, vapor-phase synthesis method at 600°C. TCO films, Al-doped ZnO films, were deposited on quartz substrates by magnetron sputtering. In order to study the effect of the growth duration on the morphological and optical properties of NRAs, the growth duration was changed from 3 to 12min. The results show that the electrical performance of the TCO films does not degrade after the growth of NRAs and the nanorods are highly crystalline. As the growth duration increases from 3 to 8min, the diffuse transmittance of the samples decreases, while the total transmittance and UV emission enhance. Two possible nanorod self-attraction models were proposed to interpret the phenomena in the sample with 9-min growth duration. The sample with 8-min growth duration has the highest total transmittance of 87.0%, proper density about 75μm^{-2}, diameter about 26nm, and length about 500nm, indicating that it can be used in hybrid solar cells.

Keywords: ZnO Nanorod, Al-Doped ZnO Films, Catalyst-Free Growth, Optical Properties

1. Introduction

ZnO, one of the most important metal oxides, has a wide bandgap of 3.37eV and a high exciton binding energy of 60 meV at room temperature. One-di- mensional nanostructures have a high aspect ratio and surface area, and can provide a direct conduction path for electrons. Accordingly, a wide range of ZnO nanostructures[1] such as nanowires (NWs), nanorods (NRs), and nanonails are extensively studied for their applications in various optoelectronic devices, e.g., gas sensors[2], UV photodetectors[3][4], lasers[5][6], electron field emitters[7], solar cells[8]-[12], and nanogenerators[13].

For most photovoltaic devices, the light is coupled in devices through transparent conductive oxide (TCO) substrate, so tailored well-aligned ZnO nanorod arrays (NRAs) grown on TCO substrate are of particular interest because they can improve the device performance[14]. Previously, ZnO NRAs and NWs on different TCO substrates have been synthesized by various growth methods including chemical bath deposition[8][10][11], electrochemical deposition[9][12][14], and thermal vaporphase deposition[15][16]. Among these methods, the vapor- phase growth method has many advantages such as excellent crystalline quality of the nanostructures[15], low cost, and simplicity[17]. Generally, ZnO NRs in dye- sensitized solar cells or hybrid solar cells are used to extract the carriers from an organic material and transfer the carriers toward the electrode[15]. Moreover, the density, diameter, length, and crystalline performance of NRs have a significant influence on the efficiency of solar cells[9][15][16]. A larger nanorod diameter will reduce spacing between NRs, which contributes to a reduction in the amount of solar absorber. Longer ZnO NRs do not improve the solar efficiency due to the lower short-circuit current[9]. Therefore, it is important to synthesize ZnO NRAs on TCO substrate with the suitable nanorod diameter, length, and density for their applications in hybrid solar cells. However, there are few reports on the growth and optical properties of ZnO NRAs on a TCO substrate by the vapor-phase deposition[15][16].

In this paper, we focus on the growth and optical properties of ZnO NRAs, which were grown by a solution-free, catalyst-free, vapor-phase synthesis method at a temperature of 600°C. This method can grow ZnO NRAs on Al-doped ZnO (AZO) films, and the performance of AZO does not degrade after the growth of NRAs. AZO has the advantage of being indium free and can be produced on a large scale. The effect of growth duration on the morphology and optical properties of NRAs has been investigated.

2. Methods

AZO films were deposited on quartz substrates using a radio-frequency (RF) magnetron sputtering system at room temperature. The quartz substrates, 0.5mm thick, 2.5cm × 2.5cm, were cleaned in acetone and ethanol several times before deposition. The target, 60-mm diameter, was a commercial ZnO and Al_2O_3 mixture (97:3 wt.%) of ≥99.99% purity. The sputtering was performed in an Ar atmosphere with a target-to-substrate distance of 5cm. The base pressure in the chamber was 4.0×10^{-4}Pa. The Ar flux determined using a mass flow-controlled regulator was maintained at 50.0sccm, and the sputtering pressure was 0.5Pa. The RF power was 300W, and deposition time was typically 10min. A typical sheet resistance of AZO film, about 480nm thick, was about 60Ω/sq.

ZnO NRAs were grown by a vapor-phase method in a horizontal tube furnace[18]. The substrates, polycrystalline AZO films on quartz substrates, were cleaned in acetone and ethanol before the NRA growth. Commercial zinc (99.99% purity) powder in a ceramic boat was used as the zinc vapor source. The ceramic boat and AZO substrate were placed in a long quartz tube, and the quartz tube was then put into the furnace. An AZO substrate was placed 5cm downstream from the sources at the heat center of the furnace. After evacuating the system to a base pressure of 12Pa, the furnace temperature was ramped to 600°C at 20°Cmin^{-1}. A 100-sccm Ar and 10-sccm oxygen mixed gas was introduced into the furnace only when the maximum temperature was reached. The growth pressure was 110Pa. The temperature was kept at 600°C for several minutes, and then the furnace was cooled down to room temperature. Changing the growth duration, several samples had been synthesized. For simplicity, the samples with growth durations of 3, 6, 8, 9, and 12min were defined as samples S1, S2, S3, S4, and S5, respectively.

Morphological and structural properties of the grown nanostructures were analyzed using a JSM-7500LV scanning electron microscope (SEM) and a

JEM-2010 high-resolution transmission electron microscope (TEM) (JEOL Ltd., Akishima-shi, Japan). For the latter, the samples were prepared by mechanically scraping NRs from the substrate, dispersing them in ethanol, and depositing a drop of the dispersion on a circular copper grid covered by a thin holey carbon film. The crystal structure and orientation were investigated using an X-ray diffractometer (XRD; Y-2000, Rigaku Corporation, Shibuya-ku, Japan) with monochromated Cu Kα irradiation (λ = 1.5418Å). The surface morphology of the AZO film was observed using an atomic force microscope (AFM; CSPM 4000, Benyuan Co. Ltd., Guandong, China) under ambient conditions. The sheet resistance was measured by the van der Pauw method[19].

Room-temperature photoluminescence (PL) spectra of the samples were obtained on a Fluorolog 3–22 fluorescence spectrophotometer (Horiba Ltd., Kyoto, Japan) using a Xe lamp with an excitation wavelength of 325nm. The total transmittance and diffuse transmittance of the samples were measured using a double-beam spectrophotometer (PerkinElmer Lambda 950, Waltham, MA, USA) equipped with an integrating sphere. In the measurement, the light propagation path was air/quartz/AZO/air or air/quartz/AZO/NRAs/air, and the reflection at the quartz/air interface was not removed.

3. Results and discussion

The top-view SEM images of samples S1 to S5 are shown in **Figures 1(a)-1(e)**, respectively, and the insets are the high-magnification images of the corresponding samples. **Figure 1(f)**, **Figure 1(g)** presents the cross-sectional SEM images of samples S2 and S5, respectively. The ZnO NR growth mechanism is the catalyst-free vapor-solid growth due to the absence of metal catalysts on NR tips[20]. Moreover, **Figure 1(f)**, **Figure 1(g)** clearly indicates a ZnO buffer layer between NRAs and AZO film, which is used as a seed layer[21]. The density and average NR dimensions of samples S1 to S4 are tabulated in **Table 1**. Sample S1 has a relatively low NR density, and its NR lengths are between 200 and 300nm. As the growth duration increases to 8min, sample S3 has a NR density of 75μm^{-2}, an average NR diameter of 26nm, and an average length of 500nm, indicating that the density, length, and aspect ratio of NR increase with the increase of growth duration. The average NR diameter, however, does not obviously change.

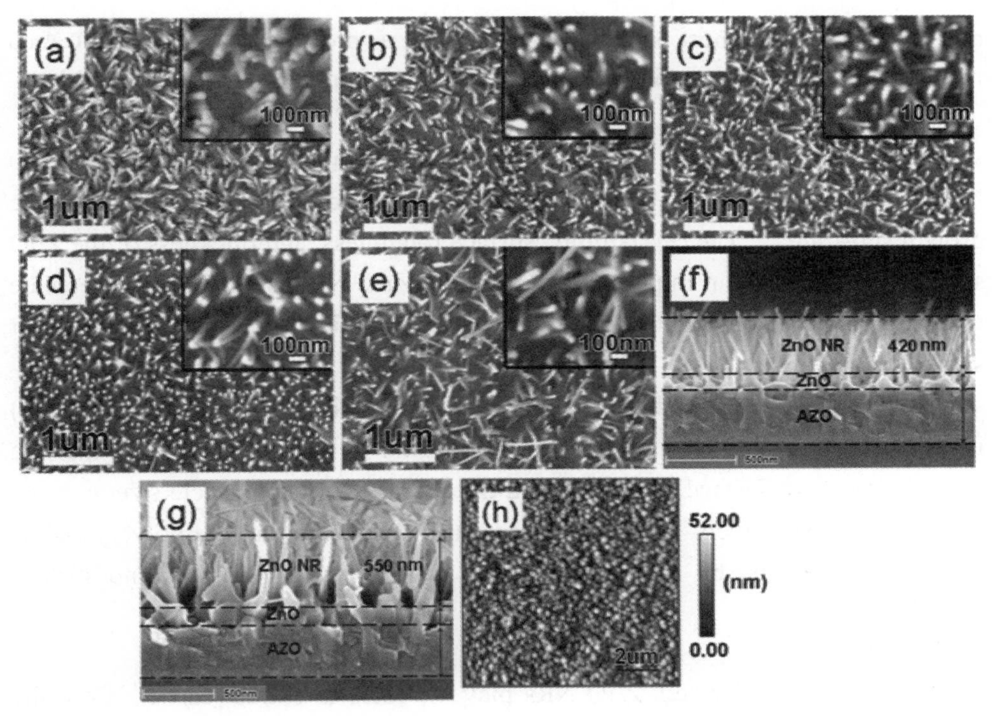

Figure 1. SEM images of ZnO NRs grown with different durations and AFM surface image of AZO film. (a to e) Top-view and (f, g) cross-sectional SEM images of ZnO NRs grown with different durations: (a) S1 - 3min, (b, f) S2 - 6min, (c) S3 - 8min, (d) S4 - 9min, and (e,g) S5 - 12min; insets are the high-magnification images of the corresponding samples. (h) AFM surface image of AZO film.

Table 1. Density and average NR dimensions (diameter, length, and aspect ratio) of the samples.

Sample	Density (per μm^2)	Average NR diameter $2r$ (nm)	Average NR length L (nm)	Aspecv ratq L/r
S3	40 ± 8	28 ± 7	250 ± 50	17.8
S4	61 ± 6	25 ± 6	420 ± 40	33.6
S3	75 ± 2	26 ± 4	500 ± 20	38.5
S6	82 ± 2	28 ± 4	550 ± 20	39.3

Moreover, as shown in **Figure 1(d)**, the phenomenon of two or three NRs self-attracting in sample S4 with 9-min growth duration can be seen clearly. NRs in sample S5 are out of order because more NRs touch each other and the new NRs grow at NR self-attraction positions. The newly grown NRs are more disordered, and some NRs are almost parallel to the substrate as presented in **Figure 1(e)**. As a result, the density and length of the NRs on sample S5 are not calculated in **Table 1**.

In previous research reports, it was found that the characteristic of ZnO NWs strongly depends on the crystallinity, type, and surface roughness of the growth substrate[20]. The crystallinity, surface roughness, and thickness of the ZnO seed layer also have an important influence on ZnO NR growth[21]. We speculate that two main reasons contribute to the not well vertically aligned NRAs in our samples. First is that the AZO film was deposited on the amorphous quartz substrate, which results in a polycrystalline AZO film as discussed below. **Figure 1(h)** is a typical AFM surface image of an AZO film. AFM results indicate that the root-mean-square surface roughness and the average surface particle size are 10.2 and 140nm, respectively. The second reason, therefore, is that the polycrystalline AZO film deposited by RF sputtering has large surface roughness and surface particle size.

In a hybrid solar cell, ZnO NRs play the roles to extract carriers from the absorber and provide a fast and direct path for these carriers. The efficiency of a solar cell strongly relies on the crystallinity, density, diameter, and length of ZnO NR[9][15]. Conradt *et al.*[15] have reported that short NRs in the range of 100 to 500 nm are of particular interest for hybrid solar cells. A smaller NR diameter will enhance the spacing between NRs and increase the solar absorber amount and the efficiency of a solar cell[9]. NR in sample S3 has a suitable length about 500nm and a small diameter about 26nm. Accordingly, we suggest that sample S3 is interesting for application in hybrid solar cells.

Most NRs in sample S4 are well aligned, as shown in **Figure 1(d)**. However, the phenomenon of two or three NRs self-attracting can be seen obviously in the inset of **Figure 1(d)**. Han *et al.*[22] and Wang *et al.*[23] had reported self-attraction among aligned ZnO NRs under an electron beam, while Liu *et al.*[24] have observed the self-attraction of ZnO NWs after the second-time growth. In our samples, NRs with a relatively small diameter are slightly oblique and easily bent, which results in NR self-attraction, given that the NRs are long enough. According to the experimental observation, we propose two possible NR self-attraction models, as presented in **Figure 2**. The insets in **Figure 2** are topview images of sample

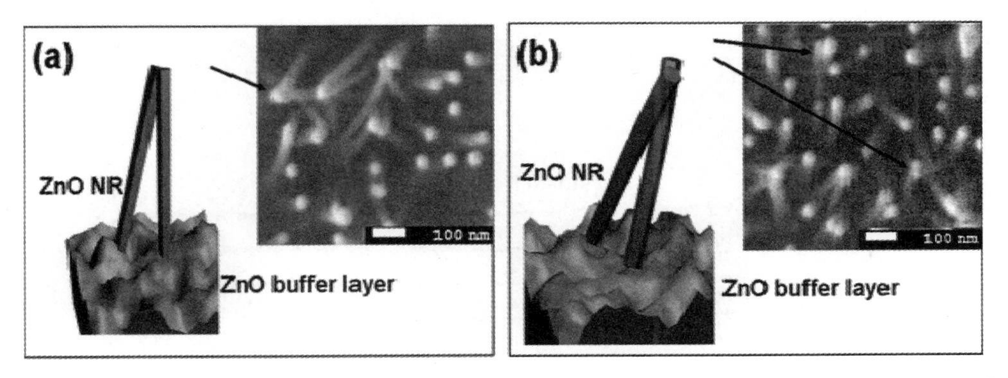

Figure 2. Schematic diagrams of two possible NR self-attraction models. (a) The tips of two NRs touch each other, (b) two NRs touch each other at the crossed position. Insets are top-view images of sample S4.

S4, and the arrows in the insets denote the examples of the self-attraction models. In the first case, in **Figure 2(a)**, NRs randomly grow and are slightly tilted, so the tips of two NRs may just touch each other when the NRs are long enough. In the second case, a NR body may slightly bend due to the oblique growth, which causes the side surfaces to be either positively or negatively charged because of the piezoelectric properties of ZnO NRs[13][24]. As a result, as indicated in **Figure 2(b)**, when two bending NRs cross, the opposite charges will lead to the attraction at the crossed position due to the large electrostatic force.

Figure 3 presents XRD patterns of an AZO film along with the samples. It clearly shows that the AZO film has a preferential orientation along the (001) axis, which is orthogonal to the substrate surface. Additionally, a weak (101) peak indicates that the AZO film is a polycrystalline structure. ZnO NRs grow coherently with the bottom AZO film, maintaining the preferential orientation of the [001] axis. For samples S1 to S4, the intensity of the (002) peak enhances with the increase of growth duration, suggesting that sample S4 has better crystallinity. The reduction of the (002) peak intensity for sample S5 is because the NRs are disordered and have more defects after the new NRs grow at NR self-attraction positions.

In order to cross-check the crystalline quality of the NRs, a TEM image of a ZnO NR is shown in **Figure 4(a)** and clearly indicates the absence of metal catalysts on the ending. In a high-resolution TEM image, **Figure 4(b)**, continuous crystal planes can be seen, which are perpendicular to the growth direction and exhibit

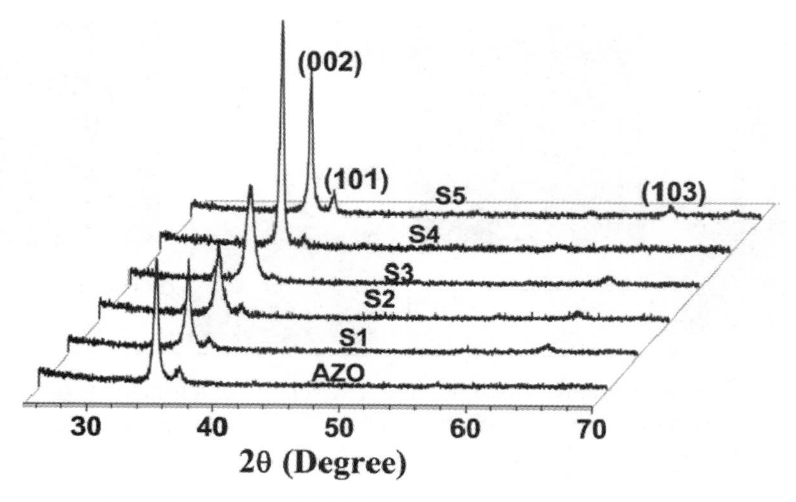

Figure 3. XRD patterns of AZO film and samples S1 to S5.

Figure 4. TEM images of a ZnO NR in sample S3. (a) TEM image of a ZnO NR in sample S3, (b) HRTEM image taken at the circle position in (a), inset is the corresponding selected-area electron diffraction pattern.

an interplanar distance of 0.26nm. The inset in **Figure 4(b)** presents the selected-area electron diffraction pattern from this NR, which suggests that NR is the single-crystal ZnO with wurtzite structure.

Room-temperature PL properties of ZnO NRAs of samples S1 to S5 are shown in **Figure 5**. There are two emission peaks in the PL spectra. One peak located at about 377nm is the near-band-edge emission or UV emission, and the other green band peak at about 500 nm is the deep-level emission[3]. The relative PL peak intensity ratio (R = IUV/IDLE) is defined as a figure of merit. R is 0.5, 1.6, 1.6, 5.1, and 1.7 for samples S1, S2, S3, S4, and S5, respectively. Comparing

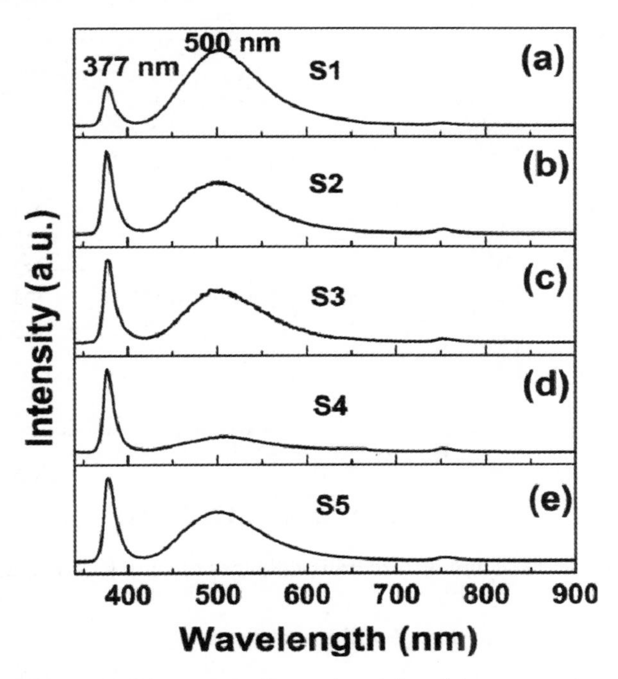

Figure 5. PL spectra of samples. (a) to (e) are samples S1 to S5.

samples S1 to S4, it is found that R enhances with the increase of growth duration, which is due to the decrease of oxygen vacancies[18]. Sample S1 has the strongest deep-level emission because it has the most oxygen vacancies and the shortest oxidation time. Although sample S5, however, has the longest growth duration, its deep-level-emission is relatively strong. This is because the new NRs grown at NR self-attraction positions have worse crystallinity, as shown in **Figure 3**, shorter growth duration, and more oxygen vacancies.

Semiconductor nanostructures offer a powerful tool to efficiently manage the light in photovoltaic devices, and the morphology of NWs or NRs has a significant effect on their transmittance and reflectance[14][25][26]. The total and diffuse transmittance spectra of the samples were measured, and the results are presented in **Figure 6**. The average total transmittance (ATT) and average diffuse transmittance (ADT) in the wavelength range of 400 to 1,100nm are shown in **Table 2**. ATT and ADT of the AZO film are 88.6% and 0.4%, respectively, indicating that the AZO film has good transparence. ATTs of samples S1 to S5 are higher than 80%. The highest diffuse transmittance of sample S5 is 44% at 416-nm

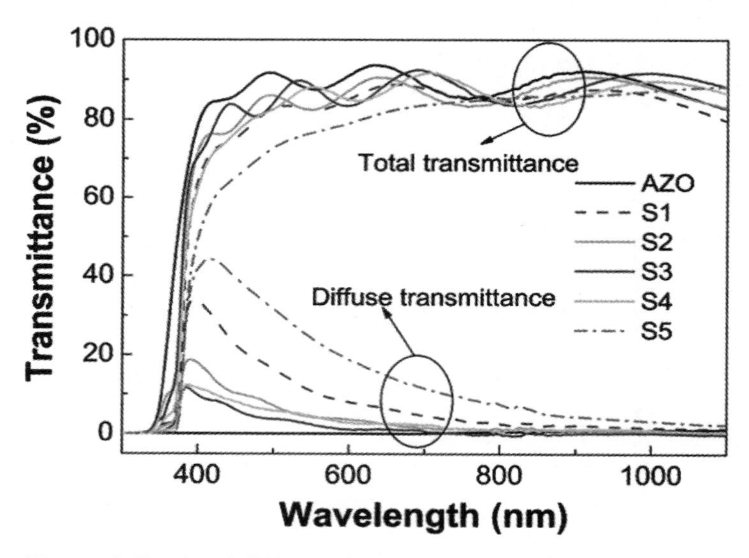

Figure 6. Total and diffuse transmittances of samples S1 to S5.

Table 2. ATT, ADT, and SR of the AZO film and samples.

Sample	AZQ	S1	S2	S3	S4	S5
ATV (%)	88.6	84.0	85.7	87.0	85.5	81.0
ADV (%)	0.4	7.3	3.2	1.5	2.8	14.2
ST (Ω/Sq)	60	19	33	48	44	36

wavelength. The diffuse transmittance decreases and total transmittance increases with increasing wavelength when the wavelength is larger than 416 nm. Sample S3 has the highest ATT and the lowest ADT because its NRs are more vertically aligned, as shown in **Figure 1**. NRs in sample S5 are disordered [**Figure 1(e)**] and have more oxygen vacancies, as discussed in the PL spectra, which results in the lowest ATT and the highest ADT of sample S5. For sample S1, although the NRs are relatively ordered, the low NR density and short NR length [**Figure 1(a)**] strongly enhance the optical surface scattering[27]. As a result, sample S1 has a large diffuse transmittance.

An AZO film must have a low resistance for use as a transparent conductive electrode in optoelectronic devices[16]. The electrical properties of an AZO film may be changed after thermal treatment at high temperature, and especially our NR growth temperature is 600°C. So, the sheet resistance (SR) of the sample was measured. The NRs at electrode positions were removed to enable good contact of the electrodes before the resistance measurement, and the results are shown in **Table 2**. All the sheet resistances of the samples are lower than that of the AZO film (60Ω/sq), indicating that the electrical performance of the AZO film does not degenerate after the NR growth. We speculate that there are two mechanisms that induce the reduction of the sheet resistances. One is that the resistance of the AZO film after the thermal treatment declines, which had been confirmed experimentally[16][28]. The other is, as indicated in **Figure 1(f)**, **Figure 1(g)**, the result of a ZnO buffer layer between NRAs and AZO film after NR growth. ZnO is naturally an n-type semiconductor due to the presence of intrinsic defects such as oxygen vacancies and zinc interstitials[29]. The resistance of a ZnO film will decline as the oxygen vacancies increase because each oxygen vacancy can generate two conductive electrons. The NRAs and ZnO buffer layer in sample S1 have the most oxygen vacancies, as confirmed by PL measurement, so it has the lowest sheet resistance (17Ω/sq).

4. Conclusions

A solution-free, catalyst-free, vapor-phase growth method was used to synthesize ZnO nanorod arrays on AZO films, which were deposited on quartz substrates by RF magnetron sputtering. The sheet resistance of the sample declines after ZnO NRA growth at 600˚C. TEM results show that the NRs are the single-crystal ZnO with wurtzite structure. As the growth duration increases from 3 to 8min, the oxygen vacancies and diffuse transmittance of the samples decrease, while the crystallinity, aspect ratio, near-band-edge emission, and total transmittance enhance. ZnO NR self-attraction in the sample with 9-min growth duration has been observed, and two possible NR self-attraction models are proposed. NRs in the sample with 12-min growth duration are disordered, which has the largest diffuse transmittance and a relatively strong deep-level emission. The sample with 8-min growth duration has a density about 75μm^{-2}, diameter about 26nm, and length about 500nm, which can be used in a hybrid solar cell.

Competing Interests

The authors declare that they have no competing interests.

Authors' Contributions

SL grew the nanorods and performed the tests on the samples. HH participated in the analysis of TEM results. WZ deposited the Al-doped ZnO films. YQ participated in the test of the samples. FL designed the study and drafted the manuscript. All authors read and approved the final manuscript.

Acknowledgements

This work was financially supported by the Natural Science Foundation of China (no. 11074041) and the Natural Science Foundation of Fujian Province of China (2012J01256).

Source: Lin S, Hu H, Zheng W, *et al.* Growth and optical properties of ZnO nanorod arrays on Al-doped ZnO transparent conductive film[J]. Nanoscale Research Letters, 2013, 8(1):1–6.

References

[1] Jiang P, Zhou JJ, Fang HF, Wang CY, Wang ZL, Xie SS: Hierarchical shelled ZnO structures made of bunched nanowire arrays. Adv Funct Mater 2007, 17:1303–1310.

[2] Chien FSS, Wang CR, Chan YL, Lin HL, Chen MH, Wu RJ: Fast-response ozone sensor with ZnO nanorods grown by chemical vapor deposition. Sens Actuators B: Chem 2010, 144:120–125.

[3] Zhang X, Han X, Su J, Zhang Q, Gao Y: Well vertically aligned ZnO nanowire arrays with an ultra-fast recovery time for UV photodetector. Appl Phys A 2012, 107:255–260.

[4] Dhara S, Giri PK: Enhanced UV photosensitivity from rapid thermal annealed vertically aligned ZnO nanowires. Nanoscale Res Lett 2011, 6:504.

[5] Liu XY, Shan CX, Wang SP, Zhang ZZ, Shen DZ: Electrically pumped random lasers fabricated from ZnO nanowire arrays. Nanoscale 2012, 4:2843–2846.

[6] Huang MH, Mao S, Feick H, Yan H, Wu Y, Kind H, Weber E, Russo R, Yang P: Room-temperature ultraviolet nanowire nanolasers. Science 2001, 292:1897–1899.

[7] Chen ZH, Tang YB, Liu Y, Yuan GD, Zhang WF, Zapien JA, Bello I, Zhang WJ, Lee

CS, Lee ST: ZnO nanowire arrays grown on Al:ZnO buffer layers and their enhanced electron field emission. J Appl Phys 2009, 106:064303.

[8] Seol M, Ramasamy E, Lee J, Yong K: Highly efficient and durable quantum dot sensitized ZnO nanowire solar cell using noble-metal-free counter electrode. J Phys Chem 2011, 115:22018–22024.

[9] Chen JW, Perng DC, Fang JF: Nano-structured Cu2O solar cells fabricated on sparse ZnO nanorods. Sol Energy Mater Sol Cells 2011, 95:2471–247.

[10] Zhang J, Que W, Shen F, Liao Y: CuInSe2 nanocrystals/CdS quantum dots/ ZnO nanowire arrays heterojunction for photovoltaic applications. Sol Energy Mater Sol Cells 2012, 103:30–34.

[11] Lee SH, Han SH, Jung HS, Shin H, Lee J, Noh JH, Lee S, Cho IS, Lee JK, Kim J, Shin H: Al-doped ZnO thin film: a new transparent conducting layer for ZnO nanowire-based dye-sensitized solar cells. J Phys Chem C 2010, 114:7185–7189.

[12] Wang L, Zhao D, Su Z, Shen D: Hybrid polymer/ZnO solar cells sensitized by PbS quantum dots. Nanoscale Res Lett 2012, 7:106.

[13] Wang ZL, Song J: Piezoelectric nanogenerators based on zinc oxide nanowire arrays. Science 2006, 312:242–246.

[14] Lee HK, Kim MS, Yu JS: Effect of AZO seed layer on electrochemical growth and optical properties of ZnO nanorod arrays on ITO glass. Nanotechnology 2011, 22:445602.

[15] Conradt J, Sartor J, Thiele C, Flaig FM, Fallert J, Kalt H, Schneider R, Fotouhi M, Pfundstein P, Zibat V, Gerthsen D: Catalyst-free growth of zinc oxide nanorod arrays on sputtered aluminum-doped zinc oxide for photovoltaic applications. J Phys Chem C 2011, 115:3539–3543.

[16] Calestani D, Pattini F, Bissoli F, Gilioli E, Villani M, Zappettini A: Solution-free and catalyst-free synthesis of ZnO-based nanostructured TCOs by PED and vapor phase growth techniques. Nanotechnology 2012, 23:194008.

[17] Liu P, Li Y, Guo Y, Zhang Z: Growth of catalyst-free high-quality ZnO nanowires by thermal evaporation under air ambient. Nanoscale Res Lett 2012, 7:220.

[18] Zhuang B, Lai F, Lin L, Lin M, Qu Y, Huang Z: ZnO nanobelts and hollow microspheres grown on Cu foil. Chin J Chem Phys 2010, 23:79–83.

[19] Lai F, Lin L, Gai R, Lin Y, Huang Z: Determination of optical constants and thickness of In_2O_3: Sn films from transmittance data. Thin Solid Films 2007, 515:7387–7392.

[20] Ho ST, Chen KC, Chen HA, Lin HY, Cheng CY, Lin HN: Catalyst-free surface-roughness-assisted growth of large-scale vertically aligned zinc oxide nanowires by thermal evaporation. Chem Mater 2007, 19:4083–4086.

[21] Li C, Fang G, Li J, Ai L, Dong B, Zhao X: Effect of seed layer on structural proper-

ties of ZnO nanorod arrays grown by vapor-phase transport. J Phys Chem C 2008, 112:990–995.

[22] Han X, Wang G, Zhou L, Hou JG: Crystal orientation-ordered ZnO nanorod bundles on hexagonal heads of ZnO microcones: epitaxial growth and self-attraction. Chem Commun 2006, 212:212–214.

[23] Wang X, Summers CJ, Wang ZL: Self-attraction among aligned Au/ZnO nanorods under electron beam. Appl Phys Lett 2005, 86:013111.

[24] Liu J, Xie S, Chen Y, Wang X, Cheng H, Liu F, Yang J: Homoepitaxial regrowth habits of ZnO nanowire arrays. Nanoscale Res Lett 2011, 6:619.

[25] Convertino A, Cuscunà M, Rubini S, Martelli F: Optical reflectivity of GaAs nanowire arrays: experiment and model. J Appl Phys 2012, 111:114302.

[26] Versteegh MAM, Van der Wel REC, Dijkhuis JI: Measurement of light diffusion in ZnO nanowire forests. Appl Phys Lett 2012, 100:101108.

[27] Lai F, Li M, Wang H, Hu H, Wang X, Hou JG, Song Y, Jiang Y: Optical scattering characteristic of annealed niobium-oxide films. Thin Solid Films 2005, 488:314–320.

[28] Wimmer M, Ruske F, Scherf S, Rech B: Improving the electrical and optical properties of DC-sputtered ZnO:Al by thermal post deposition treatments. Thin Solid Films 2012, 520:4203–4207.

[29] Hwang DK, Oh MS, Lim JH, Park SJ: ZnO thin films and light-emitting diodes. J Phys D: Appl Phys 2007, 40:R387–R412.

Chapter 24

Preparation of 1,4-Bis (4-Methylstyryl)Benzene Nanocrystals by a Wet Process and Evaluation of Their Optical Properties

Koichi Baba[1], **Kohji Nishida**[2]

[1]Department of Visual Regenerative Medicine, Osaka University Graduate School of Medicine, 2-2 Yamadaoka, Suita, Osaka 565-0871, Japan
[2]Department of Ophthalmology, Osaka University Graduate School of Medicine, 2-2 Yamadaoka, Suita, Osaka 565-0871, Japan

Abstract: Single-crystal 1,4-bis(4-methylstyryl)benzene is a promising material for optoelectronic device applications. We demonstrate the preparation of 1,4-bis (4-methylstyryl)benzene nanocrystals by a wet process using a bottom-up reprecipitation technique. Scanning electron microscopy revealed the morphology of the nanocrystals to be sphere-like with an average particle size of about 60nm. An aqueous dispersion of the nanocrystals was monodisperse and stable with a ζ-potential of −41mV. The peak wavelengths of the absorption and emission spectra of the nanocrystal dispersion were blue and red shifted, respectively, compared with those of tetrahydrofuran solution. Powder X-ray diffraction analysis confirmed the crystallinity of the nanocrystals. The presented 1,4-bis(4-methylstyryl) benzene nanocrystals are expected to be a candidate for a new class of optoelectronic material.

Keywords: Organic Nanocrystal, 1,4-Bis(4-Methylstyryl)Benzene, Water Dispersion, Bottom-up Fabrication Technique, Wet Process, Optical Properties

1. Introduction

Recently, organic single crystals have attracted considerable attention for optoelectronic device applications because of their high stimulated cross-sections, broad and high-speed nonlinear optical responses, and broad tuning wavelength[1]. In addition, single crystals have the great advantages of high electronic transport and excellent optical properties compared with those of amorphous or polycrystalline thin films[1]. Optoelectronic devices using organic single crystals such as organic field-effect transistors, light-emitting transistors, optically pumped organic semiconductor lasers, and upconversion lasers have therefore been successfully demonstrated[1]-[5]. Styrylbenzene derivatives are particularly promising candidates for organic transistor and laser oscillation materials. Kabe *et al.* demonstrated an amplified spontaneous emission from single-crystal 1,4-bis(4-methylstyryl)benzene (BSB-Me) and also studied an organic light-emitting diode using BSB-Me single nanocrystals (the molecular structure of BSB-Me is shown in **Figure 1**)[6][7]. Yang *et al.* prepared high-quality, large organic crystals of BSB-Me using an improved physical vapor growth technique and investigated their optical gain properties[1].

In contrast, we have investigated the preparation and evaluated the properties of nano-sized organic crystals, *i.e.*, organic nanocrystals[8]-[11]. Organic nanocrystals show unique physicochemical properties different from those of the molecular and bulk crystal states[12]-[15]. Organic nanocrystals have been broadly used as optoelectronic materials as well as biomedical materials[16]-[22]. Recently, Fang *et al.* demonstrated the preparation of BSB-Me nanocrystals using a femtosecond laser-induced forward transfer method[23][24]. The BSB-Me nanocrystals

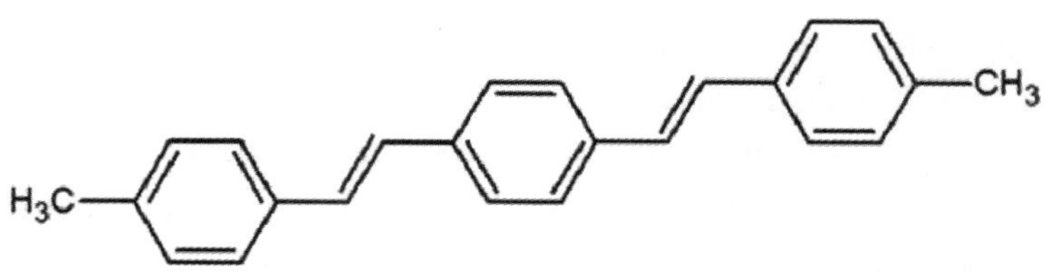

H_3C—[structure]—CH_3

Figure 1. Molecular structure of BSB-Me.

were directly deposited on a substrate to form a nanocrystal film, and their size and morphology were investigated as functions of applied laser fluence. The use of BSB-Me nanocrystals will be a promising approach for organic crystal device applications in the near future. However, according to Fang's report, the morphology of the prepared BSB-Me nanocrystals were multifarious, *i.e.*, while most nanoparticles were cubic in geometry, others were tetrahedral shaped, truncated cubes, and truncated tetrahedra[23]. To fabricate high-quality optical devices, such nanocrystals should ideally be homogenous in shape and in size because their optical properties are strongly affected by the crystal morphology. Additionally, there is a serious problem that the yields of nanoparticles prepared by laser ablation are smaller than those obtained by other nanoparticle synthesis methods because the nanocrystals are formed only in the small laser-irradiated spot[25]. This is a weak point when considering mass production for device fabrication. Furthermore, the output power of laser ablation is not suitable for organic compounds because the high energy may degrade them[26][27]. Wet processes using bottom-up techniques overcome these disadvantages. The solvent exchange method, known as the reprecipitation method, is especially suitable for preparing organic nanocrystals[18][28]. Unlike laser ablation, no excess energy is necessary to form the organic nanocrystals, and bulk production is possible[29]. Following a previously reported study, it is possible to prepare a nanocrystal-layered thin film for optical devices using the reprecipitation method[30]. Instead of top-down laser ablation, the alternative approach of this bottom-up wet process is an attractive prospect for preparing BSB-Me nanocrystals.

The aim of this study is to demonstrate the preparation of BSB-Me nanocrystals having narrow size distribution with singular morphology by means of a bottom-up, wet process using the reprecipitation method. This method makes it possible to control the particle size and morphology of the nanocrystals. We prepared BSB-Me nanocrystal dispersions in water, and investigated the size, morphology, optical properties, and powder X-ray diffraction pattern of the nano-crystals.

2. Methods

2.1. Materials

BSB-Me (>98.0%) was purchased from Tokyo Chemical Industry Co., Ltd.

(Tokyo, Japan) and used without further purification. Tetrahydrofuran (THF) (>99.5%) was purchased from Wako Pure Chemical Industries, Ltd. (Tokyo, Japan). Purified water (18.2MΩ) was obtained from a Milli-Q A-10 (Millipore, Tokyo, Japan).

2.2. Nanocrystal Preparation

BSB-Me was dissolved in THF (2mM) at 50°C, and 100μl of the solution was injected into vigorously stirred (1,500rpm) poor solvent water (10ml at 24°C) using a microsyringe. As a result, the BSB-Me suddenly precipitated to form dispersed nanocrystals. Syringe filter (pore size 1.2μm; Minisart®, Sartorius Stedim Biotech, NY, USA) was used to remove small degree of aggregates from the nanocrystal dispersion.

2.3. Evaluation

The particle size and morphology of the BSB-Me nanocrystals were evaluated using scanning electron microscopy (SEM; JSM-6510LA, JEOL, Tokyo, Japan). To prepare specimens for imaging, the nanocrystals were collected from the water dispersion using suction filtration with a membrane filter (0.05-μm pore size), followed by platinum sputter coating (JFC-1600, JEOL). The average particle size, size distribution, and ζ-potential of the nanocrystal dispersion were evaluated using an ELSZ-1000 zeta-potential and particle size analyzer (Otsuka Electronics Co., Ltd., Osaka, Japan). Ultraviolet-visible (UV-vis) absorption spectra and fluorescence spectra were measured using a V-550 UV/vis spectrophotometer (JASCO, Tokyo, Japan) and F-2500 fluorescence spectrophotometer (Hitachi, Tokyo, Japan), respectively.

3. Results and discussion

The morphology and particle size of the BSB-Me nanocrystals were investigated using SEM. The nanocrystals were found to be sphere-like and had an apparent average particle size with standard deviation of 67 ± 19nm. The average particle size was obtained by measuring the particle sizes using the ruler from the

SEM picture (the counted particle number was $n = 211$) [**Figure 2(a)**, **Figure 2(b)**]. The actual particle size, size distribution, and ζ-potential of the nanocrystals in the dispersion were investigated using the ELSZ-1000ZS analyzer (**Figure 3**). The average particle size was 60.9nm, which was analyzed by cumulant analysis method, in good agreement with that observed by SEM. The ζ-potential was −41.62mV, negative enough to make a stable dispersion. Thus, we succeeded in preparing the BSB-Me nanocrystals stable in aqueous dispersion and with homogenous particle size and morphology.

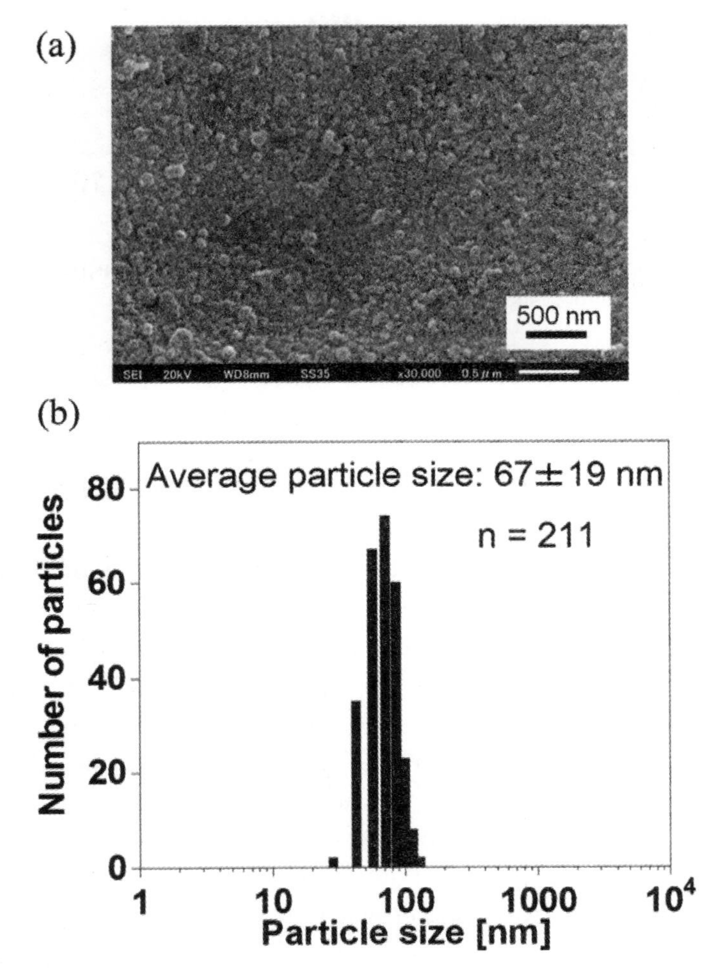

Figure 2. SEM image of the BSB-Me nanocrystals and their average particle size. SEM image of BSB-Me nanocrystals (a) and average particle size obtained by measuring the size of particles from SEM picture (b). The counted number of particles was n = 211. The average particle size was 67 ± 19nm.

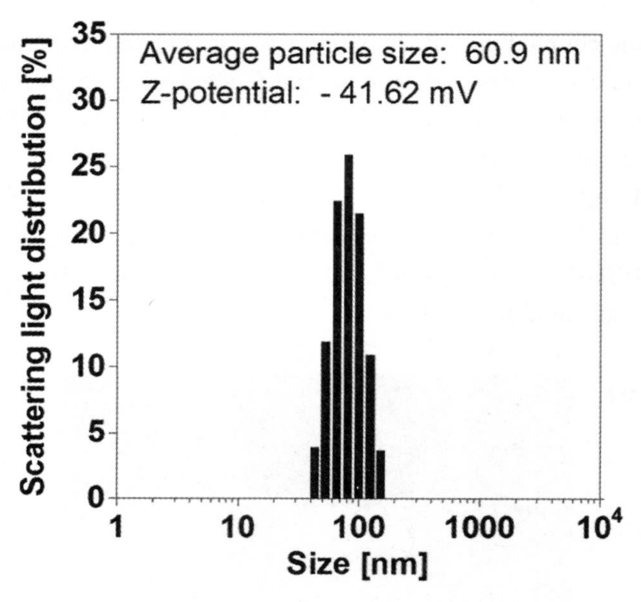

Figure 3. Average particle size and ζ-potential of BSB-Me nanocrystal water dispersion.

Photographic images of the BSB-Me nanocrystal dispersion with and without fluorescence are shown in **Figure 4**. Blue-green fluorescence was observed in the nanocrystal dispersion when it was excited at 365nm using a UV lamp (SPECTROLINE®, Spectronics Corp., Westbury, NY, USA). Absorption spectra measurements of the BSB-Me THF solution and the aqueous BSB-Me nanocrystal dispersion revealed a blue shift of the maximum absorption peak of the nanocrystal dispersion (λ_{max} = 307nm) compared with that of the THF solution (λ_{max} = 359nm) (**Figure 5**). Varghese *et al.* reported that the absorption blue shift in distyrylbenzene single crystals occurs in H-aggregates of herringbone-forming systems, where the long molecular axes are oriented in parallel. However, the short axes are inclined to each other, thus minimizing $\pi - \pi$ overlap. Hence, this side-by-side intralayer orients the transition dipole moments that constitute the main optical absorption band of distyrylbenzene (S0 → S1), leading to a blue shift compared with in solution[31]. The blue shift of the BSB-Me nanocrystal may occur by the same mechanism. Kabe *et al.* also reported that BSB-Me single crystals have a quasi-planar conformation because of a lack of steric repulsion. This planar structure induces strong supramolecular interactions, which cause the molecules to arrange layer by layer into the well-known herringbone structure[6]. This herringbone forming should affect the emission from the nanocrystals. The emission spectrum

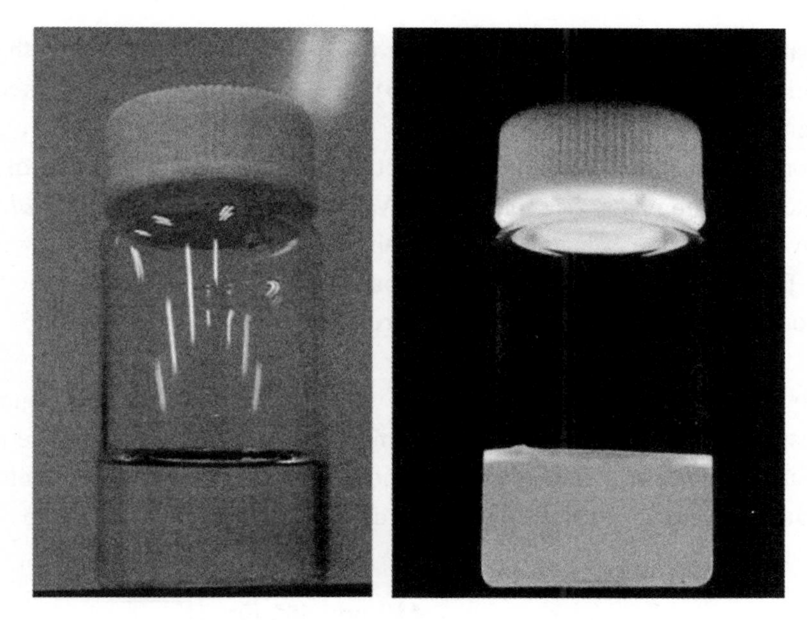

Figure 4. Imaging pictures of BSB-Me nanocrystal water dispersion with (a) and without (b) fluorescence.

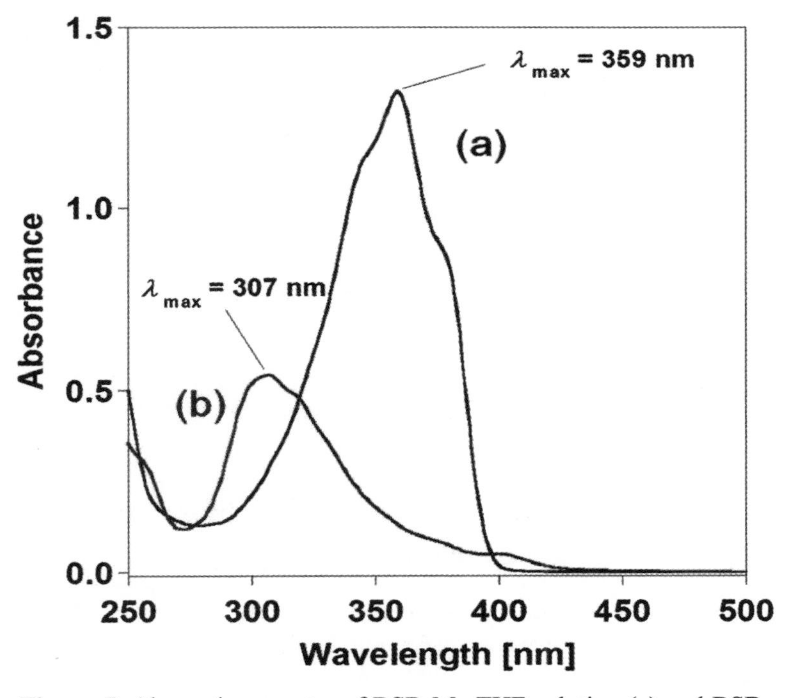

Figure 5. Absorption spectra of BSB-Me THF solution (a) and BSB-Me nanocrystal water dispersion (b).

of the nanocrystal state showed a red shift (λ_{max} = 466nm) compared with that of the solution state (λ_{max} = 415nm) (**Figure 6**). This means that the red shifted emission occurred with suppressed high-energy features and a small radiative rate, in other words, indicating the presence of intermolecular interaction in the solid-state aggregated environments, as explained by Varghese *et al.*[31] and Kabe *et al.*[6]. The peak wavelength of the excitation spectra of the nanocrystal dispersion (λ_{max} = 308nm) and the THF solution (λ_{max} = 359nm) almost corresponded to those of the respective absorption spectra (**Figure 5** and **Figure 6**).

To investigate the photoluminescence efficiency of the BSB-Me nanocrystal water dispersion, we estimated its photoluminescence quantum yield. The manner to estimate the quantum yield of a fluorophore is by comparison with standards of known quantum yield. We used the standard of BSB-Me dichloromethane solution

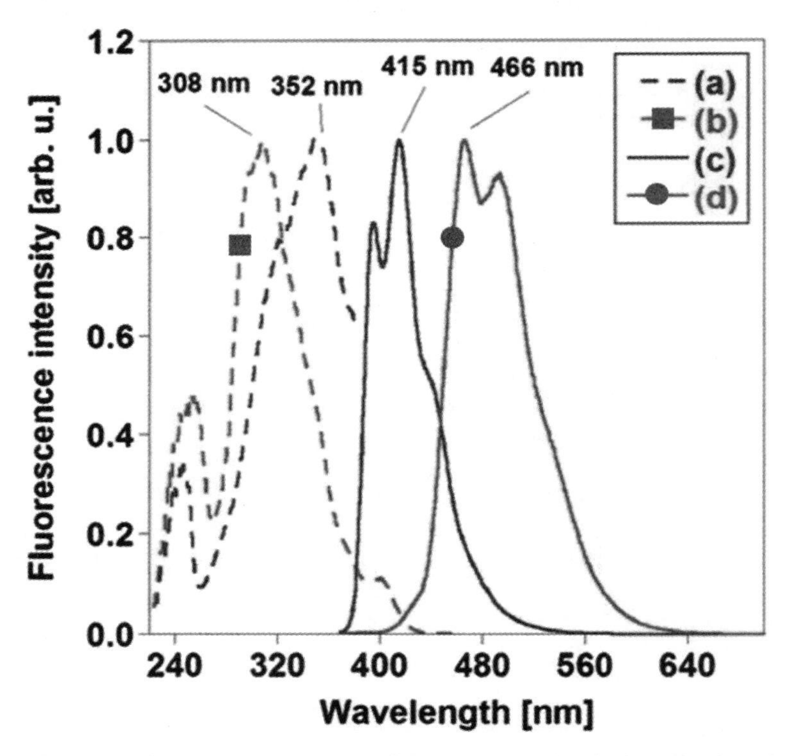

Figure 6. Fluorescence spectra of BSB-Me THF solution (a, c) and BSB-Me nanocrystal water dispersion (b, d). Excitation spectra are (a) and (b), which were measured at 395 and 465nm, respectively. Emission spectra are (c) and (d), which were excited at 350 and 310nm, respectively.

referred in the literature[6], in which the BSB-Me dichloromethane solution had an absolute photoluminescence quantum yield of 95% ± 1%. The quantum yields of the standards are mostly independent of excitation wave-length, so the standards can be used wherever they display useful absorption[32][33]. Determination of the quantum yield is generally accomplished by comparison of the wavelength integrated intensity of the unknown to that of the standard. The optical density is kept below 0.05 to avoid inner filter effects, or the optical densities of the sample and reference (*r*) are matched at the excitation wavelength. The quantum yield of the unknown is calculated using Equation (1):

$$Q = Q_R \frac{IOD_R n^2}{I_R OD n^2_R} \tag{1}$$

where Q is the quantum yield, I is the integrated intensity (areas) of spectra, OD is the optical density, and n is the refractive index. The subscripted R refers to the reference fluorophore of known quantum yield. The data of I and OD were obtained from **Figure 7**. The quantum yield of the BSB-Me nanocrystal water dispersion, which was calculated using Equation (1), was estimated to be 9.2% ± 0.1% (**Table 1**).

The crystallinity of the BSB-Me nanocrystals was confirmed using powder X-ray diffraction analysis (**Figure 8**). Two strong peaks were observed at $2\theta = 9.0$ and 13.6, corresponding with those previously reported for single bulk crystals[6]. However, interestingly, there were another four strong peaks at $2\theta = 20.9$, 23.6, 28.4, and 29.4 which did not correspond with any previously observed peaks for single crystals[6]. There may be a possibility that a different molecular arrangement to that previously reported for bulk single crystal state was formed in the nanocrystal state. Because the powder X-ray diffraction pattern of the nanocrystals showed (001) refractions as shown in (004) in $2\theta = 9.0$ and (006) in $2\theta = 13.6$, the nanocrystals basically had planar structure, supporting the occurrence of H-aggregation according to the work of Kabe *et al.*[6]. H-aggregation was also supported by the observed blue shift and red shift in the absorption and emission spectra, respectively, of the nanocrystals. However, because other refractions were observed at $2\theta = 20.9$, 23.6, 28.4, and 29.4, the nanocrystals may have had slightly different crystal structure than the bulk single crystal. Actually, we have previously reported the existence of a softened crystal lattice in nanocrystals[34][35]. A similar

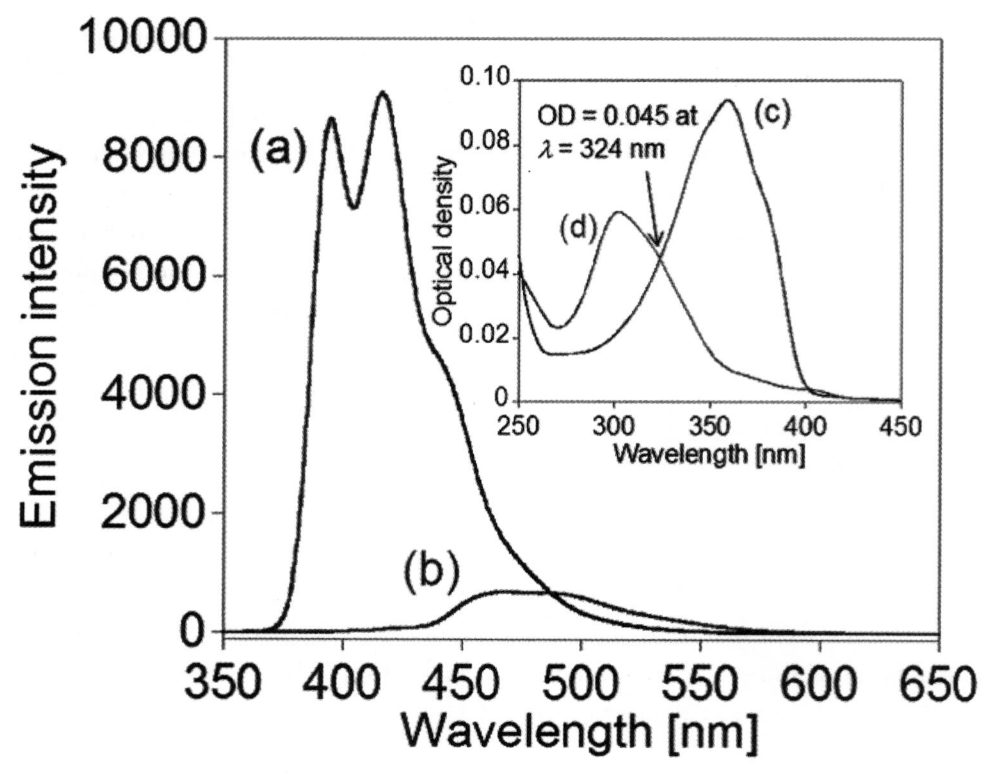

Figure 7. Emission and absorption spectra of BSB-Me dichloromethane solution and BSB-Me nanocrystal water dispersion. Emission spectra of BSB-Me dichloromethane solution (a) and BSB-Me nanocrystal water dispersion (b). The excitation wavelength was 324nm for each spectrum. The integrated intensity (areas) of the spectra was calculated as 528,826 for (a) and 58,884 for (b). Inset: the absorption spectra of the BSB-Me dichloromethane solution (c) and BSB-Me nanocrystal water dispersion (d), where both samples had the same optical density of 0.045 at 324-nm wavelength.

Table 1. Quantum yield, integrated intensity, optical density, and refractive index of the BSB-Me.

Quantum yield (Q), %		Integrated intensity (I)[b]	Optical density (OD) at $\lambda = 324$nm[c]	Refractive index (n) at 20°C
BSB-Me dissolved in dichloromethane (1μM)	95 ± 1[a]	528,826	0.045	1.42
BSB-Me nanocrystal water dispersion (2μM)	9.2 ± 0.1	58,884	0.045	1.33

[a]The data was obtained from Table one of reference[6]. [b]The data was obtained from **Figure 7(a)** and **Figure 7(b)**. [c]The data was obtained from **Figure 7** inset **(c)** and **(d)**.

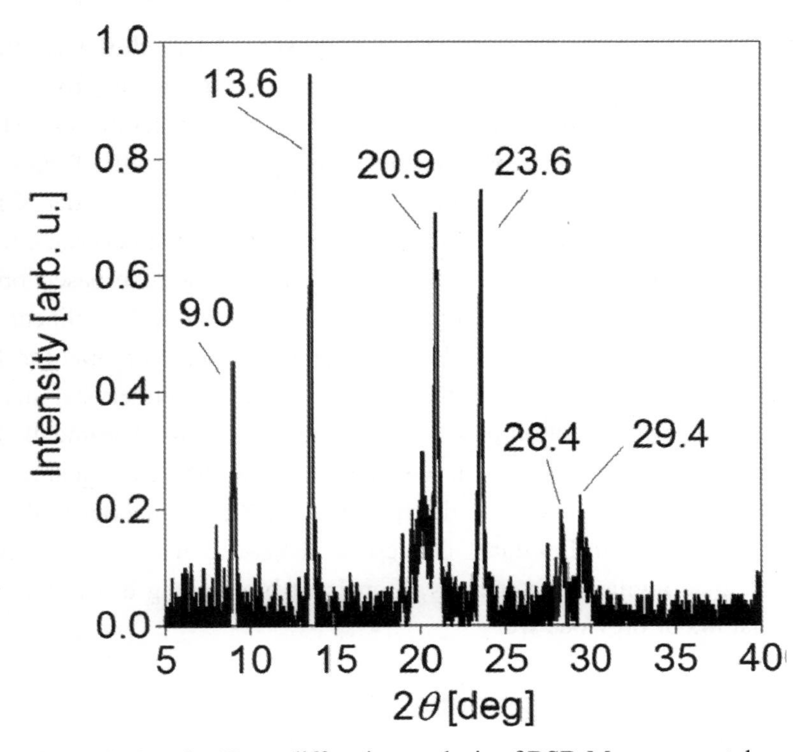

Figure 8. Powder X-ray diffraction analysis of BSB-Me nanocrystals.

softness of the crystal lattice may occur in nanocrystalline BSB-Me. Additionally, in our previous study, there were instances where the crystal structure of the nanocrystal was different from that of bulk crystal[22][36]. That unique optoelectronic properties may occur in nanocrystals compared with bulk single crystals caused by differences in crystal structure is quite interesting, but further investigation is necessary in future work.

4. Conclusions

We demonstrated the preparation of a BSB-Me nanocrystal dispersion in water by the reprecipitation method, which is a bottom-up, wet process for preparing organic nanocrystals. SEM observations revealed that the nanocrystals had a sphere-like morphology. The average particle size was 60.9nm, measured using an ELSZ-1000 zeta-potential and particle size analyzer. The nanocrystal dispersion was stable with a measured ζ-potential of −41.62mV using ELSZ-1000. The blue shift and red shift of maximum peak wavelength were observed in absorption and emission spectra, respectively. This optical feature may have arisen from supramolecular interactions like those caused by the herringbone structure, *i.e.*, H-aggregation, in the nanocrystals. The photoluminescence quantum yield of the BSB-Me nanocrystal water dispersion was estimated to be 9.2% ± 0.1%. Powder X-ray diffraction analysis confirmed the crystallinity of the BSB-Me nanocrystals. In future work, these BSB-Me nanocrystals will be applied to crystalline-based optoelectronic devices. Measuring amplified spontaneous emission and nonlinear optical properties of single nanocrystals will be a particularly interesting topic for the near future. We will also investigate and discuss elsewhere the nanocrystal size distribution using Scherrer's equation based on the data of XRD measurements. Further detailed optical properties such as an absolute photoluminescence quantum yield, fluorescence lifetime, and radiative decay rates of BSB-Me nanocrystals will be discussed elsewhere. Furthermore, fluorescent BSB-Me nanocrystals could be used in biological applications such as fluorescent bioimaging of cells and tissue similar to that in our previous work.

Competing Interests

The authors declare that they have no competing interests.

Authors' Contributions

KB contributed to the conception of the study, carried out all the experiments, and drafted the manuscript. KN contributed to the interpretation of the data and revision of the manuscript. Both authors read and approved the final manuscript.

Authors' Information

KB is an Endowed Chair Associate Professor at the Department of Visual Regenerative Medicine, Osaka University Graduate School of Medicine, Japan, and KN is a Professor and a medical doctor at the Department of Ophthalmology, Osaka University Graduate School of Medicine, Japan.

Acknowledgements

This study was partially supported by a Challenging Exploratory Research (no. 25560223) and Grant-in-Aid for Young Scientists (A) (no. 24680054) from the Japan Society for the Promotion of Science. We thank Dr. Yasunobu Wada for his technical support to the experiments.

Source: Baba K, Nishida K. Preparation of 1,4-bis(4-methylstyryl)benzene nano-crystals by a wet process and evaluation of their optical properties [J]. Nanoscale Research Letters, 2014, 9(1):1–8.

References

[1] Yang J, Fang HH, Ding R, Lu SY, Zhang YL, Chen QD, Sun HB: High-quality large-size organic crystals prepared by improved physical vapor growth technique and their optical gain properties. J Phy Chem C 2011, 115:9171–9175.

[2] Liu SH, Wang WCM, Briseno AL, Mannsfeld SCE, Bao ZN: Controlled deposition of crystalline organic semiconductors for field-effect-transistor applications. Adv Mater 2009, 21:1217–1232.

[3] Nakanotani H, Saito M, Nakamura H, Adachi C: Emission color tuning in ambipolar organic single-crystal field-effect transistors by dye-doping. Adv Funct Mater 2010, 20:1610–1615.

[4] Sasaki F, Kobayashi S, Haraichi S, Fujiwara S, Bando K, Masumoto Y, Hotta S: Microdisk and microring lasers of thiophene-phenylene co-oligomers embedded in Si/SiO2 substrates. Adv Mater 2007, 19:3653–3655.

[5] Fang HH, Yang J, Ding R, Chen QD, Wang L, Xia H, Feng J, Ma YG, Sun HB: Polarization dependent two-photon properties in an organic crystal. Appl Phys Lett 2010, 97:101101.

[6] Kabe R, Nakanotani H, Sakanoue T, Yahiro M, Adachi C: Effect of molecular morphology on amplified spontaneous emission of bis-styrylbenzene derivatives. Adv Mater 2009, 21:4034–4038.

[7] Nakanotani H, Adachi C: Organic light-emitting diodes containing multilayers of organic single crystals. Appl Phys Lett 2010, 96:053301.

[8] Baba K, Kasai H, Nishida K, Nakanishi H: Poly(N-isopropylacrylamide)-based thermoresponsive behavior of fluorescent organic nanocrystals. Jpn J Appl Phys 2011, 50:010202.

[9] Baba K, Nishida K: Calpain inhibitor nanocrystals prepared using Nano Spray Dryer

B-90. Nanoscale Res Lett 2012, 7:436.

[10] Baba K, Nishida K: Steroid nanocrystals prepared using the Nano Spray Dryer B-90. Pharmaceutics 2013, 5:107–114.

[11] Baba K, Kasai H, Okada S, Oikawa H, Nakanishi H: Novel fabrication process of organic microcrystals using microwave-irradiation. Jpn J Appl Phys 2000, 39: L1256–L1258.

[12] Katagi H, Kasai H, Okada S, Oikawa H, Komatsu K, Matsuda H, Liu ZF, Nakanishi H: Size control of polydiacetylene microcrystals. Jpn J Appl Phys 1996, 35: L1364–L1366.

[13] Nakanishi H, Katagi H: Microcrystals of polydiacetylene derivatives and their linear and nonlinear optical properties. Supramol Sci 1998, 5:289–295.

[14] Kasai H, Kamatani H, Okada S, Oikawa H, Matsuda H, Nakanishi H: Size-dependent colors and luminescences of organic microcrystals. Jpn J Appl Phys 1996, 35:L221–L223.

[15] Oikawa H, Mitsui T, Onodera T, Kasai H, Nakanishi H, Sekiguchi T: Crystal size dependence of fluorescence spectra from perylene nanocrystals evaluated by scanning near-field optical microspectroscopy. Jpn J Appl Phys 2003, 42:L111–L113.

[16] Oikawa H: Hybridized organic nanocrystals for optically functional materials. B Chem Soc Jpn 2011, 84:233–250.

[17] Onodera T, Oikawa H, Masuhara A, Kasai H, Sekiguchi T, Nakanishi H: Silver-deposited polydiacetylene nanocrystals produced by visible-light-driven photocatalytic reduction. Jpn J Appl Phys 2007, 46:L336–L338.

[18] Baba K, Pudavar HE, Roy I, Ohulchanskyy TY, Chen YH, Pandey RK, Prasad PN: New method for delivering a hydrophobic drug for photodynamic therapy using pure nanocrystal form of the drug. Mol Pharmaceut 2007, 4:289–297.

[19] Baba K, Kasai H, Masuhara A, Oikawa H, Nakanishi H: Organic solvent-free fluorescence confocal imaging of living cells using pure nanocrystal forms of fluorescent dyes. Jpn J Appl Phys 2009, 48:117002.

[20] Baba K, Tanaka Y, Kubota A, Kasai H, Yokokura S, Nakanishi H, Nishida K: A method for enhancing the ocular penetration of eye drops using nanoparticles of hydrolyzable dye. J Control Release 2011, 153:278–287.

[21] Kasai H, Murakami T, Ikuta Y, Koseki Y, Baba K, Oikawa H, Nakanishi H, Okada M, Shoji M, Ueda M, Imahori H, Hashida M: Creation of pure nanodrugs and their anticancer properties. Angewe Chem Int Edit 2012, 51:10315–10318.

[22] Baba K, Konta S, Oliveira D, Sugai K, Onodera T, Masuhara A, Kasai H, Oikawa H, Nakanishi H: Perylene and perylene-derivative nano-cocrystals: preparation and physicochemical property. Jpn J Appl Phys 2012, 51:125201.

[23] Fang HH, Yang J, Ding R, Feng J, Chena Q-D, Sun H-B: Top down fabrication of organic nanocrystals by femtosecond laser induced transfer method. Cryst Eng

Comm 2012, 14:4596–4600.

[24] Fang HH, Ding R, Lu SY, Wang L, Feng J, Chen QD, Sun HB: Direct laser interference ablating nanostructures on organic crystals. Opt Lett 2012, 37:686–688.

[25] Nishi T, Takeichi A, Azuma H, Suzuki N, Hioki T, Motohiro T: Fabrication of palladium nanoparticles by laser ablation in liquid. J Laser Micro Nanoeng 2010, 5: 192–196.

[26] Kenth S, Sylvestre JP, Fuhrmann K, Meunier M, Leroux JC: Fabrication of paclitaxel nanocrystals by femtosecond laser ablation and fragmentation. J Pharm Sci 2011, 100:1022–1030.

[27] Hobley J, Nakamori T, Kajimoto S, Kasuya M, Hatanaka K, Fukumura H, Nishio S: Formation of 3,4,9,10-perylenetetracarboxylicdianhydride nanoparticles with perylene and polyyne byproducts by 355nm nanosecond pulsed laser ablation of microcrystal suspensions. J Photoch Photobio A 2007, 189:105–113.

[28] Kasai H, Nalwa HS, Oikawa H, Okada S, Matsuda H, Minami N, Kakuta A, Ono K, Mukoh A, Nakanishi H: A novel preparation method of organic microcrystals. Jpn J Appl Phys 1992, 31:L1132–L1134.

[29] Ujiiye-Ishii K, Baba K, Wei Z, Kasai H, Nakanishi H, Okada S, Oikawa H: Mass-production of pigment nanocrystals by the reprecipitation method and their encapsulation. Mol Cryst Liq Cryst 2006, 445:177–183.

[30] Oikawa H, Onodera T, Masuhara A, Kasai H, Nakanishi H: New class materials of organic-inorganic hybridized nanocrystals/nanoparticles, and their assembled micro- and nano-structure toward photonics. Polym Mat Adv Polym Sci 2010, 231:147–190.

[31] Varghese S, Park SK, Casado S, Fischer RC, Resel R, Milian-Medina B, Wannemacher R, Park SY, Gierschner J: Stimulated emission properties of sterically modified distyrylbenzene-based H-aggregate single crystals. J Phys Chem Lett 2013, 4:1597–1602.

[32] Lakowicz JR: Principles of Fluorescence Spectroscopy. New York: Springer; 2006.

[33] Brouwer AM: Standards for photoluminescence quantum yield measurements in solution (IUPAC Technical Report). Pure Appl Chem 2011, 83:2213–2228.

[34] Baba K, Kasai H, Okada S, Nakanishi H, Oikawa H: Fabrication of diacetylene nanofibers and their dynamic behavior in the course of solid-state polymerization. Mol Cryst Liq Cryst 2006, 445:161–166.

[35] Takahashi S, Miura H, Kasai H, Okada S, Oikawa H, Nakanishi H: Single-crystal-to-single-crystal transformation of diolefin derivatives in nanocrystals. J Am Cheml Soc 2002, 124:10944–10945.

[36] Baba K, Kasai H, Okada S, Oikawa H, Nakanishi H: Fabrication of organic nanocrystals using microwave irradiation and their optical properties. Opt Mater 2003, 21: 591–594.

Chapter 25

AlGaInP LED with Low-Speed Spin-Coating Silver Nanowires as Transparent Conductive Layer

Xia Guo[1], Chun Wei Guo[1], Cheng Wang[2], Chong Li[1], Xiao Ming Sun[2]

[1]Photonic Research Lab, Beijing University of Technology, Beijing 100124, China
[2]Beijing University of Chemical Technology, Beijing 100029, China

Abstract: The low-speed spin-coating method was developed to prepare uniform and interconnected silver nanowires (AgNWs) film with the transmittance of 95% and sheet resistance of 20Ω/sq on glass, which was comparable to ITO. The fitting value of σ_{dc}/σ_{op} of 299.3 was attributed to the spin-coating process. Advantages of this solution-processed AgNW film on AlGaInP light-emitting diodes (LEDs) as transparent conductive layer were explored. The optical output power enhanced 100%, and the wavelength redshift decreased from 12 to 3nm, which indicated the AgNW films prepared by low-speed spin-coating possessed attractive features for large-scale TCL applications in optoelectronic devices.

Keywords: AlGaInP, Light-Emitting Diodes (LEDs), Low-Speed Spin-Coating, Silver Nanowires, Transparent Conductive Layer

1. Background

Transparent conductive layer (TCL) is crucial for lightemitting diodes (LEDs) which spread the carriers far away from the opaque electrodes to enhance the quantum efficiency and improve the efficiency droop effect[1]-[3]. Indium-doped tin oxide (ITO) material is widely used in LED field with the sheet resistance Rs of 10 to 30Ω/sq and optical transmittance T of 90%, which are two important figures of merit (FoM) to facilitate to describe the performance of TCLs[4]. However, due to the scarcity of the element indium in the earth and consequently the soaring prices, recently, nanomaterials, such as carbon nanotubes (CNTs)[5], graphene[6], metal grids[7], and metallic nanowires[8], have attracted great attention as candidates of TCL due to their unique electrical properties, good transparency, and mechanical flexibility. Due to the large inter-junction resistance of CNT film caused by mixture of metallic and semiconducting properties, the sheet resistance of CNT film is 200 to 1,000Ω/sq[9], which is relatively high compared with that of the ITO film. Graphene has high mobility as well as high transmittance[10][11]. However, large sheet resistance and obvious degradation of graphene layer under several milliampere current injections restricted its actual application[12].

Random and sparse silver nanowire (AgNW) film[13], which demonstrated superior FoM performances, was regarded as the most promising candidate to replace ITO, due to its low inter-wire junction resistance and low absorption loss[14]. Yi's group demonstrated solution-processed AgNW films with T550nm of 80% and Rs of 20Ω/sq[9]. J. N. Coleman's group sprayed the AgNWs over large areas with T550nm of 90% and Rs of 50Ω/sq[15]. Such AgNW films as TCLs applied to organic optoelectronic devices were reported. For example, AgNWs film as top electrodes of the organic solar cells or organic LEDs were recently established as a serious alternative to ITO[16]-[18].

One of the most significant challenges for the AgNW film was to obtain the low sheet resistance and high transmittance at the same time[19]. Besides controlling the diameter and length of the wires, uniform distribution of the wires was another important factor. Conventional drop-coating, spray-coating, and bar-coating processes inevitably caused the solution-based AgNWs self-aggregation after solvent drying process. Poor adhesion to the substrate made the uniform distribution of AgNWs more difficult. Thus, new film coating method as well as substrate material to improve the uniformity of nanowire distribution was required. Filtration coating was developed for preparing AgNW films with T of 88% at 550nm and Rs of 12Ω/sq on cellulose nanopaper[20]. Exfoliated clays were utilized for reducing the self-aggregation of nanowires with high solution viscosity on PETs by roll-to-roll coating process with T of 97.9% and Rs of 91.3Ω/sq[16].

In this paper, a low-speed spin-coating method was developed for uniform AgNW film with the T of 94% and Rs of 20Ω/sq on glass. The optical output power increased about 100% for AlGaInP LEDs with AgNW film as TCL, and the wavelength redshift decreased from 12 to 3nm under the current injection of 100mA due to the uniform carrier injection in the active region.

2. Methods

The Ag nanowires were prepared following the reported procedure[21]. In brief, 0.5g of glucose and 0.1g of polyvinyl pyrrolidone (PVP) were dissolved in 35ml of deionized water to form a clear solution. Then, 0.5ml of freshly prepared 0.1M aqueous $AgNO_3$ solution was added under vigorous stirring. The mixture was transferred into a 40-ml Teflon-sealed autoclave and heated at 140°C for 10h. After the reaction, the autoclave was allowed to cool in air and the product was purified by 3 to 5 centrifugation/rinsing/redispersion circles. Then, the AgNWs re-dispersed in isopropyl alcohol due to better dispersibility for different concentrations. The AgNW film was fabricated through spin-coating at a speed of 270rpm, which was much lower than the spin-coating speed of photoresist which was used in the microelectronics field due to the poor adhesion of AgNW solution. The nanowires were stuck on the substrate surface after the solution was spanned out of the substrate. After the spin-coating process, samples with AgNW film was put on the hot plate for 10min with a temperature of 200°C in order to decrease the nanowire-nanowire contact resistance[22].

3. Results and Discussion

Figure 1(a) presented the scanning electron microscopy (SEM) images of AgNW films on glass prepared by spin-coating, which was the mostly commonly

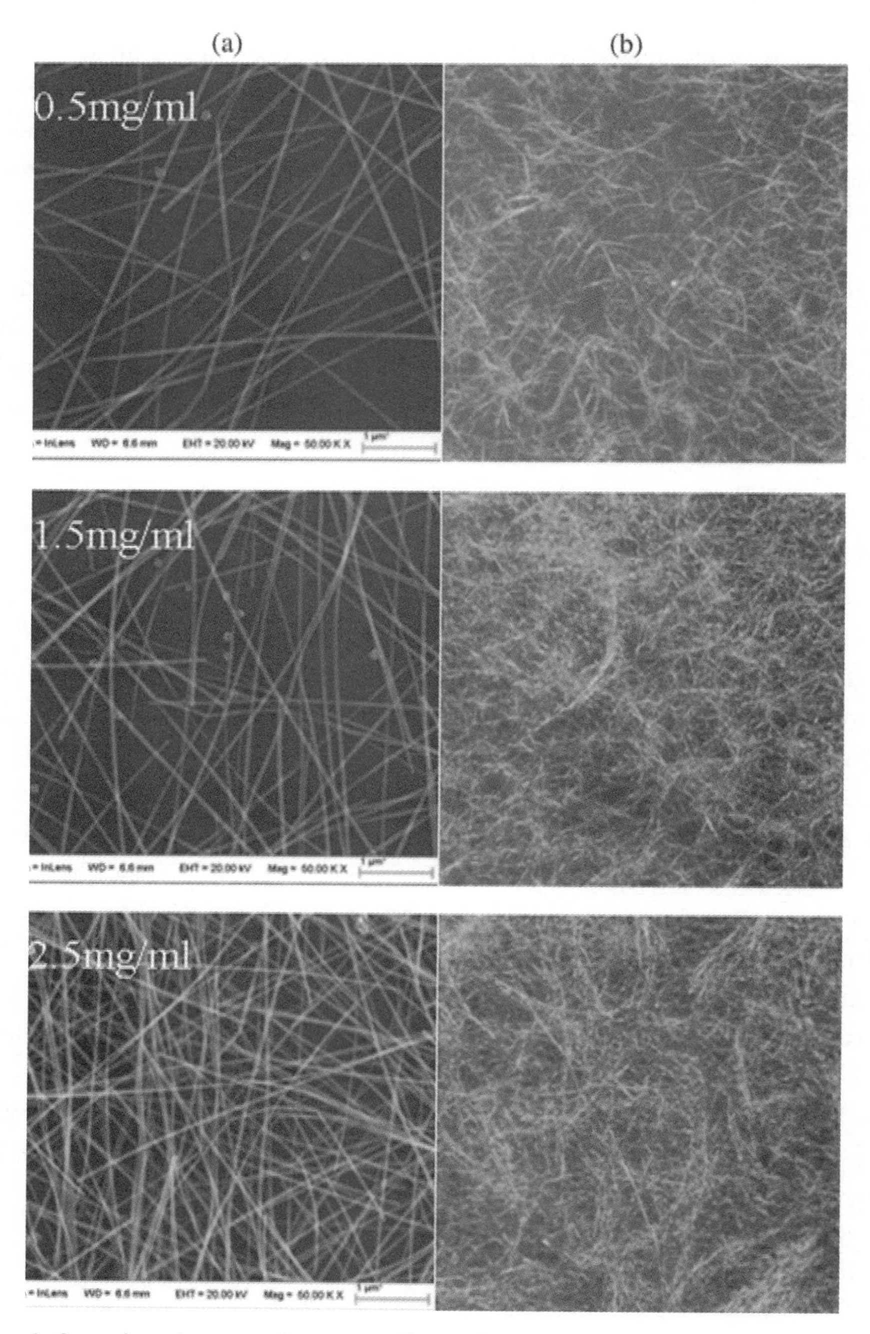

Figure 1. Scanning electron microscopy (SEM) images (a) and microscopic photographs (b). Of AgNW films with AgNW concentrations of 0.5, 1.5, and 2.5mg/ml, respectively. The microscopic photographs, with magnification of ×1,000, displayed the AgNWs were connected with each other over large areas.

used in preparing photoresist with large scale in the field of integrated circuits. The concentrations of AgNW solution were 0.5, 1.5, and 2.5mg/ml, respectively. The rotation speed was closely related with the FoM and was optimized to be 270rpm, which was far lower than that in the photolithography process due to the poor adhesion of wires to the substrate. From the SEM images, the typical diameter and length of the AgNWs can be measured about 40nm and 30−50μm, respectively. But most importantly, the wires connected were with each other and uniformly distributed, which guaranteed low resistance of the AgNW film. **Figure 2(b)** showed the corresponding microscopic photographs with magnification of ×1,000, which also displayed the uniform distribution of the AgNW films. By increasing the AgNW concentrations, the AgNW area coverage increased. For the concentration of 2.5mg/ml, it is hard to find the bare space of substrate under the microscope, which indicated low transmittance.

Figure 2(a) showed the transparent conductive performances of AgNW film on glass without any pressing treatment after spin-coating. The transmittance was evaluated using a piece of bare glass as reference. The sheet resistances of AgNW film on glass prepared by spin-coating at 270rpm showed 20Ω/sq with T550nm of about 95%. It should be noted that the value (Rs, T550nm) of (20Ω/sq, 95%) was comparable with the performance of ITO and was much superior to that of graphene and CNT, which indicated that the AgNW film as transparent conductive

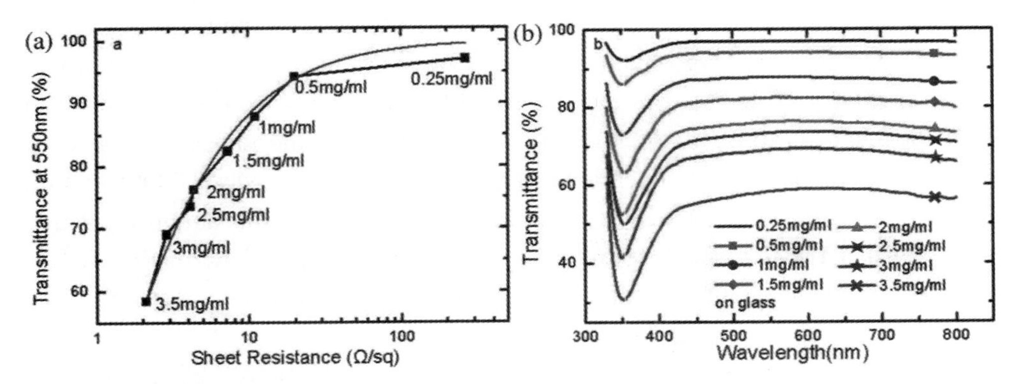

Figure 2. Optical and electrical performance of AgNW film on glass. (a) Spectral transmittance at 550nm as a function of sheet resistance for AgNW films on glass with different concentrations by spin- and drop-coating processes. The fitting curve according to Equation (1) fits the data prepared by spin-coating process well. In each case, the transmittance was evaluated with the bare glass as a reference. (b) Spectral transmittance results of AgNW films prepared by spin-coating at 270rpm with AgNW concentrations from 0.25 to 3.5mg/ml.

layer was capable of applications in the optoelectronic devices, such as LEDs and solar cells, whose performance was sensitive to the power conversion efficiency. **Figure 2(b)** showed the spectral transmittance of AgNW film on glass with the AgNW concentrations from 0.25 to 3.5mg/ml. The transmittance was kept almost flat from about 420 to 800nm for all the curves, which indicated wide applications in the visible wavelength range. The transmittance decreased with the concentration of AgNW solution due to the nanowire coverage area.

The transmittance and sheet resistance of nanowire film could be expressed as[23]

$$T(\lambda)=\left(1+\frac{Z_0}{2R_s}\frac{\sigma_{op}}{\sigma_{dc}}\right)^{-2} \tag{1}$$

in which Z_0 was the impedance of free space which was equal to 377Ω. Rs was the sheet resistance of the nanowire film. σ_{op} and σ_{dc} were the optical and DC conductivity of the film, respectively. The optical and electrical performance of the film could be evaluated by the ratio of σ_{dc}/σ_{op}. High transmittance and low sheet resistance means the large ratio of σ_{dc}/σ_{op}. The first criterion for high-performance TCL required the ratio of $\sigma_{dc}/\sigma_{op} \geq 35$ to achieve the target of $T \geq 90\%$ and $Rs \leq 100\Omega/sq$[14]. In our experiment, the σ_{dc}/σ_{op} of the film prepared by spin-coating was fitted to be 299.3, as shown in **Figure 2(a)**, which was close to that of ITO and much larger than that of CNT and graphene[14][24]. Also, the theoretical prediction fitted all the experimental data very well except the only data with the concentration of 0.25mg/ml. For comparison, the sheet resistance of AgNW film prepared by conventional drop-coating on glass was 1,000Ω/sq with T550nm of about 95%, and the fitting ratio of σ_{dc}/σ_{op} was only about 31, which indicated that the low-speed spin-coating could improve the FoM of the AgNW film greatly by enhancing the uniformity of nanowire distribution.

To demonstrate its potential of transmittance conductive properties, AlGaInP LEDs with AgNW film as current-spreading layer were fabricated. The reason that demonstration on the AlGaInP LEDs, not GaN-based LEDs, was smaller work function difference of no more than 1.5eV between the AgNW and GaP, which indicated the feasibility of ohmic contact between metal nanowire and p-doped GaP.

The AlGaInP LEDs were grown on n-GaAs substrate by metal-organic chemical vapor deposition. The details could be found in ref.[25]. In order to study the current-spreading effect of AgNW film, only 500-nm-thick Mgdoped p-GaP window layer with the doping density of 5×10^{18}cm^{-3} was grown on top. The 50-, 150-, or 200-nm-thick Au/BeAu/Au with 100μm diameter was first deposited and then patterned by wet etching as p-type electrode. The AgNW solution with the concentration of 0.5mg/ml was applied and then stuck on the surface of the LED wafer by Vander Waals force. The chip size was 300μm × 300μm in this work.

Figure 3 showed the current-voltage (I-V) curves of AlGaInP LED with and without AgNWs as current-spreading layer with the voltage drops of 2.08 and 2.18V at current injection of 20mA, which indicated better current spreading. The inset showed the microscope photographs of LED wafers before dicing under the current injection of 5mA under the probe station. It was obvious that the current-spreading effect was totally different, which echoed the I-V measurement results. For the devices without AgNWs, the emission was localized around the electrode, which indicated the carriers transport laterally with limited distance. While for the devices with AgNW film, the whole wafer was lighting up, which demonstrated the excellent capability of lateral carrier transport of AgNW film.

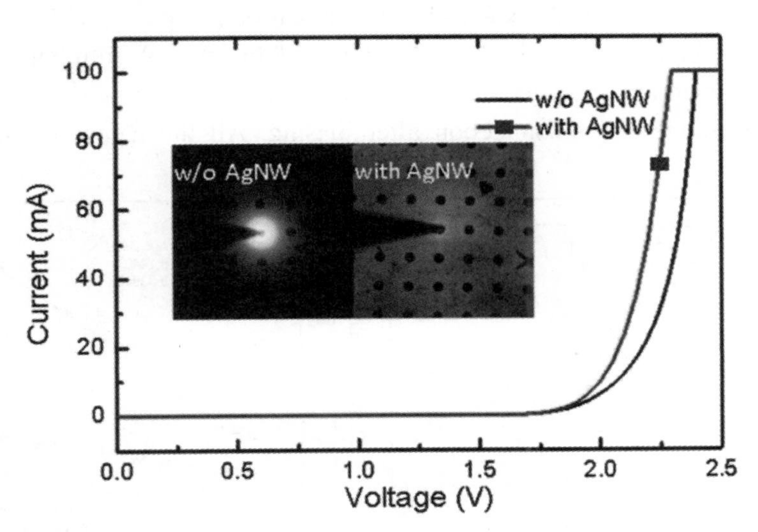

Figure 3. Current-voltage (*I-V*) curves of AlGaInP LED with and without AgNWs as current-spreading layer. The inset showed microscope photographs of LED wafers before dicing under the current injection of 5mA under the probe station.

Figure 4(a) demonstrated the optical output power on the current injection of LED with and without AgNW film as TCL. The optical output power and the linearity of LED with AgNW film were much better than that of without AgNW film. At 20mA, the optical output power of LED with AgNW film was two times of that of without AgNW film. As we known, the optical output power improved only 30% if ITO as TCL on LEDs[26][27]. Nano or microstructures, such as photonic crystal and surface roughness, could only improve the optical output power about 10% to 30%[28]. The current value corresponding to the maximum optical output power was 60 and 40mA, respectively, with and without AgNW film, which indicated the better thermal performance. The peak wavelength was 630 and 635nm, respectively, according to the electroluminescence spectra of LEDs with and without AgNW film at 20mA. The wavelength redshift was another important criterion to characterize the current-spreading effect, and AlGaInP material was very sensitive to the temperature. **Figure 4(b)** demonstrated the wavelength redshift measurement results, in which the dots were the measurement data and the line was the linear fitting of the data. The wavelength redshift was 3 and 12nm for LEDs with and without AgNW film, respectively, which verified the optical output power results.

The obvious improvement of LED's optical output power and thermal performance, we believe, not only due to the high FoM of AgNW film but also due to the current injection in different ways. The network of nanowires on the LED formed an equipotential connection after biasing. All the nanowires uniformly

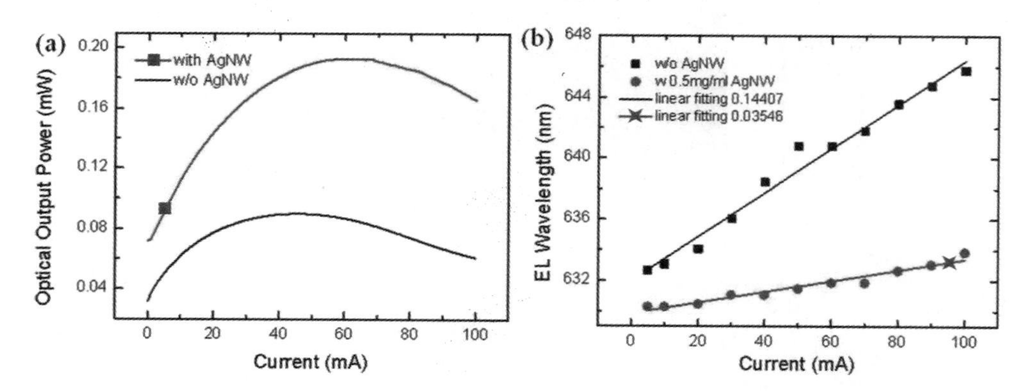

Figure 4. Dependence of optical output power and peak wavelength. (a) The dependence of optical output power on the current injection of LED chips with and without AgNW film as current-spreading layer. (b) The dependence of peak wavelength of AlGaInP LEDs with and without AgNW film on the current injection.

distributed on the surface of the LED injected the carriers at the same time with lowered current density, just like water from a shower head. Compared with the current injection from the ohmic contact electrode which usually located at the center of the device, the current density distribution in the quantum wells from the nanowire film will be more uniform, which decreased the current crowding and heat generation.

4. Conclusions

In summary, low-speed spin-coating method was demonstrated to prepare uniform and interconnected AgNW film with the transmittance of 95% and sheet resistance of 20Ω/sq on glass, which was comparable to ITO. The fitting value of $\sigma dc/\sigma op$ of around 300 was attributed to the spin-coating process. Advantages of this solution-processed AgNW film on AlGaInP LEDs as TCL were explored. The optical output power enhanced 100% and the wavelength redshift decreased four times, which indicated the AgNW films prepared by low-speed spin-coating-possessed attractive features for wide TCL applications in optoelectronic devices.

Competing Interests

The authors declare that they have no competing interests.

Authors' Contributions

GX and GCW drafted the manuscript. GX conducted the experiment design and analysis of all the experiments. GCW carried out most of the experimental work including all the measurements. WC prepared the AgNW solution. LC and SXM participated in all the discussion on this study. All authors read and approved the final manuscript.

Acknowledgements

This work was supported by the National Natural Science Foundation of

China (Grant Nos. 61222501 and 61335004) and the Specialized Research Fund for the Doctoral Program of Higher Education of China (Grant No. 20111103 110019).

Source: Guo X, Guo C W, Wang C, *et al.* AlGaInP LED with low-speed spin-coating silver nanowires as transparent conductive layer[J]. Nanoscale Research Letters, 2014, 9(1):1–6.

References

[1] Kim A, Won Y, Woo K, Jeong S, Moon J: Transparent electronics: all-solution-processed indium-free transparent composite electrodes based on Ag nanowire and metal oxide for thin-film solar cells (Adv. Funct. Mater. 17/2014). Adv Funct Mater 2014, 24:2414.

[2] Yusoff A, Bin Mohd R, Lee SJ, Shneider FK, Da Silva WJ, Jang J: High-performance semitransparent tandem solar cell of 8.02% conversion efficiency with solution-processed graphene mesh and laminated Ag nanowire top electrodes. Adv Energy Mater 2014, 12:4.

[3] Shim J, Han D, Kim H, Shin D, Lin G, Meyaard DS, Shan Q, Cho J, Schubert EF, Shim H: Efficiency droop in AlGaInP and GaInN light-emitting diodes. Appl Phys Lett 2012, 100:111106.

[4] Chae DJ, Kim DY, Kim TG, Sung YM, Kim MD: AlGaN-based ultraviolet light-emitting diodes using fluorine-doped indium tin oxide electrodes. Appl Phys Lett 2012, 100:81110.

[5] Feng C, Liu K, Wu JS, Liu L, Cheng JS, Zhang Y, Sun Y, Li Q, Fan S, Jiang K: Flexible, stretchable, transparent conducting films made from superaligned carbon nanotubes. Adv Funct Mater 2010, 20:885–891.

[6] Pang S, Hernandez Y, Feng X, Müllen K: Graphene as transparent electrode material for organic electronics. Adv Mater 2011, 23:2779–2795.

[7] Hsu P, Wang S, Wu H, Narasimhan VK, Kong D, Lee HR, Cui Y: Performance enhancement of metal nanowire transparent conducting electrodes by mesoscale metal wires. Nat Commun 2013, 4:2522.

[8] Garnett EC, Cai W, Cha JJ, Mahmood F, Connor ST, Christoforo MG, Cui Y, McGehee MD, Brongersma ML: Self-limited plasmonic welding of silver nanowire junctions. Nat Mater 2012, 11:241–249.

[9] Hu L, Kim HS, Lee J, Peumans P, Cui Y: Scalable coating and properties of transparent, flexible, silver nanowire electrodes. ACS Nano 2010, 4:2955–2963.

[10] Biswas C, Lee YH: Graphene versus carbon nanotubes in electronic devices. Adv Funct Mater 2011, 21:3806–3826.

[11] Bonaccorso F, Sun Z, Hasan T, Ferrari AC: Graphene photonics and optoelectronics. Nat Photonics 2010, 4:611–622.

[12] Youn D, Yu Y, Choi H, Kim S, Choi S, Choi C: Graphene transparent electrode for enhanced optical power and thermal stability in GaN light-emitting diodes. Nanotechnology 2013, 24:75202.

[13] Michaelson HB: The work function of the elements and its periodicity. J Appl Phys 1977, 48:4729–4733.

[14] De S, Coleman JN: The effects of percolation in nanostructured transparent conductors. MRS Bull 2011, 36:774–781.

[15] Scardaci V, Coull R, Lyons PE, Rickard D, Coleman JN: Spray deposition of highly transparent, low-resistance networks of silver nanowires over large areas. Small 2011, 7:2621–2628.

[16] Kim T, Kim YW, Lee HS, Kim H, Yang WS, Suh KS: Uniformly interconnected silver-nanowire networks for transparent film heaters. Adv Funct Mater 2013, 23:1250–1255.

[17] Krantz J, Stubhan T, Richter M, Spallek S, Litzov I, Matt GJ, Spiecker E, Brabec CJ: Spray-coated silver nanowires as top electrode layer in semitransparent P3HT: PCBM-based organic solar cell devices. Adv Funct Mater 2013, 23:1711–1717.

[18] Krantz J, Richter M, Spallek S, Spiecker E, Brabec CJ: Solution-processed metallic nanowire electrodes as indium tin oxide replacement for thin-film solar cells. Adv Funct Mater 2011, 21:4784–4787.

[19] Song M, You DS, Lim K, Park S, Jung S, Kim CS, Kim DH, Kim DG, Kim JK, Park J: Highly efficient and bendable organic solar cells with solution-processed silver nanowire electrodes. Adv Funct Mater 2013, 23:4177–4184.

[20] Koga H, Nogi M, Komoda N, Nge TT, Sugahara T, Suganuma K: Uniformly connected conductive networks on cellulose nanofiber paper for transparent paper electronics. NPG Asia Mater 2014, 6:e93. 1–7.

[21] Sun XM, Li YD: Cylindrical silver nanowires: preparation, structure, and optical properties. Adv Mater 2005, 17:2626–2630.

[22] Coskun S, Ates ES, Unalan HE: Optimization of silver nanowire networks for polymer light emitting diode electrodes. Nanotechnology 2013, 24:125202.

[23] Khanarian G, Joo J, Liu X, Eastman P, Werner D, O'Connell K, Trefonas P: The optical and electrical properties of silver nanowire mesh films. J Appl Phys 2013, 114:24302.

[24] Jeong J, Kim H: Ag nanowire percolating network embedded in indium tin oxide nanoparticles for printable transparent conducting electrodes. Appl Phys Lett 2014,

104:71906.

[25] Guo X, Guo CW, Jin YH, Chen Y, Li QQ, Fan SS: AlGaInP light-emitting diodes with SACNTs as current-spreading layer. Nanoscale Res Lett 2014, 9:171.

[26] Yen C, Liu Y, Yu K, Lin P, Chen T, Chen L, Tsai T, Huang N, Lee C, Liu W: On an AlGaInP-based light-emitting diode with an ITO direct ohmic contact structure. IEEE Electron Device Lett 2009, 30:359–361.

[27] Hsu S, Wuu D, Lee C, Su J, Horng R: High-efficiency 1-mm 2 AlGaInP LEDs sandwiched by ITO omni-directional reflector and current-spreading layer. IEEE Photonics Technol Lett 2007, 19:492–494.

[28] Ryu SW, Park J, Oh JK, Long DH, Kwon KW, Kim YH, Lee JK, Kim JH: Analysis of improved efficiency of InGaN light-emitting diode with bottom photonic crystal fabricated by anodized aluminum oxidxe. Adv Funct Mater 2009, 19:1650–1655.

Chapter 26

Cu-Doped ZnO Nanorod Arrays: The Effects of Copper Precursor and Concentration

Musbah Babikier, Dunbo Wang, Jinzhong Wang, Qian Li, Jianming Sun, Yuan Yan, Qingjiang Yu and Shujie Jiao

Department of Opto-Electric Information Science, School of Materials Science and Engineering, Harbin Institute of Technology, Harbin 150001, China

Abstract: Cu-doped ZnO nanorods have been grown at 90°C for 90min onto a quartz substrate pre-coated with a ZnO seed layer using a hydrothermal method. The influence of copper (Cu) precursor and concentration on the structural, morphological, and optical properties of ZnO nanorods was investigated. X-ray diffraction analysis revealed that the nanorods grown are highly crystalline with a hexagonal wurtzite crystal structure grown along the c-axis. The lattice strain is found to be compressive for all samples, where a minimum compressive strain of −0.114% was obtained when 1at.% Cu was added from $Cu(NO_3)_2$. Scanning electron microscopy was used to investigate morphologies and the diameters of the grown nanorods. The morphological properties of the Cu-doped ZnO nanorods were influenced significantly by the presence of Cu impurities. Near-band edge (NBE) and a broad blue-green emission bands at around 378 and 545nm, respectively, were observed in the photoluminescence spectra for all samples. The transmittance characteristics showed a slight increase in the visible range, where the total transmittance increased from approximately 80% for the nanorods doped with $Cu(CH_3COO)_2$ to approximately 90% for the nanorods that were doped with $Cu(NO_3)_2$.

Keywords: Zinc Oxide, Nanostructures, Doping, Hydrothermal Crystal Growth, Photoluminescence

1. Background

ZnO semiconductor attracted considerable research attention in the last decades due to its excellent properties in a wide range of applications. ZnO is inherently an n-type semiconductor and has a wide bandgap of approximately 3.37eV and a large exciton binding energy of approximately 60meV at room temperature. As mentioned above, ZnO is a promising semiconductor for various applications such as UV emitters and photodetectors, light-emitting diodes (LEDs), gas sensors, field-effect transistors, and solar cells[1]-[6]. Additionally, ZnO resists radiation, and hence, it is a suitable semiconductor for space technology applications. Recently, ZnO nanostructures have been used to produce short-wavelength optoelectronic devices due to their ideal optoelectronic, physical, and chemical properties that arise from a high surface-to-volume ratio and quantum confinement effect[6]-[8]. Among the ZnO nanostructures, ZnO nanorods showed excellent properties in different applications and acted as a main component for various nanodevices[1][2][9]-[11]. Previous research showed that the optical and structural properties of ZnO nanorods can be modified by doping with a suitable element to meet pre-determined needs[12][13]. The most commonly investigated metallic dopants are Cu and Al[13]-[15]. Specifically, copper is known as a prominent luminescence activator, which can enhance the green luminescence band by creating localized states in the bandgap of ZnO[16]-[19]. Previous research showed that Cu has high ionization energy and low formation energy, which speedup the incorporation of Cu into the ZnO lattice[16][20]. Experimentally, it was observed that the addition of Cu into ZnO-based systems has led to the appearance of two defective states at +0.45eV (above the valence band maximum) and −0.17eV (below the conduction band minimum)[21][22]. Currently, a green emission band was observed for many Cu-doped ZnO nanostructures grown by different techniques[23][24]. Moreover, Cu as a dopant gained more attention due to its room-temperature ferromagnetism, deep acceptor level, some similar properties to those of Zn, gas sensitivity, and enhanced green luminescence[15]-[17]. However, there are several points that have to be analyzed such as the effect of the copper source on the structural, morphological, and optical properties of Cu-doped ZnO. Moreover, the luminescence and the structural properties of Cu-doped ZnO nanorods are affected by different parameters such as growth conditions, growth mechanism, post growth treatments, and Cu concentration. Despite the promising properties, research on the influence of Cu precursors on Cu-doped ZnO na-norod properties remains low.

ZnO nanostructures can be synthesized by a variety of techniques including vapor-phase transport, chemical vapor deposition, sol-gel, condensation, spray pyrolysis, and hydrothermal method. Among these methods, the hydrothermal method is used to prepare ZnO nanorods due to its low cost and simplicity[16][25][26].

In order to improve the structural and optical properties of Cu-doped ZnO nanorods, the effect of the Cu precursor is worth clarification. In the study reported here, we have synthesized pure and Cu-doped ZnO nanorods onto a quartz substrate pre-coated with a ZnO seed layer using the hydrothermal method. The main focus has been put on the effect of the copper precursor on the morphology, structural, transmittance, and photoluminescence properties of the synthesized ZnO nanorods.

2. Methods

The nanorod growth was accomplished in two steps: (1) the sputtering of ZnO seed layer to achieve highly aligned Cu-doped ZnO nanorods[27] and (2) the nanorod growth using the hydrothermal method.

2.1. Sputtering of ZnO Seed Layer

Prior to the nanorod growth, a 120-nm-thick seed layer of undoped ZnO was deposited onto a quartz substrate using RF magnetron sputtering at room temperature. Before the deposition of the ZnO seed layer, a surface treatment of the quartz substrate was conducted using acetone, ethanol, and deionized water for 10min for each at RT and then dried in air. Pure ZnO (99.999%) with a 50-mm diameter and 5-mm thickness was used as the ZnO target. The seed layer sputtering was accomplished in a mixture of O and Ar gas atmosphere with the gases' flow rates of 2.5

and 35sccm, respectively. The base pressure attained was 10^{-4}Pa, and the working pressure was 1Pa during sputtering. The sputtering power was 100W. In order to remove the contaminants from the ZnO target, pre-sputtering for 10min was performed. Finally, the ZnO-sputtered seed layer thin films were annealed at 500°C for 30min.

2.2. Nanorod Growth

Undoped and Cu-doped ZnO nanorods were grown by the hydrothermal method on a quartz substrate seeded with the ZnO thin film using hexamethylenetetramine (HMT) $[(CH_2)_6N_4]$, zinc acetate dihydrate meV $[Zn(CH_3COO)_2·2H_2O]$, and either cupric acetate $[Cu(CH_3COO)_2·H_2O]$ or cupric nitrate $[Cu(NO_3)_2·3H_2O]$ as hydroxide, zinc (Zn), and copper (Cu) precursors, respectively. The nanorod growth was accomplished by suspending the substrates in a conical flask containing the aqueous solution that was prepared from zinc acetate (0.025M) and HMT (0.025M). Before suspending the samples, the aqueous solution was magnetically stirred for 30min. The flask that contains the equimolar aqueous solution was placed in a combusting waterbath deposition system at 90°C for 90min. After the nanorods were grown, the samples were removed from the beakers, rinsed in deionized water several times to remove the unreacted materials, and then finally dried in an oven at 60°C for 2h. In order to introduce the Cu dopants, either cupric acetate (0.025M) or cupric nitrate (0.025M) was added directly to the reaction path. To study the effects of Cu concentration and precursor on the Cu-doped ZnO nanorods, five samples (S1 to S5) were prepared. For simplicity, the undoped ZnO nanorod (sample S1) was used as a reference sample. Samples S2 and S3 were doped with 1 and 2at.% of Cu, respectively, from $Cu(CH_3COO)_2$. Samples S4 and S5 were doped with 1 and 2at.% of Cu, respectively, from Cu $(NO_3)_2$. For more details, see **Table 1** to clarify the concentrations and precursors for each sample.

2.3. Characterization and Measurements

In order to characterize the structure of the grown nanorods, X-ray diffraction (XRD) measurements were performed using a MiniFlex-D/MAX-rb with

Table 1. Precursors, concentrations, and crystal parameters of undoped and Cu-doped ZnO nanorods.

	S1	S2	S3	S4	S5
Zn precursor	Zn ACT	Zn ACT	Zn ACT	Zn ACT	Zn ACT
OH precursor	HMT	HMT	HMT	HMT	HMT
Cu precursor	-	Cu acetate	Cu acetate	Cu nitrate	Cu nitrate
Cu (at.%)	-	1	2	1	2
FWHM (degrees)	0.096	0.087	0.087	0.099	0.134
c (Å)	5.186	5.192	5.200	5.201	5.184

CuKα radiation. The morphology of the hydrothermally grown nanorods was investigated by field emission scanning electron microscope (SEM) using SEM Helios Nanolab 600i (Hillsboro, OR, USA). Photoluminescence (PL) spectra were measured at room temperature with an excitation source of 325-nm wavelength using a He-Cd laser. Transmittance measurements were recorded by a UV-vis spectrophotometer (Phenix-1700 PC, Shanghai, China).

3. Results and Discussion

3.1. Crystal Structure

Figure 1 shows the XRD patterns of the undoped and Cu-doped ZnO nanorod samples grown with varied concentrations and doped from two different Cu precursors. Clearly, a strong and narrow peak corresponding to ZnO (002) is observed, indicating that all samples possess a hexagonal wurtzite crystal structure with highly preferred growth direction along the c-axis perpendicular to the substrate. Additionally, there were two weak diffraction peaks observed at around 63.2° and 72.8°, which correspond to ZnO (103) and ZnO (004), respectively. For the Cu-doped ZnO nanorod samples, no other diffraction peaks are observed, only ZnO-related peaks, which is consistent with previous results[6][16][18][28]. It may be seen that the diffraction intensity from the (002) plane is more pronounced for the undoped ZnO nanorods (sample S1) and decreases with the increase of Cu concentration regardless of the Cu precursor, indicating that the incorporation of Cu dopants into the ZnO lattice induces more crystallographic defects and hence

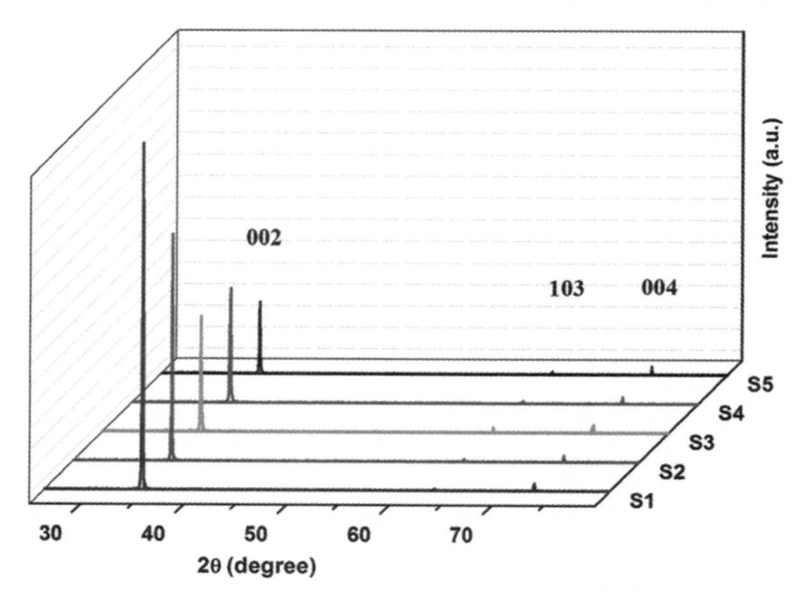

Figure 1. XRD patterns of undoped and Cu-doped ZnO nanorods.

degrades the crystal quality[16][28]. In terms of Cu precursor, the samples doped with 1 and 2at.% of Cu from $Cu(CH_3COO)_2$ (samples S2 and S3) exhibited strong diffraction intensities from the (002) plane compared to the samples doped with 1 and 2at.% of Cu from $Cu(NO_3)_2$ (samples S4 and S5). This result suggests that the samples doped with $Cu(CH_3COO)_2$ (S2 and S3) have a low concentration of crystallographic defects. The decrease in the crystal quality of the samples doped with $Cu(NO_3)_2$ (S4 and S5) might be attributed to (i) the formation of $[Cu_{Zn}\text{–}Zn_i]^{\chi}$ complexes and/or (ii) the lack of hydrolysis process in NO_3^-, which could increase the anion vacancies in the ZnO lattice[29][30]. However, the strong (002) peaks' positions of the Cu-doped nanorods showed a slight shift toward a lower angle relative to the undoped nanorods. This shift is more significant for sample S3. On the other hand, previous research showed that at low concentrations (<1.5at.%) of Cu, the peak position is not significantly affected by Cu doping, while at high concentration, a slight shift towards higher angles is reported due to the substitution of Zn^{2+} (ionic radii = 0.074nm) by Cu^{2+} (ionic radii = 0.057nm)[30][31]. Additionally, these changes in crystallinity might be due to the changes in the atomic environment as a result of Cu incorporation into the ZnO lattice. It is evident that there is a slight lattice deformation in the Cu-ZnO lattice, which may be assigned to the diminishing CuZn-O bonds[32]. In this study, with up to 2% Cu concentration from the two precursors, neither the

Cu nor CuO phases are observed in the XRD measurements, which indicates that the Cu impurities are dissolved completely in the ZnO crystal lattice[26][30].

To explore more details about the influence of Cu precursors and the concentration on the crystal structure of the grown nanorods, Scherrer's equation[33] was used to estimate the crystallite size (D) of the nanorods along the (002) peak. From **Figure 2(a)**, the nanorods doped with 1 and 2at.% from $Cu(CH_3COO)_2$ (S2

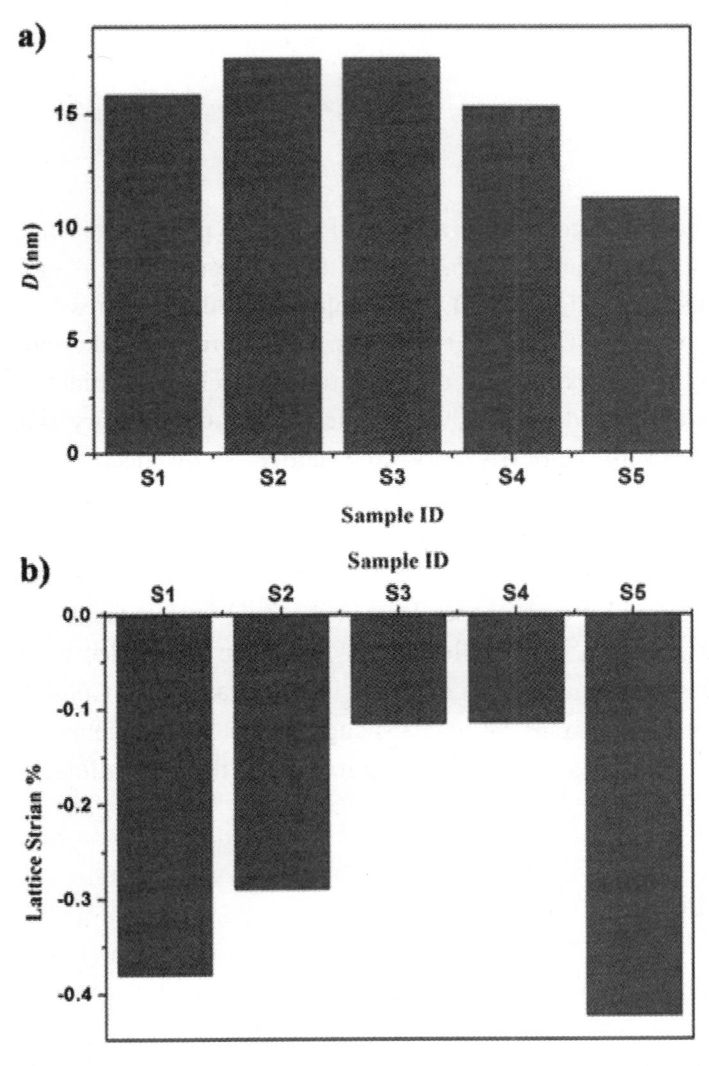

Figure 2. Crystallite size (a) and lattice strain (b) of undoped and Cu-doped ZnO nanorods.

and S3, respectively) showed higher crystallite size (D = 17.4nm) compared to the undoped nanorod (S1) (D = 15.8nm). When we use $Cu(NO_3)_2$ as the Cu precursor instead of $Cu(CH_3COO)_2$, the crystallite size decreases from 15.8nm (for the undoped nanorods) to 11.3nm (for sample S5). Clearly, the nanorods doped using $Cu(NO_3)_2$ (S4 and S5) had slightly smaller crystallite sizes relative to the ZnO nanorods doped using $Cu(CH_3COO)_2$ (S2 and S3). Such variations in the crystallite size might be the result of the changes in the host lattice parameters due to Cu incorporation[16][27]. The lattice strain of the undoped ZnO nanorods and the Cu-doped ZnO nanorods was calculated using Equation (1).

$$\varepsilon_z = \frac{c - c^{\circ}}{c^{\circ}} \times 100, \quad (1)$$

where c is the lattice constant (**Table 1**) of the ZnO nanorods calculated from the XRD measurements, and $c^{\circ} = 5.206$Å is the lattice constant of the standard unstrained ZnO. From **Figure 2(b)**, all samples showed a compressive strain. It appears that when $Cu(CH_3COO)_2$ is used as the Cu precursor, the lattice strain decreases with the increase in the Cu concentration, reaching its minimum (−0.115%) for the nanorods doped with 2at.% (sample S3). On the contrary, when $Cu(NO_3)_2$ is used instead of $Cu(CH_3COO)_2$, the lattice strain decreased significantly (−0.114%) for 1at.% Cu (S4) and increased to maximum when 2at.% is added (sample S5). It is evident that sample S4 [doped with 1at.% from $Cu(NO_3)_2$] showed a minimum lattice strain [**Figure 2(b)**]. This result suggests that the Cu dopants in sample S4 took proper sites in the ZnO lattice. Generally, the substitution of Zn^{2+} by Cu^{2+} would lead to a change in the lattice parameters[18][27]. However, the pronounced changes in the lattice strain when $Cu(NO_3)_2$ is used as the Cu precursor (samples S4 and S5) suggest that the concentration of OH^- in the aqueous solution plays an important role in the crystalline quality of the grown nanorods.

3.2. Morphology

The morphology of the nanorods was investigated by scanning electron microscopy. The top-view SEM images for the undoped and Cu-doped ZnO nanorods are shown in **Figure 3**. The density and diameters of the nanorods showed dependency on Cu precursor and concentration. It can be seen that the average rod

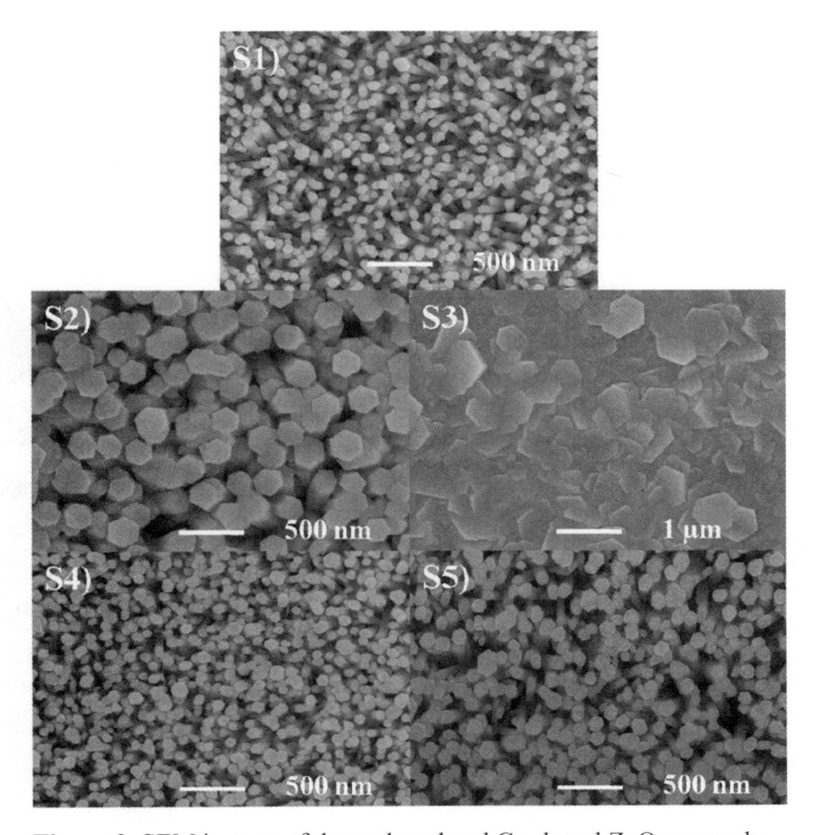

Figure 3. SEM images of the undoped and Cu-doped ZnO nanorods.

diameter increases from approximately 75nm for undoped nanorods (sample S1) to approximately 210nm when 1at.% Cu is added from $Cu(CH_3COO)_2$ (sample S2), while when 2at.% (sample S3) is added from the same precursor, the nanorods aggregated and the structure becomes compact. On the other hand, when 1at.% of Cu (sample S4) is added from $Cu(NO_3)_2$, the average nanorod diameter increases slightly relative to the undoped nanorods. Increasing the Cu content to 2at.% (sample S5) from $Cu(NO_3)_2$, the average nanorod diameter increases to approximately 120nm.

The variations in the nanorod diameters and densities as functions of Cu concentration and precursors are explained in **Figure 4(a)**, **Figure 4(b)**. The ZnO unit cell is shown in **Figure 4(a)**, where the cations (zinc ions) and the anions (oxygen ions) are arranged alternatively along the c-axis perpendicular to the substrate. Basically, the nanorod diameter and density are highly affected by the density

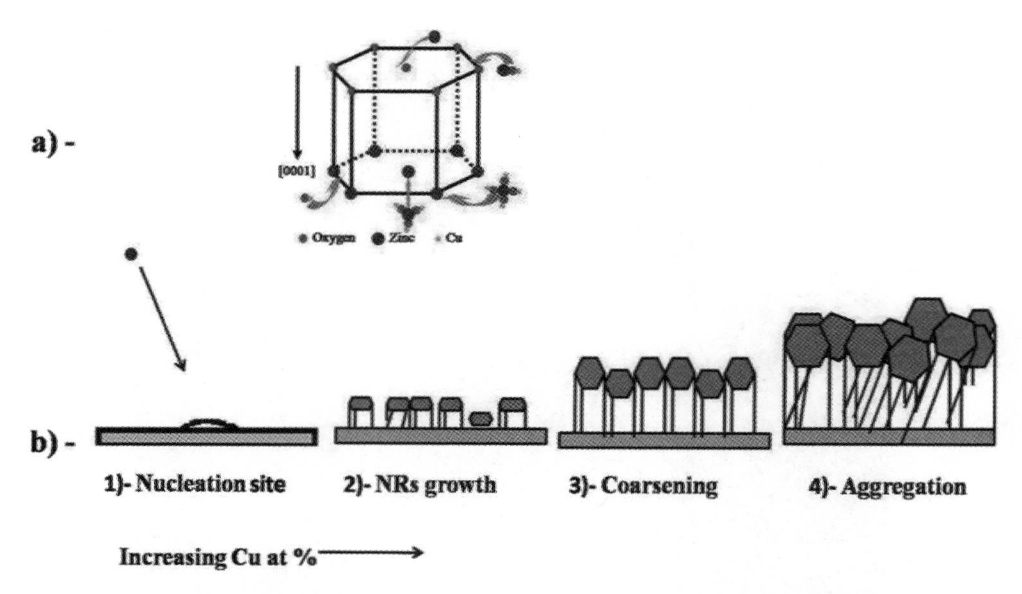

Figure 4. Schematics of ZnO unit cell (a) and nanorod growth and aggregation (b).

of the nucleation sites and the pH value of the aqueous solution. Therefore, introducing Cu dopants into the reaction path would increase the nucleation density and hence enhance the growth rate, which in turn, results in a coarsening and lateral aggregation of the nanorods.

The reason why the nanorods doped with $Cu(CH_3COO)_2$ exhibited a larger diameter compared to the nanorods doped with the same concentration of $Cu(NO_3)_2$ is that as shown in Equations (2) and (3), both $Cu(CH_3COO)_2$ and Cu $(NO_3)_2$ release the same concentration of Cu^{2+}. Therefore, the anion concentration is a determinant factor.

$$Cu(CH_3COO)_2 \rightarrow Cu^{2+} + CH_3COO^- \quad (2)$$

$$Cu(NO_3)_2 \rightarrow Cu^{2+} + 2NO_3 \quad (3)$$

The two different anions CH_3COO^- and NO_3^- will affect the nanorod growth process in different ways. In the hydrolysis process of CH_3COO^-, more OH^- will be released when the amount of OH^- in the aqueous solution decreases [Equation (4)]. Accordingly, both lateral nd vertical growth rates will increase with the increase of $Cu(CH_3COO)_2$.

$$CH_3COO^- + H_2O \leftrightarrow CH_3COOH + OH^- \quad (4)$$

Conversely, the lack of hydrolysis process in $Cu(NO_3)_2$ would lead to a low concentration of OH^-, which may slowdown the growth rate[34].

3.3. Photoluminescence

Room-temperature photoluminescence spectra of all the samples are shown in **Figure 5(a)**. All samples exhibited two dominant peaks. The first and sharpest

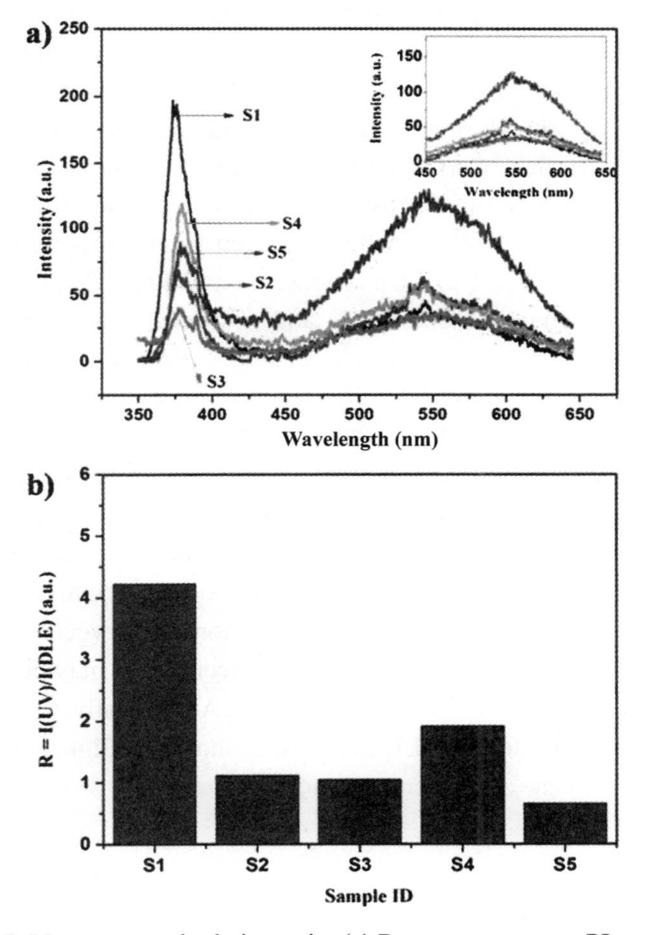

Figure 5. PL spectra and relative ratio. (a) Room-temperature PL spectra of undoped and Cu-doped ZnO nanorods; the inset shows the blue-green emission bands. (b) The relative ratio of PL intensity ($R = I$(UV)/I(DLE)).

peak is centered on 378nm and was assigned to the near-band edge (NBE) emission or to the free exciton emission. The intensity of the NBE emission decreases with the increase of Cu concentration for both precursors $Cu(CH_3COO)_2$ and $Cu(NO_3)_2$. This may have resulted from the formation of the nonradiative centers in the Cudoped samples[28]. In comparison between the two precursors, the nanorods doped with $Cu(NO_3)_2$ (samples S4 and S5) showed a higher NBE emission compared to the nanorods doped with $Cu(CH_3COO)_2$ (samples S2 and S3). This observation could be due to the higher anion concentration in samples S2 and S3[35]. The UV emission peak of the Cu-doped samples showed a small redshift (approximately 6nm) relative to the undoped ZnO, where the shift is clearer for the samples doped with $Cu(NO_3)$ (S4 and S5). This may be attributed to the rigid shift in the valence and the conduction bands due to the coupling of the band electrons and the localized Cu^{2+} impurity spin[16]. It can be observed that there is a small shoulder at around 390nm, and it becomes pronounced for sample 3, which is doped with 2at.% Cu from $Cu(CH_3COO)_2$, and this shoulder is ascribed to the free electron-shallow acceptor transitions[25][26]. Additionally, there is a luminescence peak at around 544nm, which is called the deep-level emission (DLE) or blue-green emission band. When 1at.% Cu is added from $Cu(CH_3COO)_2$, the intensity of this peak increased slightly (sample S2) and decreased again when 2at.% Cu is added from the same precursor (sample S3), becoming nearly identical with the undoped ZnO nanorods (sample S1). This result suggests that the green emission is independent of Cu concentration. On the other hand, when we use $Cu(NO_3)_2$ as the Cu source (samples S4 and S5), the green emission enhanced significantly for sample S5 (doped with 2at.%). Interestingly, the origin of the green emission is questionable because it has been observed in both un-doped and Cu-doped ZnO nanorod samples. Vanheusden *et al.*[36] attributed the green emission to the transitions between the photoexcited holes and singly ionized oxygen vacancies. Based on these arguments, the high oxygen vacancy concentration may be responsible for the higher green emission intensity of sample S5. Additionally, the ratio (R) of the NBE emission intensity to the DLE intensity is shown in **Figure 5(b)**. The R decreases with the increase of Cu concentration.

3.4. Transmittance

Figure 6 shows the total transmittance spectra for the undoped and Cu-doped ZnO nanorods, where all samples are found to be transparent in the visible

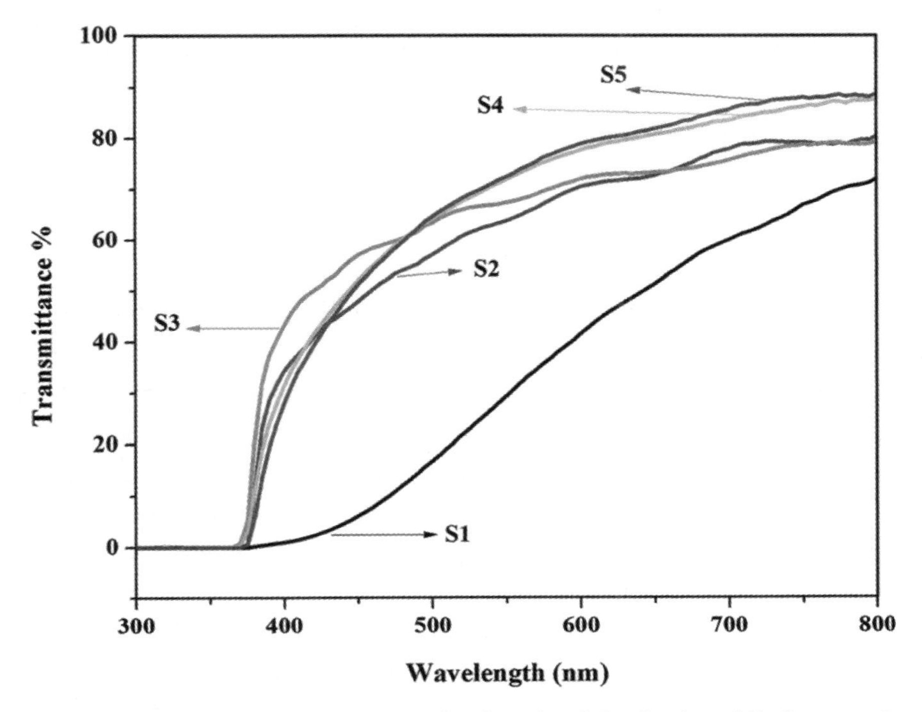

Figure 6. Total transmittance spectra of undoped and the Cu-doped ZnO nanorods.

region. It is evident that the rise of the absorption edge near the band edge for the pure ZnO nanorods (sample S1) increased gradually, while it becomes sharper for the Cu-doped ZnO nanorods (samples S2 to S5), indicating the presence of localized states within the bandgap. The un-doped ZnO nanorods (sample S1) showed lower transmittance (approximately 70%) compared to the Cu-doped ZnO nanorods. This could be attributed to the scattering either from the unfilled inter-columnar volume and voids or from the inclined nanorods. Using $Cu(CH_3COO)_2$ as the Cu source (samples S2 and S3), the total transmittance increased, reaching approximately 80%, and was found to be independent on the amount of Cu dopants. Comparatively, using $Cu(NO_3)_2$ as the Cu precursor (samples S4 and S5), the total transmittance increased further, reaching approximately 90%. Lin *et al.*[37] related the presence of oxygen vacancies to the transmittance ratio, where lower transmittance indicates that there are more oxygen vacancies and vice versa. However, in the study reported here, we can attribute the reduction in the total transmittance to the increase in the rod diameter for the samples doped with $Cu(CH_3COO)_2$. It can be seen that at the absorption edge for Cu-doped ZnO nanorods, the slight blue-shift indicates that the bandgap was tuned by the incorporation of the Cu dopants.

It may be observed that there are obvious interference fluctuations in the transmission spectra when $Cu(CH_3COO)_2$ was used as the Cu precursor (samples S2 and S3). These fluctuations can be attributed to the presence of scattering centers[36].

4. Conclusions

In conclusion, we explored the effect of Cu precursors ($Cu(CH_3COO)_2$ and $Cu(NO_3)_2$) and concentration on the structural, morphological, and optical properties of the hydrothermally synthesized Cu-doped ZnO nanorods. The XRD results revealed that the slight changes in the lattice parameters have occurred due to the substitution of Zn^{2+} by Cu^{2+} and the formation of defect complexes. The nanorods doped with $Cu(NO_3)_2$ had less crystallinity than the nanorods doped with $Cu(CH_3COO)_2$, where the maximum compressive lattice strain (−0.423%) was obtained when 2at.% of Cu was added from $Cu(NO_3)_2$. From the SEM studies, $Cu(CH_3COO)_2$ was found to be an effective precursor for the formation of Cu-doped ZnO nanorods with large diameter. Conversely, Cu-doped ZnO nanorods with a small diameter (approximately 120nm when 2at.% was added) can be obtained when $Cu(NO_3)_2$ is used as a Cu precursor due to the lack of hydrolysis process. UV and green emission peaks at 378 and 544nm were observed for all samples and are attributed to the near-band edge UV emission and the defect-related emission, respectively. A redshift of approximately 6nm in the UV emission band was seen for the Cu-doped ZnO nanorods and was attributed to the rigid shift in the valence and the conduction bands due to the coupling of the band electrons and the localized Cu^{2+} impurity spin. Irrespective of Cu concentration, the nanorods doped with Cu $(CH_3COO)_2$ showed a transmittance of approximately 80% in the visible range, while the nanorods doped with $Cu(NO_3)_2$ showed a rather high transmittance (approximately 90%). The obtained results are comparable with the previous results. In conclusion, by choosing a suitable Cu precursor and concentration, we can control the diameter of Cu-doped ZnO nanorods, which is important for the fabrication of nano-optoelectronic devices.

Competing Interests

The authors declare that they have no competing interests.

Authors' Contributions

MB fabricated all the samples, performed the XRD and transmission measurements, and wrote the manuscript. DW performed the PL and FESEM measurements. JW participated in the discussion and manuscript writing. JS and QL contributed in the preparation of some samples. YY, QY, and SJ contributed with valuable discussions. All authors read and approved the final manuscript.

Authors' Information

MB obtained his MSc degree in nanoscience from Lund University, Sweden. He is currently a Ph.D. student in Harbin Institute of Technology. His research interests include fabrication and properties of metal-doped ZnO nanostructures.

DW is an MSc student in Harbin Institute of Technology. His research interests include fabrication and properties of ZnO thin films. JW obtained his Ph.D. degree from Jilin University. He is currently a full professor at Harbin Institute of Technology. His research interests cover pure and doped ZnO nanomaterials, solar cell, and optoelectronic devices. QL is an MSc student at Harbin Institute of Technology. Her research interests include fabrication and properties of p-type ZnO thin films. JS is an MSc student in Harbin Institute of Technology. His research interests include fabrication and properties of ZnO UV detectors. YY obtained his MSc degree in engineering from Harbin Institute of Technology. He is currently a Ph.D. student in Harbin Institute of Technology. His research interests include fabrication and properties of metal oxide solar cells. QY is currently a full professor at Harbin Institute of Technology. His research interests cover metal oxide nanomaterials, solar cell, and gas sensors. SJ is currently a full professor at Harbin Institute of Technology. Her research interests cover pure and doped ZnO nanomaterials.

Acknowledgements

This work has been partly supported by the Program for New Century Excellent Talents in University (NCET-10-0066), an 863 project grant (2013AA031502), and Project No. 2011RFLXG006.

Source: Babikier M, Wang D, Wang J, *et al*. Cu-doped ZnO nanorod arrays: the effects of copper precursor and concentration[J]. Nanoscale Research Letters, 2014, 9(1):1–40.

References

[1] Li Y, Gong J, Deng Y: Hierarchical structured ZnO nanorods on ZnO nanofibers and their photoresponse to UV and visible lights. Sensor Actuat A: Phys 2010, 158: 176–182.

[2] Lao CS, Liu J, Gao P, Zhang L, Davidovic D, Tummala R, Wang ZL: ZnO nanobelt/nanowire Schottky diodes formed by dielectrophoresis alignment across Au electrodes. Nano Lett 2006, 6:263–266.

[3] Bender M, Fortunato E, Nunes P, Ferreira I, Marques A, Martins R, Katsarakis N, Cimalla V, Kiriakidis G: Highly sensitive ZnO ozone detectors at room temperature. Jpn J Appl Phys 2003, 42:435–437.

[4] Fortunato E, Gonçalves A, Pimentel A, Barquinha P, Gonçalves G, Pereira L, Ferreira I, Martins R: Zinc oxide, a multifunctional material: from material to device applications. Appl Phys A 2009, 96:197–205.

[5] Stolt L, Hedstrom J, Kessler J, Ruckh M, Velthaus KO, Schock HW: ZnO/CdS/CuInSe_2 thin film solar cells with improved performance. Appl Phys Lett 1993, 62: 597–599.

[6] Lupan O, Pauporté T, Le Bahers T, Viana B, Ciofini I: Wavelength emission tuning of ZnO nanowire based light emitting diodes by Cu doping: experimental and computational insights. Adv Funct Mater 2011, 21:3564–3572.

[7] Jiang S, Ren Z, Gong S, Yin S, Yu Y, Li X, Xu G, Shen G, Han G: Tunable photoluminescence properties of well-aligned ZnO nanorod array by oxygen plasma post-treatment. Appl Surf Sci 2014, 289:252–256.

[8] Lin K-F, Cheng H-M, Hsu H-C, Hsieh W-F: Band gap engineering and spatial confinement of optical phonon in ZnO quantum dots. Appl Phys Lett 2006, 88: 263113–263117.

[9] Wang ZL: Zinc oxide nanostructures: growth, properties and applications. Condens Matter Phys 2004, 16:829–857.

[10] Choi MY, Choi D, Jin MJ, Kim I, Kim SH, Choi JY, Lee SY, Kim JM, Kim SW: Mechanically powered transparent flexible charge generating nanodevices with piezoelectric ZnO nanorods. Adv Mater 2009, 21:2185–2189.

[11] Huo K, Hu Y, Fu J, Wang X, Chu PK, Hu Z, Chen Y: Direct and large-area growth of one-dimensional ZnO nanostructures from and on a brass substrate. J Phys Chem C 2007, 111:5876–5881.

[12] Snure M, Tiwari A: Band-gap engineering of Zn1-xGaxO nanopowders: synthesis, structural and optical characterizations. J Appl Phys 2008, 104:073707–5.

[13] Wang X, Song C, Geng K, Zeng F, Pan F: Photoluminescence and Raman scattering of Cu-doped ZnO films prepared by magnetron sputtering. Appl Surf Sci 2007, 253:6905–6909.

[14] Zhang Z, Yi JB, Ding J, Wong LM, Seng HL, Wang SJ, Tao JG, Li GP, Xing GZ, Sum TC: Cu-doped ZnO nanoneedles and nanonails: morphological evolution and physical properties. J Phys Chem C 2008, 112:9579–9585.

[15] Ding J, Chen H, Zhao X, Ma S: Effect of substrate and annealing on the structural and optical properties of ZnO:Al films. J Phys Chem Solids 2010, 71:346–350.

[16] Muthukumaran S, Gopalakrishnan R: Structural, FTIR and photoluminescence studies of Cu doped ZnO nanopowders by co-precipitation method. Opt Mater 2012, 34:1946–1953.

[17] Yamada T, Miyake A, Kishimoto S, Makino H, Yamamoto N, Yamamoto T: Effects of substrate temperature on crystallinity and electrical properties of Ga-doped ZnO films prepared on glass substrate by ion-plating method using DC arc discharge. Surf Coat Technol 2007, 202:973–976.

[18] Lupan O, Pauporté T, Viana B, Aschehoug P: Electrodeposition of Cu-doped ZnO nanowire arrays and heterojunction formation with p-GaN for color tunable light emitting diode applications. Electrochim Acta 2011, 56:10543–10549.

[19] Dingle R: Luminescent transitions associated with divalent copper impurities and the green emission from semiconducting zinc oxide. Phys Rev Lett 1969, 23:579–581.

[20] Yan Y, Al-Jassim M, Wei S-H: Doping of ZnO by group-IB elements. Appl Phys Lett 2006, 89:181912–181913.

[21] Kanai Y: Admittance spectroscopy of Cu-doped ZnO crystals. J Appl Phys 1991, 30:703–707.

[22] Xu C, Sun X, Zhang X, Ke L, Chua S: Photoluminescent properties of copper-doped zinc oxide nanowires. Nanotechnology 2004, 15:856–861.

[23] Kim J, Byun D, Ie S, Park D, Choi W, Choi J-W, Angadi B: Cu-doped ZnO-based p–n hetero-junction light emitting diode. Semicond Sci Technol 2008, 23:095004.

[24] Herng T, Lau S, Yu S, Tsang S, Teng K, Chen J: Ferromagnetic Cu doped ZnO as an electron injector in heterojunction light emitting diodes. Appl Phys 2008, 104: 103104–103106.

[25] Yang J, Fei L, Liu H, Liu Y, Gao M, Zhang Y, Yang L: A study of structural, optical and magnetic properties of Zn0.97−xCuxCr0.03O diluted magnetic semiconductors.

J Alloys Compd 2011, 509:3672–3676.

[26] Aravind A, Jayaraj M, Kumar M, Chandra R: Optical and magnetic properties of copper doped ZnO nanorods prepared by hydrothermal method. J Mater Sci: Mater Electron 2013, 24:106–112.

[27] Wang S-F, Tseng T-Y, Wang Y-R, Wang C-Y, Lu H-C: Effect of ZnO seed layers on the solution chemical growth of ZnO nanorod arrays. Ceram Int 2009, 35:1255–1260.

[28] Chow L, Lupan O, Chai G, Khallaf H, Ono L, Roldan Cuenya B, Tiginyanu I, Ursaki V, Sontea V, Schulte A: Synthesis and characterization of Cu-doped ZnO one-dimensional structures for miniaturized sensor applications with faster response. Sensor Actuat A: Phys 2013, 189:399–408.

[29] West C, Robbins D, Dean P, Hayes W: The luminescence of copper in zinc oxide. Physica B+C 1983, 116:492–499.

[30] Kim AR, Lee J-Y, Jang BR, Lee JY, Kim HS, Jang NW: Effect of Zn2+ source concentration on hydrothermally grown ZnO nanorods. J Nanosci Nanotechnol 2011, 11:6395–6399.

[31] Kumar S, Koo B, Lee C, Gautam S, Chae K, Sharma S, Knobel M: Room temperature ferromagnetism in pure and Cu doped ZnO nanorods: role of copper or defects. Func Mater Lett 2011, 4:17–20.

[32] Gao D, Xue D, Xu Y, Yan Z, Zhang Z: Synthesis and magnetic properties of Cu-doped ZnO nanowire arrays. Electrochim Acta 2009, 54:2392–2395.

[33] Ma Q, Buchholz DB, Chang RP: Local structures of copper-doped ZnO films. Phys Rev B 2008, 78:214429.

[34] Amin G, Asif M, Zainelabdin A, Zaman S, Nur O, Willander M: Influence of pH, precursor concentration, growth time, and temperature on the morphology of ZnO nanostructures grown by the hydrothermal method. J Nanomater 2011, 2011:269692.

[35] Sanon G, Rup R, Mansingh A: Growth and characterization of tin oxide films prepared by chemical vapour deposition. Thin Solid Films 1989, 190:287–301.

[36] Vanheusden K, Warren W, Seager C, Tallant D, Voigt J, Gnade B: Mechanisms behind green photoluminescence in ZnO phosphor powders. J Appl Phys 1996, 79:7983–7990.

[37] Lin S, Hu H, Zheng W, Qu Y, Lai F: Growth and optical properties of ZnO nanorod arrays on Al-doped ZnO transparent conductive film. Nanoscale Res Lett 2013, 8: 158–163.

Chapter 27

Electromagnetic Enhancement of Graphene Raman Spectroscopy by Ordered and Size-Tunable Au Nanostructures

Shuguang Zhang[1], Xingwang Zhang[2], Xin Liu[2]

[1]State Key Laboratory of Luminescent Materials and Devices, South China University of Technology, Guangzhou 510641, China
[2]Key Lab of Semiconductor Materials Science, Institute of Semiconductors, CAS, Beijing 100083, China

Abstract: The size-controllable and ordered Au nanostructures were achieved by applying the self-assembled monolayer of polystyrene microspheres. Few-layer graphene was transferred directly on top of Au nanostructures, and the coupling between graphene and the localized surface plasmons (LSPs) of Au was investigated. We found that the LSP resonance spectra of ordered Au exhibited a redshift of ~20nm and broadening simultaneously by the presence of graphene. Meanwhile, the surface-enhanced Raman spectroscopy (SERS) of graphene was distinctly observed; both the graphene G and 2D peaks increased induced by local electric fields of plasmonic Au nanostructures, and the enhancement factor of graphene increased with the particle size, which can be ascribed to the plasmonic coupling between the ordered Au LSPs and graphene.

Keywords: Surface Plasmons, Graphene, Ordered Au Nanostructures, Raman Scattering

1. Background

Graphene is the first two-dimensional carbon atomic crystal which is constructed by several layers of honeycomb-barrayed carbon atoms. This promising material is highly attractive for the fabrication of high-frequency nano-electronic and optoelectronic devices due to its exceptional optical and electrical properties, such as extreme mechanical strength, ultrahigh electrical carrier mobility, and very high light transmittance[1]–[3]. Unfortunately, the graphene of only one-atomic-layer thickness exhibits lower light absorption (only ~2.3% for a single layer) originating from the weak light-graphene interaction, which is unfavorable for high-performance graphene-based optoelectronic devices. Several approaches have been proposed to enhance the absorption of graphene, including using one-dimensional photonic crystal and localized surface plasmons (LSPs)[4][5]. LSPs in conventional systems are the collective oscillations of conduction electrons in the metal nanoparticles when illuminated and excited by light with appropriate wavelength, and the resonance excitation of the LSPs induces a large enhancement and confinement of the local electric field in the vicinity of the metal nano-structures. Generation of LSPs stimulates a wide range of applications such as ultratrace biochemical sensing, enhanced absorption in photovoltaic cells, surface plasmon-enhanced fluorescence, and Raman scattering. From a spectroscopic point of view, surface-enhanced Raman spectroscopy (SERS) has become a promising spectacular application of plasmonics especially for the graphene-LSP hybrid system. On the other hand, the two-dimensional nature of graphene and its well-known Raman spectrum make it a favorable test bed for investigating the mechanisms of SERS, and various nanoparticle geometries have proven to deliver a considerable Raman enhancement in the case of graphene[6]–[11]. A Raman enhancement of 103 times had been detected for graphene from the dimer cavity between two closely packed Au nanodisks. However, fabrication and space control of the Au nanodisks are very complex and costly[6]. Sun *et al.* deposited Ag on the surface of a graphene film, and distinct Raman enhancement had been achieved. However, the quality of graphene significantly deteriorated after deposition of metal nanoparticles which limits the further application of graphene[8].

Except for enhancing intensity of Raman scattering of graphene by LSPs, graphene had also been adopted to tune the surface plasmons resonance wavelength of metal nanostructures. For instance, the plasmonic behavior of Au nanoparticles can be tuned by varying the thickness of the Al_2O_3 spacer layer inserted between the graphene and nanoparticles[12]. Nevertheless, the Au nanoparticles are randomly distributed on the surface of the Al_2O_3 layer which is unfavorable for the precise controllability and investigation of the inter-coupling of the graphene-metal hybrid system. Obviously, combination of enhanced near-fields of ordered plasmonic nanostructures with unusual optoelectronic properties of graphene will provide a more promising application for novel graphene-based optoelectronic devices, and thus, research on the plasmonic coupling between graphene and plasmonic ordered nanoparticles is highly desirable.

In this paper, ordered and size-controlled Au nanostructures were fabricated using the inverted self-assembled monolayer template of polystyrene microspheres. A chemical vapor deposition (CVD) graphene was transferred directly on top of Au nanostructures, and the interaction between graphene and LSPs of Au has been systematically investigated. We found that the SERS of graphene was apparently observed and Raman intensities of both the graphene G and 2D peaks increased with the size of Au which was induced by the local electric field of plasmonic Au nanostructures. On the other hand, the absorption spectra of Au nanostructures exhibited a redshift of ~20nm and a slight broadening by the presence of graphene, which was due to the inter-coupling between Au LSPs and graphene.

2. Methods

2.1. Sample Preparation

Colloidal microspheres of polystyrene (PS) (2.5wt.%) with a diameter of 500nm were purchased from Alfa Aesar and self-assembled to form a hexagonal close-packed (hcp) monolayer on the SiO_2/Si or quartz substrates. Prior to the hcp alignment of the PS microspheres, the target substrates were firstly immersed in piranha solution (98% H_2SO_4:37% H_2O_2 = 7:3) for 3h to achieve a completely hydrophilic surface, which contributed to the adhesion of PS microspheres to the substrates surface. Two parallel hydrophilic Si wafers with a distance of ~100μm were mounted on the dip coater, and two or three drops of the PS sphere suspension were dropped into the gap of the Si wafers. With one Si substrate fixed, the other parallel Si was lifted with a constant speed of approximately 500μm/s. The monolayer PS microspheres were ultimately formed on the hydrophilic surface of Si. Then, the PS monolayer was used as a template for the deposition of the Au film. The Au films were deposited on the PS template using the ion beam-assisted deposition (IBAD) system with a Kaufmann ion source. The size and shape of Au nanostructures were adjusted by varying the nominal thicknesses of the initial Au films ranging from 15 to 40nm. After ultrasonic washing in acetone for 30min, the PS microspheres together with the upper Au on them were completely removed and the ordered and sizetunable bottom Au nanostructures were formed. The few-layer graphene (FLG) films were synthesized on the Cu foils by low-pressure CVD in a tubular quartz reactor, using methane as the carbon source under H_2 and Ar atmosphere at 1000°C. Then, they were transferred to cover the Au nanostructures after the Cu foils were removed by wet chemical etching. The schematic illustration of fabrication processes for the ordered Au nanostructures with graphene coverage is shown in **Figure 1**.

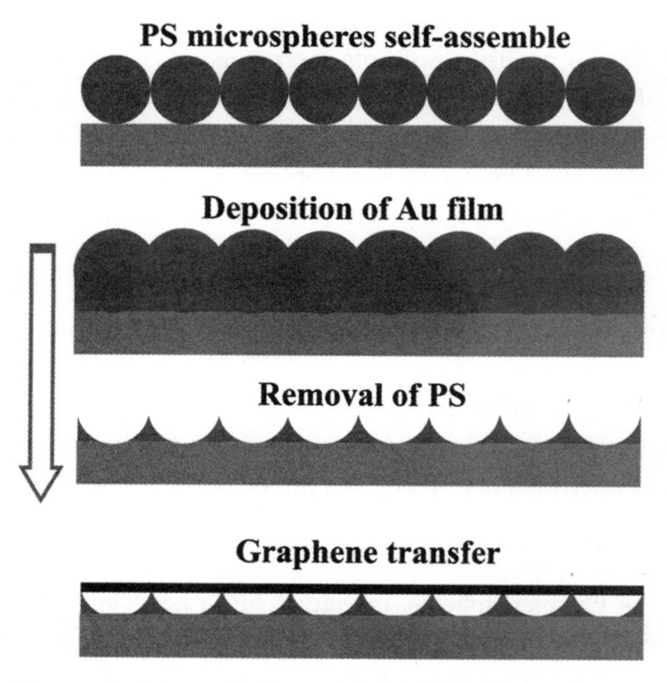

Figure 1. Schematic illustration of fabrication processes for the ordered Au nanostructures with graphene coverage.

2.2. Characterization

Surface morphologies of the ordered Au nanostructures with various shapes and sizes were characterized by atomic force microscopy (AFM, NT-MDT solver P47). The Au nanostructures with graphene coverage were investigated by field-emission scanning electron microscopy (SEM, Hitachi FE-S4800). Raman spectra of graphene were recorded with a Horiba LabRAM HR800 spectrometer using the 514-nm excitation line from an Ar ion laser. For both SEM and Raman measurements, the SiO_2/Si substrates were adopted. The ultraviolet (UV)-visible absorption spectra of the ordered Au nanostructures were measured as a function of the incident wavelength using a Varian Cary 5000 spectrophotometer in a double-beam mode. The quartz substrate was adopted for the UV-vis measurement of the Au nanostructures with and without graphene, respectively.

3. Results and Discussion

Figure 2 shows the AFM images of Au nanostructures with different sizes and shapes after removing the PS microspheres. Obviously, the long-range hexagonal order inserted from the original PS template is conserved. When the initial thickness of the Au film is 15nm, the shape of the Au nanostructures is sphere-like nanoparticles, with an average width of ~100 ± 4.5nm. Another point we have to notice from the SEM images is that the shape of the Au nanostructures gradually changed from more sphere-like nanodots to sharp triangles with the increase of the initial thickness of the Au film from 15 to 20nm. The effect of the shape changes was also quantified by evaluating the circularities of the individual nanostructures, defined as the ratio of the square of the perimeter to $4\pi A$, where A is the area of the particular nanostructure. The circularity should be 1.654 for a regular triangle and should approach 1 for a perfect circle[13]. The resultant values of the Au nanostructures with and without graphene coverage are listed in **Table 1**. From the table, we can distinctly see that with the deposition time increasing from 10min to 30min (corresponding initial thickness from 15 to 40nm), the circularity of the Au nanostructures first increases and then decreases, and the average width changes from 100 ± 4.5 to 140 ± 7.8nm. As the diameter of the PS sphere is 500nm, the gap (inscribed circle) among the PS spheres is approximately 77nm. When the initial thickness of the Au film is only 15nm, the Au film cannot cover the whole

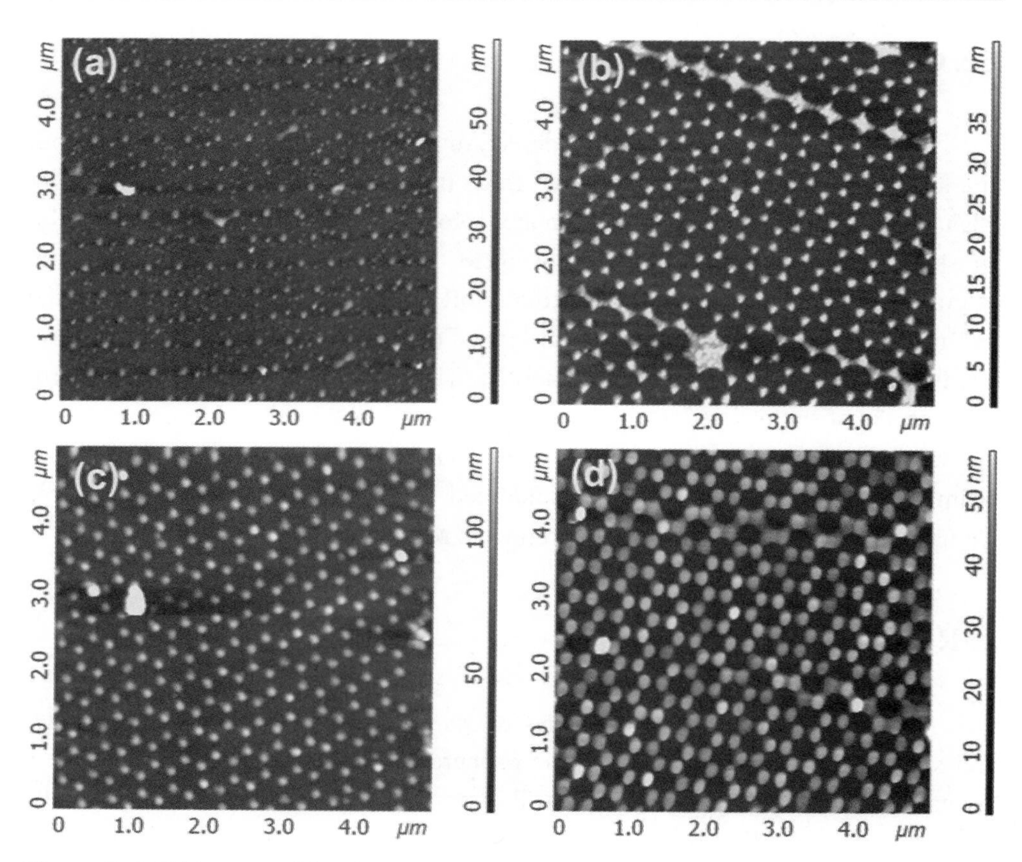

Figure 2. AFM images of the Au nanostructures with the initial thicknesses of (a) 15nm, (b) 20nm, (c) 30nm, and (d) 40nm.

Table 1. Circularity, width, and resonance position of Au LSP with and without graphene coverage as a function of the initial thicknesses of the Au film.

Initial thickness of Au film (nm)	Circularity	Width (nm)	LSPR without graphene (nm)	LSPR with graphene (nm)
15	1.09 ± 0.02	100 ± 4.5	530	554
20	1.21 ± 0.01	120 ± 5.4	542	562
30	1.13 ± 0.02	128 ± 6.7	547	564
40	1.06 ± 0.04	140 ± 7.8	553	570

gap surface. Due to the different thermal expansion coefficients between the Au films and the substrate, when the initial thickness of the Au film is ~15nm, the compressive stress induced by the Ostwald ripening mechanism would cause the

Ag films to form isolated nanoparticles. With the increase of the initial thickness of the Au film to 20nm, the whole gap among the PS nanospheres can be approximately filled, leading to shape transformation of the Au nanostructures from nanodots to triangles. When the initial thickness of Au was further increased to 40nm, the whole gap of the PS spheres can be fully filled and the shape of the Au nanostructures changes to nanospheres due to the larger thickness of Au. The typical SEM images of 15- and 20-nm Au nanostructures covered with graphene are presented in **Figure 3**. It is clear that the continuous graphene film has been successfully transferred on the surface of Au nanostructures, and the electron beam can easily penetrate through the atomically thin graphene to display the underlying Au nanostructures. The ridges and cracks formed on the graphene surface during the wet transfer processes can be also distinctly observed.

To investigate effects of graphene on the plasmonic Au nanostructures, LSP properties of Ag nanostructures with and without graphene coating were characterized by a UV-visible absorption spectrophotometer. As shown in **Figure 4**, the absorption spectra of the Au nanostructures on quartz substrates exhibit a wide plasmonic resonance peak varying from sample to sample, indicating that the LSP resonance of Au can be tuned by adjusting the size and shape of the Au nanostructures. The absorption intensity after graphene coverage increases slightly which can be ascribed to the absorption of the graphene film. Meanwhile, an obvious redshift and broadening of the Au LSP peak can also be clearly observed for all samples with graphene coating. Resonance positions of Au LSP before and after graphene coverage are also summarized in **Table 1**, and we can clearly see that the Au LSP resonances exhibit an ~20-nm redshift. On the other hand, the full width

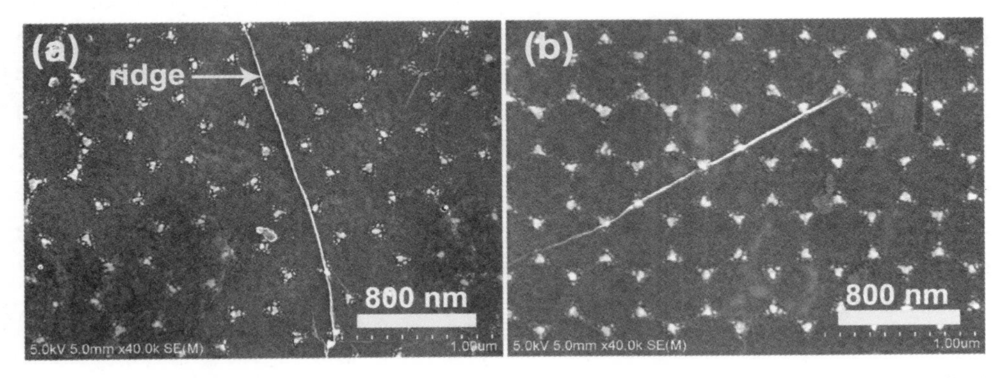

Figure 3. Typical SEM images of the Au nanostructures with initial thicknesses of (a) 20nm (b) 30nm covered by the graphene film.

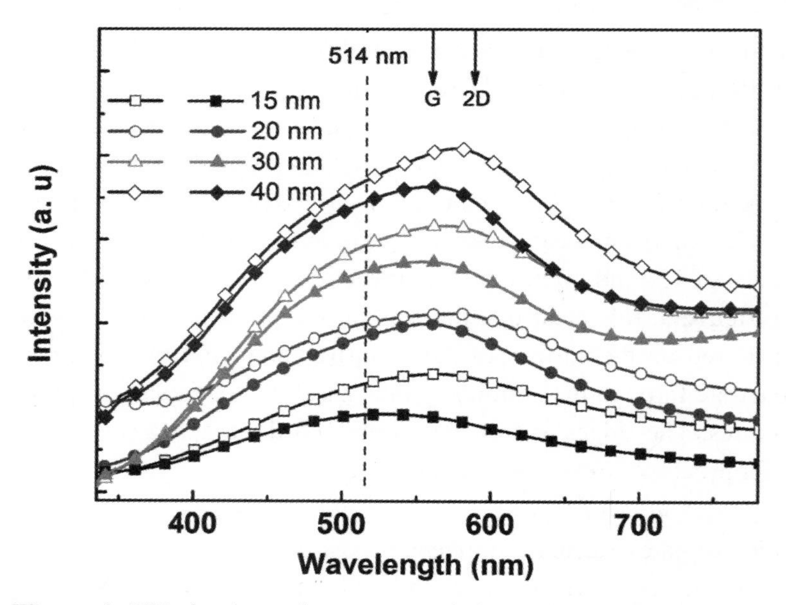

Figure 4. UV-vis absorption spectra of the various-sized Au nanostructures with (open symbols) and without (solid symbols) graphene coating. The Raman excitation wavelength of 514nm is shown as a vertical dashed line, together with the corresponding wavelength of the G and 2D modes of graphene.

at half maximum (FWHM) of Au nanostructures with graphene also shows a distinct broadening compared with their counterparts without graphene. As the sizes of the nanostructures are smaller than the wavelength of incident light, the quasistatic model can be used to describe the position of the resonance peak. When irradiated by light, the conduction electrons will move and resonate with a specific frequency which is referred to as plasmon frequency of the particle dipole. According to the quasistatic analysis, the absorption peak corresponds to the dipole surface plasmon resonance for the ordered Au nanostructures on the quartz substrates[14]. When the particle size increases, the conduction electrons cannot all move in phase anymore, that is to say, the quadrupole resonances may occur except for the dipoles. The optical spectra of Au nanoprisms display in-plane dipole and quadrupole resonances. And the interaction of the dipole and quardrupole leads to a reduction of the depolarization field, which is the origin of the redshift of Au LSP resonances[14].

The above results also indicate that the Au LSPs are also strongly affected by the presence of graphene, which can be attributed to the coupling between the graphene film and the localized electromagnetic field of the Au nanostructures. As the incident light is perpendicular to the surface of the Au nanostructures, the incident electric field is parallel to the sample surface and has no vertical component, and only the lateral electron oscillations within the Au nanostructures can be induced. When the LSPs of Au nanostructures with graphene coating are excited, the image dipoles or quadrupoles within the graphene sheet which are antiparallel to the dipoles or quadrupoles in Au will be formed[5][12]. The presence of the antiparallel image dipoles and quadrupoles can reduce the internal electric field in the Au nanostructures, which results in the redshift and broadening of LSP resonance peaks for the Au nanostructures with graphene.

Theoretical calculation based on the dipole approximation (results not shown here) has been conducted to understand the redshift of Au LSPs after graphene coverage. In the calculation, the Au nanosphere is utilized as a representative of the Au nanostructures and is placed above a transparent substrate. The thickness of the substrate is assumed to be semi-infinite, and absorption of the substrate is completely omitted. The dielectric constant of graphene is based on an assumption that the optical response of every graphene layer is given by optical sheet conductivity[5][9], and the dielectric constant of Au can be found in the literature[15]. The graphene sheet covers the surface of the Au nanosphere, which is treated as a single dipole. Considering presence of the antiparallel dipole in graphene, the polarizability α of the Au nanosphere can be written as[9][16]

$$\alpha = 4\pi a^3\left(\frac{\varepsilon_1-\varepsilon_2}{\varepsilon_1+2\varepsilon_2}\right)\left[1-\beta\left(\frac{a}{2d}\right)^3\left(\frac{\varepsilon_1-\varepsilon_2}{\varepsilon_1+2\varepsilon_2}\right)\left(\frac{\varepsilon_3-\varepsilon_2}{\varepsilon_3+\varepsilon_2}\right)\right]^{-1} \tag{1}$$

where α is the polarizability of Au nanosphere, β is 1 for the lateral electric field, and d is the distance between the center of the Au nanosphere and the graphene sheet. ε_1, ε_2, and ε_3 are the dielectric constants of Au, ambient, and graphene, respectively[16]. The absorption efficiency of Au Q_{abs} can be expressed as $Q_{abs} = [k/\pi a^2]\mathrm{Im}(\alpha)$. The calculated results of Q_{abs} for a single 30-nm Au nanosphere with and without graphene coating show that the absorption peak of the Au nanosphere exhibits a slight redshift (about 20nm) after graphene coating, which is in good agreement with the experimental data. The discrepancy of the absorption results between calculation and experimental results can be ascribed to the simplified assumptions during the calculation such as the semi-infinite substrate, the dipole approximation, the homogeneous particle size, and the crystal perfection of the Au nanosphere[5][10].

To shed light on the effects of plasmonic ordered Au nanostructures on the Raman properties of graphene, the Stokes Raman spectra of pristine graphene and grapheme transferred on differently shaped and sized Au nanostructures were measured and the corresponding results are shown in **Figure 5**. The Raman spectrum of the pristine graphene reveals the well-known D peak (1353cm^{-1}), G peak (1590cm^{-1}), and 2D peak (2690cm^{-1}). The G peak originates from the first-order Raman scattering process, while the 2D peak is due to a double-resonance inter-valley Raman scattering process[17][18]. The nearly negligible intensity of the D peak indicates few structural and crystalline defects in the CVD graphene. The Raman peak of grapheme on top of the 100-nm Au nanostructures exhibits almost the same amplitude with its counterpart on the SiO_2/Si substrate. It is apparently revealed that the intensities of graphene G and 2D peaks are significantly enhanced with increasing size of the ordered Au nanostructures from 100 to 140nm. With increasing size of the ordered Au nanostructures to 140nm, the Raman peak intensities of graphene on top of Au display an enhancement factor of ~3-fold for

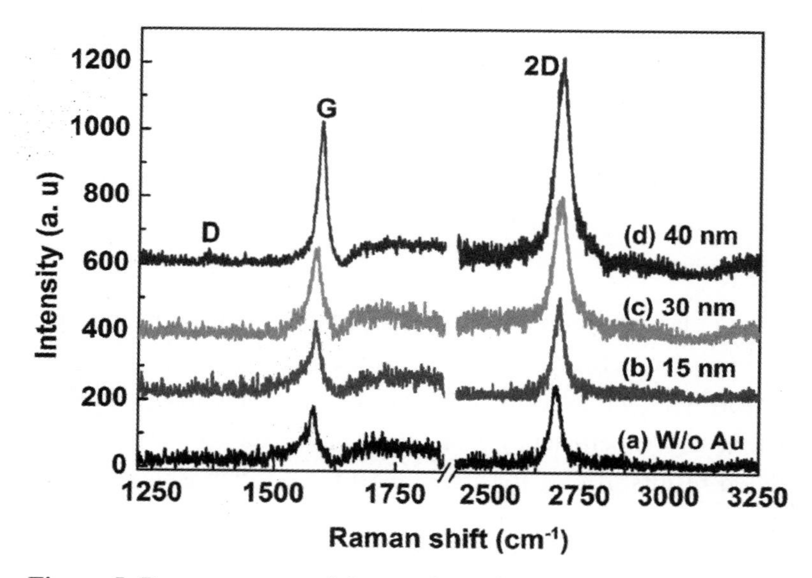

Figure 5. Raman spectra of the graphene films (a) without Au nanostructures and on the surface of the ordered Au nanostructures with initial thicknesses of (b) 15nm, (c) 30nm, and (d) 40nm.

the G peak and 2.5 for the 2D peak. Therefore, the SERS of graphene caused by the ordered Au nanostructures was distinctly observed, and the underlying mechanism will be discussed in detail later.

Generally, both the LSP and charge transfer can play a role in the SERS of graphene modes, since the SERS electromagnetic enhancement results from the Raman excitation and coupling with the LSP of nanostructures, while the metal-induced charge transfer can lead to the chemical resonance enhancement[19]. In our case, the enhanced Raman intensity of graphene/Au nanostructures is mainly attributed to the electromagnetic field enhancement induced by the plasmonic resonance of Au LSP. As illustrated in **Figure 5**, the enhancement factor of both G and 2D Raman peaks increase with the size of the Au nanostructures, which is an indication of the electromagnetic mechanism. The rather qualitative image presented in **Figure 5** coincides well with a theoretical model that has been proposed for Raman enhancement in the graphene-nanoparticle hybrid[10][20]. According to this model, the Raman enhancement due to the stand-alone nanoparticle is given by

$$\frac{\Delta I_{\mathrm{SERS}}}{I_0} \approx \frac{3}{28}\sigma Q^2(\omega) Q^2(\omega_s)\left(\frac{a}{h}\right)^{10} \tag{2}$$

where ΔI_{SERS} is the increase in Raman intensity with respect to its original intensity I_0, σ is the relative cross-sectional area of the nanoparticle, $Q(\omega)$ is the plasmonic enhancement from the Mie theory with ω_s representing the Stokes Raman frequency, α is the particle radius, and h represents the separation between the particle center and the surface of the graphene sheet. From Equation (2), we can easily deduce that the Raman enhancement scales with the cross section of the metallic nanostructures, with the fourth power of the Mie enhancement, and inversely with the tenth power of the distance between the graphene and nanoparticle center. Therefore, the Raman scattering enhancement with increasing particle size (as shown in **Figure 5**) can be mainly ascribed to the plasmonic absorption profile $Q(\omega)$ of the nanoparticle. The shapes of the particles vary with different initial thicknesses of the Au film, although majority of Au nanostructures can be treated as spheres or triangle prisms. Experiments of the absorption spectra have shown that the plasmonic resonance positions for the ordered Au nanostructures with various size ranges from 100 to 140nm lie around the excitation laser wavelength

(514nm, dotted line in **Figure 4**); therefore, the incident laser beam excites the Au LSPs which will form a strong localized electromagnetic field around the nano-structures. As the graphene sheet is in close vicinity of the Au nanostructures, the electric field will penetrate into the graphene sheet, and an enhanced electromagnetic field will be formed on the graphene surface, although the antiparallel image dipole will form and reduce the local electric field around the Au nanostructures[12]. Ultimately, the $Q(\omega)$ in Equation (2) will be significantly improved, and the SERS signal is greatly enhanced. Moreover, the enhancement factor of the G and 2D peaks gradually increases with the size of the Au nanostructures, as shown in **Figure 5**, which can be interpreted as follows: The corresponding wavelengths of the G and 2D peaks of graphene are consistent with those of the Au LSP resonances (see **Figure 4**). Therefore, the enhanced local electromagnetic field induced by Au LSPs contributes to the improved Raman signal. On the other hand, with increasing average radius of the ordered Au nanostructures, ΔI_{SERS} will be distinctly increased which can be simply estimated from Equation (2).

Now we turn towards the nature of the chemical interaction between the ordered Au nanostructures and graphene, which is another physical mechanism for the SERS especially for the graphene-nanoparticle hybrid system[3][19]. From the Raman spectrum, we find that the G peak position spectrally shifts from 1350 to 1360cm^{-1} for graphene that was transferred on the surface of the ordered Au nanostructures while the 2D peak position almost remains constant. It has been reported that the G peak of graphene is blueshifted for both electron and hole doping, while the 2D peak is redshifted for electron doping and blueshifted for hole doping[18][21]. In this case, graphene is in direct contact with the ordered Au nanostructures, and the work function of Au (5.0eV) is nearly the same with that of graphene (4.8eV). Considering that there are huge amounts of electrons for the Au nanostructures, electron transfer from Au to graphene will occur, leading to an electron doping for graphene; thus, the G peak is blueshifted and the 2D peak is slightly redshifted. On the other hand, we deduce that graphene is under compressive strain when it is directly transferred on top of Au as some ridges appear on the surface of graphene (clearly in **Figure 3**); hence, both the G and 2D peaks exhibit a blueshift trend[21]. As a result, both strain and doping effects lead to a slight blueshift for the G peak and negligible shift for the 2D peak position. However, we consider that charge transfer is not the dominant mechanism for the enhanced Raman intensity since the Raman intensity is increased with the size of Au nanostructures and the charge transfer effect should be independent on the Au size.

4. Conclusions

In summary, the coupling between graphene and LSPs of ordered and size-controllable Au nanostructures has been investigated systematically by directly transferring graphene on the surface of the Au nanostructures. The absorption spectra of Au exhibit a redshift of ~20nm after graphene coverage, which can be ascribed to the plasmonic coupling between the Au LSPs and graphene. On the other hand, the graphene SERS is significantly observed, and intensities of the G and 2D peaks increase with increasing size of the ordered Au nanostructures. The electromagnetic plasmonic effect rather than the charge transfer mechanism is considered to be the dominant mechanism for the SERS effect of graphene. We believe the results are beneficial not only for further understanding the coupling mechanism between graphene and the ordered metallic nanostructures, but also for developing plasmonic graphene-based optoelectronic devices.

Competing Interests

The authors declare that they have no competing interests.

Authors' Contributions

SZ designed and carried out the experiments, analyzed results, and drafted the manuscript. XZ supervised the research and revised the manuscript. XL participated in the experiments and offered technical supports. All authors read and approved the final manuscript.

Acknowledgements

This work was financially supported by the National Natural Science Foundation of China (Grant No. 61404051) and Fundamental Research Funds for the Central Universities (2014ZB0016).

Source: Zhang S, Zhang X, Liu X. Electromagnetic Enhancement of Graphene Raman Spectroscopy by Ordered and Size-Tunable Au Nanostructures[J]. Nanoscale Research Letters, 2015, 10(1):1–7.

References

[1] Novoselov KS, Fal'ko VI, Colombo L, Gellert PR, Schwab MG, Kim K (2012) A roadmap for graphene. Nature 490:192–200.

[2] Yan H, Xia F, Zhu W, Freitag M, Dimitrakopoulos C, Bol AA, Tulevski G, Avouris P (2011) Infrared spectroscopy of wafer-scale graphene. ACS Nano 5:9854–9860.

[3] Xu W, Mao N, Zhang J (2013) Graphene: a platform for surface-enhanced Raman spectroscopy. Small 9:1206–1224.

[4] Liu JT, Liu NH, Li J, Li XJ, Huang JH (2012) Enhanced absorption of grapheme with one-dimensional photonic crystal. Appl Phys Lett 101:052104–3.

[5] Zhang SG, Zhang XW, Liu X, Yin ZG, Wang HL, Gao HL, Zhao YJ (2014) Raman peak enhancement and shift of few-layer graphene induced by plasmonic coupling with silver nanoparticles. Appl Phys Lett 104:121109–5.

[6] Heeg S, Garcia RF, Oikonomou A, Schedin F, Narula R, Maier SA, Vijayaraghavan A, Reich S (2013) Polarized plasmonic enhancement by Au nanostructures probed through Raman scattering of suspended graphene. Nano Lett 13:301–308.

[7] Xu W, Ling X, Xiao J, Dresselhaus MS, Kong J, Xu H, Liu Z, Zhang J (2012) Surface enhanced Raman spectroscopy on a flat graphene surface. Proc Natl Acad Sci USA 109:9281–9286.

[8] Zhou H, Qiu C, Yu F, Yang H, Chen M, Hu L, Sun L (2011) Thickness-dependent morphologies and surface-enhanced Raman scattering of Ag deposited on n-layer graphenes. J Phys Chem C 115:11348–11354.

[9] Niu J, Truong VG, Huang H, Tripathy S, Qiu C, Wee ATS, Yu T, Yang H (2012) Study of electromagnetic enhancement for surface enhanced Raman spectroscopy of SiC graphene. Appl Phys Lett 100:191601–191604.

[10] Schedin F, Lidorikis E, Lombardo A, Kravets VG, Geim AK, Grigorenko AN, Novoselov KS, Ferrari AC (2010) Surface-enhanced Raman spectroscopy of graphene. ACS Nano 4:5617–5626.

[11] Liu CY, Liang KC, Chen W, Tu C, Liu CP, Tzeng Y (2011) Plasmonic coupling of silver nanoparticles covered by hydrogen-terminated graphene for surface-enhanced Raman spectroscopy. Opt Express 19:17092–17098.

[12] Niu J, Shin YJ, Lee Y, Ahn JH, Yang H (2012) Graphene induced tunability of the surface plasmon resonance. Appl Phys Lett 100:061116–4.

[13] Taguchi A, Saito Y, Watanabe K, Yijian S, Kawata S (2012) Tailoring plasmon resonances in the deep-ultraviolet by size-tunable fabrication of aluminum nanostructures. Appl Phys Lett 101:081110–081114.

[14] Kelly KL, Coronado E, Zhao LL, Schatz GC (2003) The optical properties of metal nanoparticles: the influence of size, shape, and dielectric environment. J Phys Chem B 107:668–677.

[15] Palik ED, Ghosh G (1998) Handbook of optical constants of solids. Academic, Orlando, Florida.

[16] Okamoto T, Yamaguchi I (2003) Optical absorption study of the surface plasmon resonance in gold nanoparticles immobilized onto a gold substrate by self-assembly technique. J Phys Chem B 107:20321–10324.

[17] Malard LM, Pimenta MA, Dresselhaus G, Dresselhaus MS (2009) Raman spectroscopy in graphene. Phys Rep 473:51–87.

[18] Ni Z, Wang Y, Yu T, Shen Z (2008) Raman spectroscopy and imaging of graphene. Nano Res 1:273–291.

[19] Yi C, Kim TH, Jiao W, Yang Y, Lazarides A, Hingerl K, Bruno G, Brown A, Losurdo M (2012) Evidence of plasmonic coupling in gallium nanoparticles/graphene/SiC. Small 8:2721–2730.

[20] Balasubramanian K, Zuccaro L, Kern K (2014) Tunable enhancement of Raman scattering in graphene nanoparticle hybrids. Adv Funct Mater 24:6348–6358.

[21] Qiu C, Zhou H, Cao B, Sun L, Yu T (2013) Raman spectroscopy of morphology-controlled deposition of Au on graphene. Carbon 59:487–494.

Chapter 28

Functionalized Silicon Quantum Dots by N-Vinylcarbazole: Synthesis and Spectroscopic Properties

Jianwei Ji, Guan Wang, Xiaozeng You, Xiangxing Xu

State Key Laboratory of Coordination Chemistry, Nanjing National Laboratory of Microstructures, School of Chemistry and Chemical Engineering, Nanjing University, Nanjing 210093, China

Abstract: Silicon quantum dots (Si QDs) attract increasing interest nowadays due to their excellent optical and electronic properties. However, only a few optoelectronic organic molecules were reported as ligands of colloidal Si QDs. In this report, N-vinylcarbazole—a material widely used in the optoelectronics industry was used for the modification of Si QDs as ligands. This hybrid nanomaterial exhibits different spectroscopic properties from either free ligands or Si QDs alone. Possible mechanisms were discussed. This type of new functional Si QDs may find application potentials in bioimaging, photovoltaic, or optoelectronic devices.

Keywords: Silicon Quantum Dots, N-Vinylcarbazole, Surface Modification, Spectroscopic Property

1. Background

Silicon (Si) is one of the most important semiconductor materials for the electronics industry. The energy structure of bulk Si is indirect bandgap, which is greatly changed by the quantum confinement effect for small enough Si nanocrystals (NCs) called Si quantum dots (QDs), making Si QDs fluorescent with a tunable spectrum. Excellent spectroscopic properties, such as high quantum yield, broad absorption window, and narrow fluorescent wavelength, contribute to a rapid development in Si QD research[1]. Nontoxicity to the environment and the use of an economic source material are other two merits for the application of Si QDs in optoelectronics[2][3], solar energy conversion[4][5], biology[6]–[8], splitting water[9], etc. Si QDs can be prepared using a variety of techniques such as wet chemical reduction[10]–[18], metathesis reaction[19], disproportionation reaction[20][21], thermal annealing of Si-rich SiC[22], electrochemical etching[23], plasma synthesis or plasma-enhanced chemical vapor deposition (PECVD) [24]–[27], and high-temperature hydrogen reduction method[28]–[32]. Because Si QDs are chemically active, their surface should be passivated for further use. Molecules with alkyl chains and $-CH_3$, $-COOH$, or $-NH_2$ ends have been widely employed as surface ligands to enhance the stability of Si QDs[28]–[36]. These ligands help prevent the oxidation of silicon and enhance the dispersibility of Si QDs in organic or aqueous solution. In addition to the surface protection, optoelectronic functional molecules as ligands of Si QDs are attracting increasing interest in recent years for the crucial role of the ligands to the interfacial related process in optoelectronic or light-harvesting devices. Kryschi and co-workers showed that 3-vinylthiophene ligands may act as surface-bound antennae that mediate ultrafast electron transfer or excitation energy transfer across the Si QD interface via high-energy two-photon excitation[37][38]. They also reported that for 2- and 4-vinylpyridine-terminated Si QDs, ultrafast excitation relaxation dynamics involving decay and rise dynamics faster than 1 ps were ascribed to electronic excitation energy transfer from an initially photoexcited ligand state to Si QD conduction band states[39]. Larsen and Kauzlarich and their co-workers investigated the transient dynamics of 3-aminopropenyl-terminated Si QDs[40]. A formation and decay of a charge transfer excited state between the delocalized π electrons of the carbon linker and the Si core excitons were proposed to interpret one-photon excitation. Zuilhof *et al.* reported Si QDs functionalized with a red-emitting ruthenium complex to exhibit Förster resonance energy transfer (FRET) from Si QDs to the complex[41]. The ligands on the Si surface may also induce optoelectronic interactions to other QDs such as CdSe QDs, e.g., Sudeep and Emrick found that hydrosilylation of Si QDs provides a corona of phosphine oxides that may serve as ligands for CdSe QDs[42]. This surface functionalization of the Si QDs was proved a key to the photoluminescence quenching of CdSe QDs, as conventional (alkane-covered) Si QD samples give no evidence of such optoelectronic interactions. Recently, we reported 9-ethylanthracene-modified Si QDs showing dual emission peaks that originate from the Si QD core and the ligands[43]. In this report, we demonstrate the synthesis and surface modification of Si QDs with N-ethylcarbazole, using hydrogen-terminated Si QDs and N-vinylcarbazole as the starting materials. Both anthracene and carbazole are fluorescent molecules and organic semiconductors. The main difference is that anthracene is an electron transport material while carbazole is a hole transport material. This difference is important for the structure design of optoelectronic or photovoltaic devices utilizing these Si QD-based hybrid materials. N-vinylcarbazole and its derivatives as a class of typical optoelectronic molecules show abundant attractive properties and can be applied in dye, optics, electronics, and biology[44]–[48]. N-vinylcarbazole is also the monomer precursor of poly(N-vinylcarbazole) (PVK) polymer which is widely used as a hole transport or electroluminescent material in organic optoelectronic devices[49]–[51]. The N-ethylcarbazole-modified Si QDs (referred to as 'N-ec-Si QDs' for short) exhibit photoluminescence quite different from freestanding N-vinylcarbazole- or hydrogen-modified Si QDs. This hybrid nanomaterial was characterized and investigated by powder X-ray diffraction (XRD), transmission electron microscopy (TEM), Fourier transform infrared spectroscopy (FTIR), photoluminescence (PL), and PL lifetime measurement.

2. Methods

2.1. Materials and Equipment

N-vinylcarbazole (98%), $HSiCl_3$ (99%), and mesitylene (97%) were purchased from Aladdin Reagent Co., Ltd. (Shanghai, China). Analytical-grade ethanol (99.5%) and hydrofluoric acid (40% aqueous solution) were received from Sinopharm Chemical Reagent Co., Ltd. (SCRC; Shanghai, China). All reagents were used as purchased without further purification. The XRD spectrum was performed on a Bruker D8 Advance instrument (Bruker AXS GmbH, Karlsruhe, Germany) with Cu Kα radiation (λ = 1.5418Å). TEM images were obtained on a JEM-2100 transmission electron microscope with an acceleration voltage of 200kV (JEOL, Ltd., Akishima, Tokyo, Japan). The FTIR spectra were measured by a Bruker VECTOR 22 spectrometer (Bruker, Germany) with KBr pellets. The PL and excitation spectra were collected by a Hitachi F-4600 fluorescence spectrophotometer (Hitachi, Ltd., Chiyoda-ku, Japan). The UV-vis absorption spectra were measured by a Shimadzu UV-2700 UV-vis spectrophotometer (Shimadzu Corporation, Kyoto, Japan). The PL lifetime was obtained on a Zolix Omni-λ 300 fluorescence spectrophotometer (Zolix Instruments Co., Ltd., Beijing, China).

2.2. Synthesis of Hydrogen-Terminated Si QDs

Si QDs were synthesized by reduction of $(HSiO_{1.5})_n$ powder with hydrogen[28][29]. Typically, 5 mL of $HSiCl_3$ (49.5mmol) was added to a three-neck flask equipped with a mechanical stir bar, cooled to −78°C in an ethanol bath, and kept for 10min, using standard Schlenk techniques with N_2 protection. With the injection of 20mL H_2O by a syringe, a white precipitate formed immediately. After 10min, the white $(HSiO_{1.5})_n$ was collected by centrifugation, washed by distilled water, and dried in vacuum at 60°C. In the reduction step, $(HSiO_{1.5})_n$ (1.10g) was placed in a corundum crucible and transferred to a tube furnace. The sample was heated to 1,150°C and maintained for 1.5h with a heating rate of 5°C/min under a slightly reducing atmosphere containing 5% H_2 and 95% Ar (≥99.999%). After cooling to room temperature, a light brown product of Si/SiO_2 composite was collected. The Si/SiO_2 composite (50mg) was grinded with a mortar and pestle for 10 min. Then the powder was transferred to a Teflon container (20mL) with a magnetic stir bar. A mixture of ethanol (1.5mL) and hydrofluoric acid (40%, 2.5mL) was added. The light brown mixture was stirred for 60min to dissolve the SiO_2. Finally, 5mL mesitylene was added to extract the hydrogen-terminated Si QDs into the upper organic phase, forming a brown suspension (A), which was isolated for further surface modification.

2.3. Modification of Si QDs by Functional Organic Molecules

N-vinylcarbazole (1 mmol) was dissolved in 15mL mesitylene and loaded in a 50-mL three-neck flask equipped with a reflux condenser. Then 2mL Si QDs (A) was injected by a syringe. The mixture was degassed by a vacuum pump for 10min to remove any dissolved gases from the solution. Protected by N_2, the solution was heated to 156°C and kept for 12h. After cooling to room temperature, the resulting Si QDs were purified by vacuum distillation and then washed by ethanol to remove excess solvent and organic ligands. The as-prepared brown solid product was readily re-dispersed in mesitylene to give a yellow solution.

3. Results and Discussion

The synthesis route of N-ec-Si QDs is summarized in **Figure 1**. The $HSiCl_3$ hydrolysis product $(HSiO_{1.5})_n$ was reduced by H_2 at 1,150°C for 1.5h. In this step, the temperature and time are crucial in controlling the size of Si QDs. The higher the temperature and the longer the reduction time, the bigger the sizes of Si QDs. The following HF etching procedure also plays a key role for the size tuning of the Si QDs. HF not only eliminates the SiO_2 component and liberates the free Si QDs but also etches Si QDs gradually. Another contribution of HF etching is the modification of the surface of Si QDs with hydrogen atoms in the form of Si-H bonds, which can be reacted with an ethylenic bond or acetylenic bond to form a Si-C covalent bond[28]–[32].

The hydrogen-terminated Si QDs are characterized by XRD [**Figure 2(a)**]. The XRD pattern shows broad reflections (2θ) centered at around 28°, 47°, and 56°,

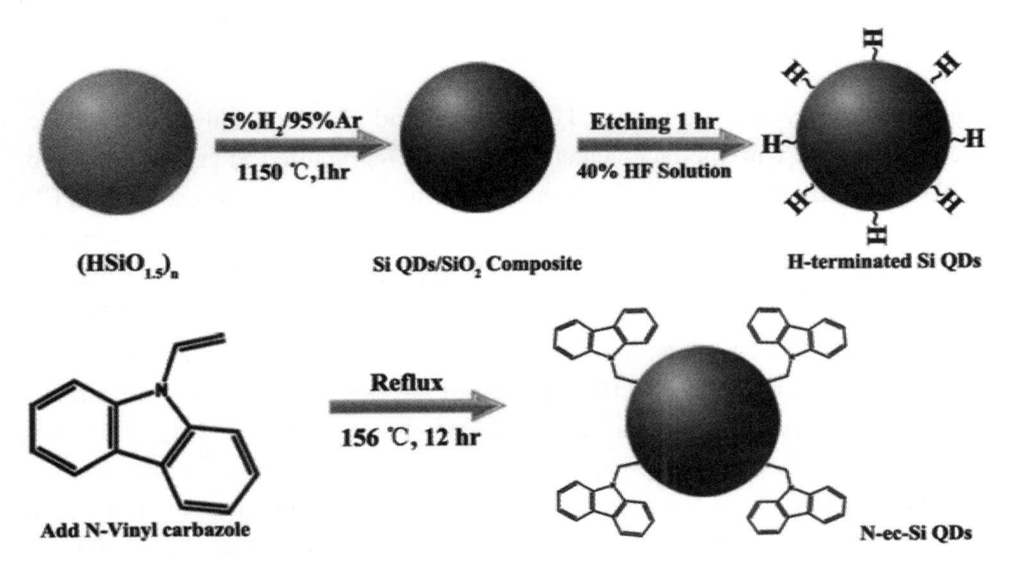

Figure 1. Synthetic strategy of N-ec-Si QDs.

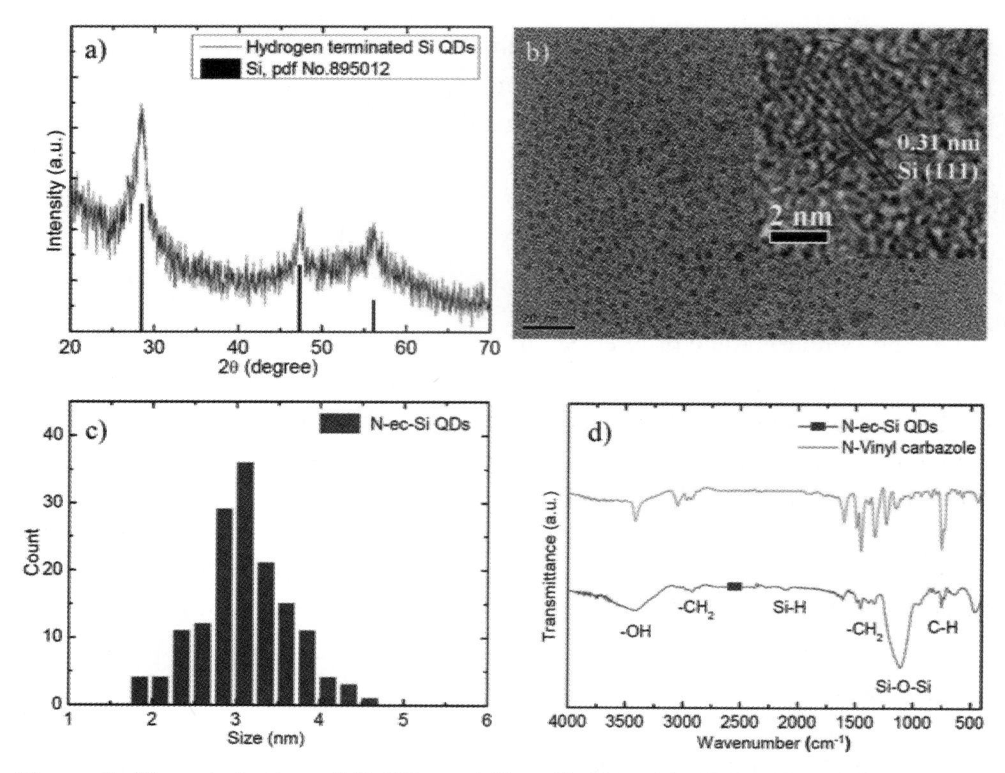

Figure 2. Characterization of Si QDs and N-ec-Si QDs. (a) XRD pattern of the hydrogen-terminated Si QDs. (b) TEM image and HRTEM image (inset) of the N-ec-Si QDs (scale bar 20nm, inset 2nm). (c) Size distribution of the N-ec-Si QDs. (d) FTIR spectra of the N-ec-Si QDs and pure N-vinylcarbazole.

which are readily indexed to the {111}, {220}, and {311} crystal planes, respectively, consistent with the face-centered cubic (fcc)-structured Si crystal (PDF No. 895012). **Figure 2(b)** and its inset show typical TEM and high-resolution TEM (HRTEM) images of N-ec-Si QDs, respectively. A d-spacing of approximately 0.31nm is observed for the Si QDs by HRTEM. It is assigned to the {111} plane of the fcc-structured Si. The size distribution of N-ec-Si QDs measured by TEM reveals that the QD sizes range from 1.5 to 4.6nm and the average diameter is about 3.1nm [**Figure 2(c)**]. In the FTIR spectrum of N-ec-Si QDs, a series of characteristic vibrations from Si QDs and carbazole are observed [**Figure 2(d)**]. The weak vibration resonance centered at 2,090cm^{-1} can be assigned to the coupled H-Si-Si-H stretching or monohydride Si-H bonds. This result shows that the Si-H bonds were only partially replaced by Si-C because of the rigid and steric effect of the N-vinylcarbazole molecule. Compared to the IR spectrum of N-vinylcarbazole, similar vibrational peaks can be found in the spectrum of N-ec-Si QDs. The CH_2 symmetric and asymmetric stretching vibrations in the range 2,920 to 2,850cm^{-1}, the CH_2 bending vibration at approximately 1,450cm^{-1}, and the aromatic group vibration bands at approximately 750cm^{-1} can be assigned to the surface-modified N-ethylcarbazole ($-NC_{14}H_{12}$) ligands. This indicates the successful modification of N-vinylcarbazole onto the Si QDs. It should be noticed that the Si-O-Si vibration band at 1,000 to 1,200cm^{-1} is recorded, suggesting possible oxidation of the Si QD surface. This may due to the steric effect of carbazole, that is, the Si QD surface cannot be fully protected by the ligand, in which some Si-H remained and encountered oxidation when exposed to air.

Figure 3(a) shows the absorption spectra of N-vinylcarbazole and N-ec-Si QDs. The absorption band at 320 to 360nm of the N-ec-Si QDs is assigned to the

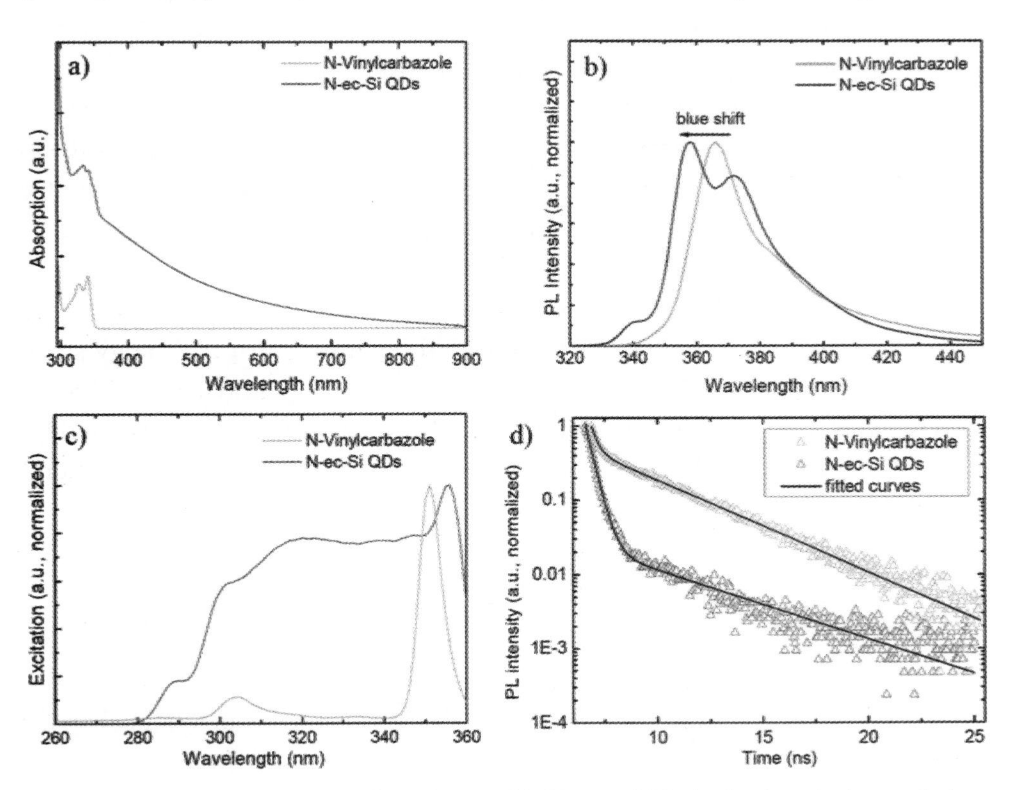

Figure 3. Spectroscopic properties of N-ec-Si QDs and N-vinylcarbazole in mesitylene solution. (a) UV spectra. (b) Photoluminescence spectra. (c) Excitation spectra. (d) PL decay curves (excitation at 302nm; emissions of 358nm for N-ec-Si QDs and 366nm for N-vinylcarbazole were adopted for the excitation spectra measurement).

carbazole ligand. It suggests that ligands can be employed to enhance the absorption of pure Si QDs, therefore providing a potential strategy to increase the light-harvesting efficiency of QDs in solar cells[52][53]. Upon excitation at 302nm, the N-ec-Si QDs and N-vinylcarbazole show intense emission bands at approximately 358nm and approximately 366nm, respectively [**Figure 3(b)**]. In comparison with N-vinylcarbazole, the emission in the 9-ea-Si QDs exhibits a blueshift of 8nm and a shoulder peak at approximately 372. When carbazole was linked to the surface of Si QDs by Si-C bond by the hydrosilylation reaction, the vinyl group in N-vinylcarbazole was transformed into an ethyl group. Therefore, the conjugate system of the molecule reduced from N-vinylcarbazole to carbazole, inducing a bigger electronic bandgap. In addition, the ligand to QD bonding would enhance the structural rigidity of the ligand. These reasons may contribute to the blueshift of the PL spectrum. Commonly, the extension of molecular conjugated orbitals of a lig- and to the attached materials would lead to a redshift. In N-ec-Si QDs, the ethyl group formed through the hydrosilylation reaction separates the conjugated part, the carbazole group, from the silicon nanocrystal, which prevents or weakens the interaction of the carbazole group with the electronic wave functions of the Si QDs. Therefore, a redshift is prohibited. A similar blueshift was also demonstrated in our recent work for 9-ethylanthracene modified on Si QDs[43].

The N-ec-Si QDs and N-vinylcarbazole show distinct excitation spectra within the range of 280 to 360nm [**Figure 3(c)**], indicating that the energy structure of N-ec-Si QDs is different from N-vinylcarbazole. PL decay curves of N-ec-Si QDs and N-vinylcarbazole were investigated at room temperature in mesitylene solution [**Figure 3(d)**]. The PL decay curves are fitted to the exponential function

$$I(t)=\sum_{i=1}^{n} A_i \exp\left(-(t-t_0)/T_i\right)$$

where T_i is the PL decay lifetime, A_i is the weighting parameter, and $n = 2$. The fitting parameters are given in **Table 1**. The average lifetime is determined by the equation[54]

$$\tau_{av}=\sum_{i=1}^{n}\left(A_i\tau_i^2\right)\Big/\sum_{i=1}^{n}\left(A_i\tau_i\right)$$

Table 1. Fitting parameters of the PL decay curves.

Sample	Emission (nm)	τ_1 (ns)	τ_2 (ns)	a_1^a	a_2^a	R^2	τ_{av} (ns)
N-vinylcarbazole	366	0.27	3.5	0.58	0.42	0.998	3.2
N-ec-Si QDs	358	0.35	4.6	0.98	0.02	0.997	1.4

[a] $a_i = \frac{A_i}{\sum_{j=1}^{n} A_j}$, $i = 1, 2, n = 2$.

The average PL decay lifetime of N-ec-Si QDs is 1.4 ns, much shorter than that of N-vinylcarbazole which is 3.2ns. The lifetime diversity may be influenced by many factors. First, the hydrosilylation reaction induces the transformation of the molecule structure. Second, the N-vinylcarbazole dispersion state in the mesitylene is not clear. Possible π–π packing of the molecules may lead to a redshift. Support can be found in the fact that N-ec-Si QDs show a more symmetric PL spectrum to the absorption spectrum than N-vinylcarbazole exhibits. Third, the interaction of the ligands with the Si-QDs and interaction between the modified ligands are inevitably encountered[55]. Additionally, the oxidation of the silicon surface may induce additional non-radiative pass-ways for the excitation. All of these factors would lead to PL lifetime shortening[56]. Unlike alkyl ligands or 9-ethylanthracene-modified Si QDs, the fluorescence from hydrogen-terminated Si QDs was quenched after the carbazole modification (**Figure 4**). It may be induced by the interaction of carbazole with the Si QDs. The fluorescence quantum yield of N-vinylcarbazole and N-ec-Si QDs was estimated to be 26.6% and 11.2%, respectively, by using Coumarin 540 dye in methanol as a reference (91%)[57]. The decrease of the quantum yield could be a result from fast non-radiative relaxation of the excited states, induced by the interaction of the ligands to Si QDs or surface states, which also could be an interpretation for the lifetime shortening. From the molecular design aspect, the functional group modified by a long alkyl tail with an ethyl or vinyl end would be an ideal ligand structure in which the Si QDs and the functional group are spatially separated. Also, the flexibility of the long alkyl chain exhibits a smaller steric effect. The surface of Si QDs could be more effectively protected, thus preserving the fluorescence of the Si QD core.

4. Conclusions

In conclusion, N-ec-Si QDs were successfully prepared and characterized.

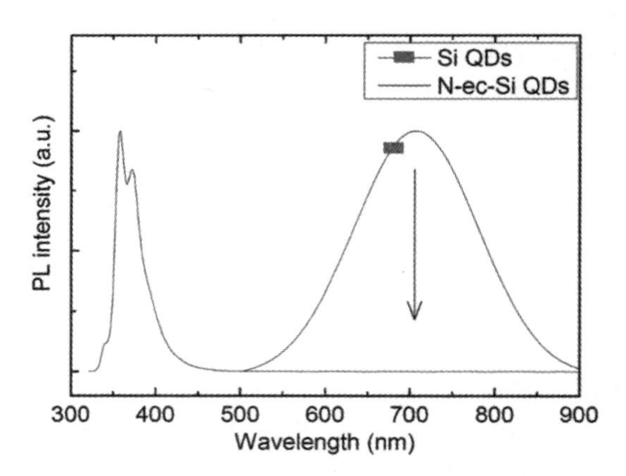

Figure 4. Photoluminescence spectra of N-ec-Si QDs (excitation 302nm) and hydrogen-modified Si QDs (excitation 360nm).

Spectroscopic properties were investigated and discussed. The absorption, excitation, PL, and PL decay properties of N-ethylcarbazole ligands on the Si QD surface are significantly different from those of N-vinylcarbazole in solution. Hopefully, the synthesis strategy could be extended for the syntheses of a series of Si QDs containing various optoelectronic functional organic ligands, with application potentials ranging from optic, electronic, and photovoltaic devices to biotechnology.

Competing Interests

The authors declare that they have no competing interests.

Authors' contributions

JWJ and GW contributed equally to the manuscript. XZY and XXX designed the research. JWJ, GW, and XXX carried out the experiments and drafted the manuscript. All authors read and approved the final manuscript.

Acknowledgements

This work was supported by the Major State Basic Research Development Program of China (Grant Nos. 2013CB922102 and 2011CB808704), the National

Natural Science Foundation of China (Grant Nos. 91022031 and 21301089), and Jiangsu Province Science Foundation for Youths (BK20130562).

Source: Ji J, Wang G, You X, *et al.* Functionalized silicon quantum dots by N-vinylcarbazole: synthesis and spectroscopic properties[J]. Nanoscale Research Letters, 2014, 9(1):1−7.

References

[1] Veinot JGC: Synthesis, surface functionalization, and properties of freestanding silicon nanocrystals. Chem Commun 2006, 40:4160.

[2] Puzzo DP, Henderson EJ, Helander MG, Wang ZB, Ozin GA, Lu ZH: Visible colloidal nanocrystal silicon light-emitting diode. Nano Lett 2011, 11:1585.

[3] Cheng KY, Anthony R, Kortshagen UR, Holmes RJ: High-efficiency silicon nanocrystal light-emitting devices. Nano Lett 2011, 11:1952.

[4] Yuan GB, Aruda K, Zhou S, Levine A, Xie J, Wang DW: Understanding the origin of the low performance of chemically grown silicon nanowires for solar energy conversion. Angew Chem Int Ed 2011, 50:2334.

[5] Liu CY, Kortshagen UR: A silicon nanocrystal Schottky junction solar cell produced from colloidal silicon nanocrystals. Nanoscale Res Lett 2010, 5:1253.

[6] Pacholski C, Sartor M, Sailor MJ, Cunin F, Miskelly GM: Biosensing using porous silicon double-layer interferometers: reflective interferometric Fourier transform spectroscopy. J Am Chem Soc 2005, 127:11636.

[7] He Y, Kang ZH, Li QS, Tsang CHA, Fan CH, Lee ST: Ultrastable, highly fluorescent, and water-dispersed silicon-based nanospheres as cellular probes. Angew Chem Int Ed 2009, 48:128.

[8] Stanca L, Petrache SN, Serban AI, Staicu AC, Sima C, Munteanu MC, Zărnescu O, Dinu D, Dinischiotu A: Interaction of silicon-based quantum dots with gibel carp liver: oxidative and structural modifications. Nanoscale Res Lett 2013, 8:254.

[9] Erogbogbo F, Lin T, Tucciarone PM, LaJoie KM, Lai L, Patki GD, Prasad PN, Swihart MT: On-demand hydrogen generation using nanosilicon: splitting water without light, heat, or electricity. Nano Lett 2013, 13:451.

[10] Heath JR: A liquid-solution-phase synthesis of crystalline silicon. Science 1992, 258:1131.

[11] Bley RA, Kauzlarich SM: A low-temperature solution phase route for the synthesis of silicon nanoclusters. J Am Chem Soc 1996, 118:12461.

[12] Dhas NA, Raj CP, Gedanken A: Preparation of luminescent silicon nanoparticles: a novel sonochemical approach. Chem Mater 1998, 10:3278.

[13] Wilcoxon JP, Samara GA: Tailorable, visible light emission from silicon nanocrystals. App Phys Lett 1999, 74:3164.

[14] Baldwin RK, Pettigrew KA, Ratai E, Augustine MP, Kauzlarich SM: Solution reduction synthesis of surface stabilized silicon nanoparticles. Chem Commun 2002, 17:1822.

[15] Warner JH, Hoshino A, Yamamoto K, Tilley RD: Water-soluble photoluminescent silicon quantum dots. Angew Chem Int Ed 2005, 44:4550.

[16] Tilley RD, Yamamoto K: The microemulsion synthesis of hydrophobic and hydrophilic silicon nanocrystals. Adv Mater 2006, 18:2053.

[17] Rosso-Vasic M, Spruijt E, van Lagen B, Cola LD, Zuilhof H: Alkyl-functionalized oxide-free silicon nanoparticles: synthesis and optical properties. Small 2008, 4:1835.

[18] Lin SW, Chen DH: Synthesis of water-soluble blue photoluminescent silicon nanocrystals with oxide surface passivation. Small 2009, 5:72.

[19] Pettigrew KA, Liu Q, Power PP, Kauzlarich SM: Solution synthesis of alkyland alkyl/alkoxy-capped silicon nanoparticles via oxidation of Mg2Si. Chem Mater 2003, 15:4005.

[20] Liu SM, Sato S, Kimura K: Synthesis of luminescent silicon nanopowders redispersible to various solvents. Langmuir 2005, 21:6324.

[21] Liu SM, Yang Y, Sato S, Kimura K: Enhanced photoluminescence from Si nano-organosols by functionalization with alkenes and their size evolution. Chem Mater 2006, 18:637.

[22] Wan ZY, Huang SJ, Green MA, Conibeer G: Rapid thermal annealing and crystallization mechanisms study of silicon nanocrystal in silicon carbide matrix. Nanoscale Res Lett 2011, 6:129.

[23] Carter RS, Harley SI, Power PP, Augustine MP: Use of NMR spectroscopy in the synthesis and characterization of air- and water-stable silicon nanoparticles from porous silicon. Chem Mater 2005, 17:2932.

[24] Jurbergs D, Rogojina E, Mangolini L, Kortshagen U: Silicon nanocrystals with ensemble quantum yields exceeding 60%. Appl Phys Lett 2006, 88:2331161.

[25] Kortshagen U, Mangolini L, Bapat A: Plasma synthesis of semiconductor nanocrystals for nanoelectronics and luminescence applications. J Nanoparticle Res 2007, 9:39.

[26] Lin GR, Lin CJ, Lin CT: Low-plasma and high-temperature PECVD grown silicon-rich SiOx film with enhanced carrier tunneling and light emission. Nanotechnol 2007, 18:395202.

[27] Lin GR, Lin CJ, Kuo HC, Lin HS, Kao CC: Anomalous microphotoluminescence of high-aspect-ratio Si nanopillars formatted by dry-etching Si substrate with self-aggregated Ni nanodot mask. Appl Phys Lett 2007, 90:143102.

[28] Henderson EJ, Veinot JGC: From phenylsiloxane polymer composition to size-controlled silicon carbide nanocrystals. J Am Chem Soc 2009, 131:809.

[29] Henderson EJ, Kelly JA, Veinot JGC: Influence of HSiO1.5 sol–gel polymer structure and composition on the size and luminescent properties of silicon nanocrystals. Chem Mater 2009, 21:5426.

[30] Mastronardi ML, Hennrich F, Henderson EJ, Maier-Flaig F, Blum C, Reichenbach J, Lemmer U, Kübel C, Wang D, Kappes MM, Ozin GA: Preparation of monodisperse silicon nanocrystals using density gradient ultracentrifugation. J Am Chem Soc 2011, 133:11928.

[31] Mastronardi ML, Maier-Flaig F, Faulkner D, Henderson EJ, Kübel C, Lemmer U, Ozin GA: Size-dependent absolute quantum yields for size-separated colloidally-stable silicon nanocrystals. Nano Lett 2012, 12:337.

[32] Hessel CM, Reid D, Panthani MG, Rasch MR, Goodfellow BW, Wei J, Fujii H, Akhavan V, Korgel BA: Synthesis of ligand-stabilized silicon nanocrystals with size-dependent photoluminescence spanning visible to near-infrared wavelengths. Chem Mater 2012, 24:393.

[33] Sieval AB, Linke R, Zuilhof H, Sudhölter EJR: High-quality alkyl monolayers on silicon surfaces. Adv Mat 2000, 12:1457.

[34] Buriak JM: Organometallic chemistry on silicon and germanium surfaces. Chem Rev 2002, 102:1271.

[35] Shirahata N, Hozumi A, Yonezawa T: Monolayer-derivative functionalization of non-oxidized silicon surfaces. Chem Rec 2005, 5:145.

[36] Boukherroub R: Chemical reactivity of hydrogen-terminated crystalline silicon surfaces. Curr Op Sol St Mat Sci 2005, 9:66.

[37] Cimpean C, Groenewegen V, Kuntermann V, Sommer A, Kryschi C: Ultrafast exciton relaxation dynamics in silicon quantum dots. Laser Photonics Rev 2009, 3:138.

[38] Groenewegen V, Kuntermann V, Haarer D, Kunz M, Kryschi C: Excited-state relaxation dynamics of 3-vinylthiophene-terminated silicon quantum dots. J Phys Chem C 2010, 114:11693.

[39] Sommer A, Cimpean C, Kunz M, Oelsner C, Kupka HJ, Kryschi C: Ultrafast excitation energy transfer in vinylpyridine terminated silicon quantum dots. J Phys Chem C 2011, 115:22781.

[40] Atkins TM, Thibert A, Larsen DS, Dey S, Browning ND, Kauzlarich SM: Femtosecond ligand/core dynamics of microwave-assisted synthesized silicon quantum dots in aqueous solution. J Am Chem Soc 2011, 133:20664.

[41] Rosso-Vasic M, Cola LD, Zuilhof H: Efficient energy transfer between silicon nanoparticles and a Ru-polypyridine complex. J Phys Chem C 2009, 113:2235.

[42] Sudeep PK, Emrick T: Functional Si and CdSe quantum dots: synthesis, conjugate formation, and photoluminescence quenching by surface interactions. ACS Nano 2009, 3:4105.

[43] Wang G, Ji JW, Xu XX: Dual-emission of silicon quantum dots modified by 9-ethylanthracene. J Mater Chem C 2014, 2:1977.

[44] Dalton LK, Demerac S, Elmes BC, Loder JW, Swan JM, Teitei T: Synthesis of the tumour-inhibitory alkaloids, ellipticine, 9-methoxyellipticine, and related pyrido[4,3-b]carbazoles. Aust J Chem 1967, 20:2715.

[45] Thomas KRJ, Lin JT, Tao YT, Ko CW: Light-emitting carbazole derivatives: potential electroluminescent materials. J Am Chem Soc 2001, 123:9404.

[46] Tsai MH, Lin HW, Su HC, Ke TH, Wu CC, Fang FC, Liao YL, Wong KT, Wu CI: Highly efficient organic blue electrophosphorescent devices based on 3,6-bis(triphenylsilyl)carbazole as the host material. Adv Mater 2006, 18:1216.

[47] Tao YT, Wang Q, Yang CL, Wang Q, Zhang ZQ, Zou TT, Qin JG, Ma DG: A simple carbazole/oxadiazole hybrid molecule: an excellent bipolar host for green and red phosphorescent OLEDs. Angew Chem Int Ed 2008, 47:8104.

[48] Gale PA: Synthetic indole, carbazole, biindole and indolocarbazole-based receptors: applications in anion complexation and sensing. Chem Commun 2008, 38:4525.

[49] Diaz-Garcia MA, Wright D, Casperson JD, Smith B, Glazer E, Moerner WE, Sukhomlinova LI, Twieg RJ: Photorefractive properties of poly(N-vinylcarbazole)-based composites for high-speed applications. Chem Mater 1999, 11:1784.

[50] Ikeda N, Miyasaka T: A solid-state dye-sensitized photovoltaic cell with a poly(N-vinyl-carbazole) hole transporter mediated by an alkali iodide. Chem Commun 2005, 14:1886.

[51] D'Angelo P, Barra M, Cassinese A, Maglione MG, Vacca P, Minarini C, Rubino A: Electrical transport properties characterization of PVK (poly N-vinylcarbazole) for electroluminescent devices applications. Solid State Electron 2007, 51:123.

[52] Liu CY, Holman ZC, Kortshagen UR: Hybrid solar cells from P3HT and silicon nanocrystals. Nano Lett 2009, 9:449.

[53] Werwie M, Xu XX, Haase M, Basché T, Paulsen H: Bio serves nano: biological light-harvesting complex as energy donor for semiconductor quantum dots. Langmuir 2012, 28:5810.

[54] Fujii T, Kodaira K, Kawauchi O, Tanaka N, Yamashita H, Anpo M: Photochromic behavior in the fluorescence spectra of 9-anthrol encapsulated in Si – Al glasses prepared by the sol–gel method. J Phys ChemB 1997, 101:10631.

[55] Xu XX, Ji JW, Wang G, You XZ: Exciton coupling of surface complexes on a nano-

crystal surface. Chem Phys Chem 2014, doi:10.1002/ cphc.201402156.

[56] Antwis L, Gwilliam R, Smith A, Homewood K, Jeynes C: Characterization of a-$FeSi_2$/c-Si heterojunctions for photovoltaic applications. Semicond Sci Technol 2012, 27:035016.

[57] Ritty JN, Thomas KJ, Jayasree VK, Girijavallabhan CP, Nampoori VPN, Radhakrishnan P: Study of solvent effect in laser emission from Coumarin 540 dye solution. Appl Optics 2007, 46:4786.

Chapter 29

Improving the Photoelectric Characteristics of MoS_2 Thin Films by Doping Rare Earth Element Erbium

Miaofei Meng, Xiying Ma

Suzhou University of Science and Technology, Kerui Road No. 1, Gaoxin section, Suzhou 215011, Jiangsu, China

Abstract: We investigated the surface morphologies, crystal structures, and optical characteristics of rare earth element erbium (Er)-doped MoS_2 (Er: MoS_2) thin films fabricated on Si substrates via chemical vapor deposition (CVD). The surface mopography, crystalline structure, light absorption property, and the photoelectronic characteristics of the Er: MoS_2 films were studied. The results indicate that doping makes the crystallinity of MoS_2 films better than that of the undoped film. Meanwhile, the electron mobility and conductivity of the Er-doped MoS_2 films increase about one order of magnitude, and the current- voltage (I-V) and the photoelectric response characteristics of the Er:MoS_2/Si heterojunction increase significantly. Moreover, Er-doped MoS_2 films exhibit strong light absorption and photoluminescence in the visible light range at room temperature; the intensity is enhanced by about twice that of the undoped film. The results indicate that the doping of MoS_2 with Er can significantly improve the photoelectric characteristics and can be used to fabricate highly efficient luminescence and optoelectronic devices.

Keywords: MoS_2 Film, Er Doping, Chemical Vapor Deposition, Photoelectric Characteristics, Light Absorption, Photoluminescence

1. Background

Layered quasi-two dimensional (2D) chalcogenide materials have attracted great interest due to their excellent optical, electrical, catalysis, and lubrication characteristics[1]-[3]. Especially, 2D molybdenum disulfide (MoS_2) has been widely studied and applied in field-effect transistors[4][5]and energy harvesting[6][7]. It has been found that a monolayer of MoS_2 has a direct bandgap of 1.8eV when it is stripped from a bulk material that has an indirect bandgap of 1.29eV[8][9]. This large change in the energy band holds great potential for applications of MoS_2 in optoelectronic fields, such as red photodiodes and photodetectors[10][11]. However, the efficiency of photoluminescence (PL) and photoelectric conversion of 2D MoS_2[12] are relatively low. Researchers have explored many avenues to improve the PL intensity and the response rate of MoS_2 films. For example, Singha *et al.* found that gold nanoparticles may impose an obvious p-doping effect in single-layer and bi-layer MoS_2 samples, resulting in enhanced PL[13].

Rare earth elements (REE) are active elements that have been widely added in optoelectronic devices to improve PL and photoelectric conversion efficiency[14][15]. So far, there have been few reports for REE-doped MoS_2. Herein, we report a study of the doping effects of rare earth element Er on the surface morphologies, crystal structures, and optical characteristics of MoS_2 thin films. Pure MoS_2 and Er: MoS_2 samples were fabricated on Si substrates by chemical vapor deposition (CVD). Additionally, we systematically analyzed the surface morphologies, structures, and optical absorption characteristics of the samples.

2. Methods

$Er(NO_3)_3 \cdot 5H_2O$ (99.9%) and MoS_2 powder (AR, 99%) reagents were used as the precursor materials. A mixed solution comprising 1-g analytical grade MoS_2 micro powder, 1-g analytical grade erbium nitrate pentahydrate ($Er(NO_3)_3 \cdot 5H_2O$) crystals, and 200mL of diluted sulfuric acid (H_2SO_4) was formed by mixing the above mentioned components for 5min, followed which the solution was maintained at 70°C via a water bath. The CVD system consisted of a horizontal quartz tube furnace, a vacuum system, an intake system, and a water bath. The Si substrates were placed in the center of the furnace, and subsequently, the pressure in the furnace was reduced to 10^{-2}Pa and the furnace was heated up to 650°C for 20min. Ar gas was introduced into the mixed solution at a flow rate of 25sccm, carrying Er^{3+} and MoS_2 molecules into the furnace. Furthermore, to investigate the material properties of MoS_2 films, some pure MoS_2 samples were deposited on the Si substrates by the same method.

The surface morphologies and crystalline structures of the thin films were characterized using atomic force microscope (AFM) and X-ray diffraction (XRD). The electrical properties of the thin films were analyzed by a Hall Effect Measurement System (HMS-3000, Ecopia, Anyang, Republic of Korea). The ultraviolet- visible (UV-vis) absorption spectra and photoluminescence properties of the samples were investigated by a UV-vis spectrophotometer (Shimadzu UV-3600) and fluorescence spectrophotometer at room temperature. Photocurrent current-voltage (I-V) curves of the doped and undoped MoS_2/Si heterojunction were investigated by a semiconductor analysis system (Keithley 4200).

3. Results and Discussion

The AFM images of the pure MoS_2 and Er: MoS_2 thin films on the Si substrates are shown in **Figure 1**. The surface of the pure MoS_2 film in **Figure 1(a)** is a continuous film with an average thickness about 25nm, and some quantum dots around 20nm are uniformly scattered on the Si substrate. The Er: MoS_2 film shown in **Figure 1(b)** is a large fluctuation film composed of compact quantum dots with a uniform color, and the average thickness is about 50nm. For the same deposition conditions and time, the density and size of the quantum dots in Er: MoS_2 film increase remarkably resulting from the catalytic action of Er^{3+} on the deposition course.

The crystal structures of the synthesized samples were characterized by using the X-ray diffraction (XRD) technique, as shown in **Figure 2**. For the pure MoS_2 sample, there are four sharp diffraction peaks located at 14.7°, 47.8°, 54.6°, and 56.4°, corresponding to the (002), (105), (106), and (110) crystal planes of

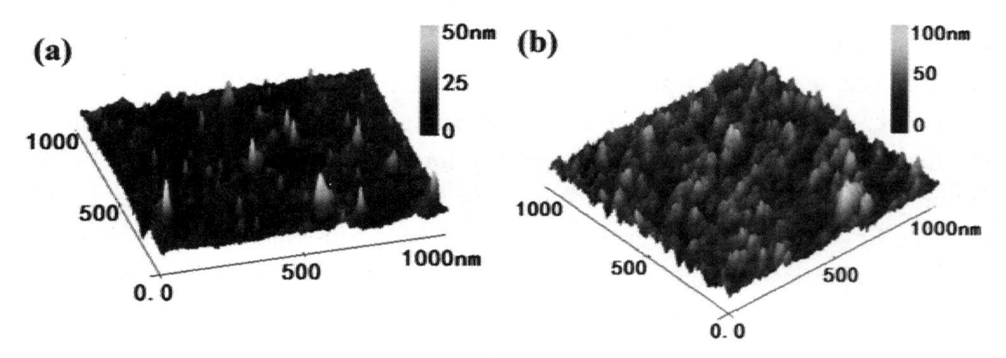

Figure 1. AFM images of MoS_2 samples. (a) The MoS_2 film. (b) The Er: MoS_2 film.

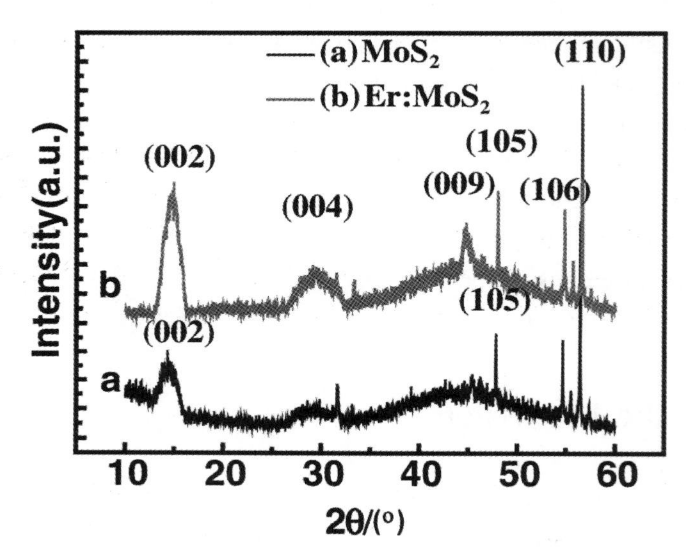

Figure 2. XRD patterns of the MoS_2 film and the Er: MoS_2 film for the diffraction angle in the range of 10°–60°.

MoS_2, respectively, showing that the film is characterized by a polycrystal structure. In the Er: MoS_2 film, the position of the above four diffraction peaks is almost the same as that of pure MoS_2. Besides, there are two more peaks at 29.5° and 44.8°, corresponding to the (004) and (009) planes, respectively. No diffraction peaks from elemental Er is observed, indicating that the Er doping does not change the crystal structure of the MoS_2 film. Er atoms were doped in MoS_2 film in the way of substitution doping, and Mo atom was replaced by Er element. By doping, the diffraction peaks of MoS_2 crystal increased and the diffraction intensity was enhanced, showing that doping improved the crystallinity of the MoS_2 films.

The surface J-V properties, carrier mobilities, and Hall coefficients of the MoS_2 and Er: MoS_2 samples were measured using a Hall Effect measurement system via the four measured points on the samples at dark condition, as shown in **Figure 3**. The currents of the samples show a linear dependency on the applied voltage, revealing that the films have a good conductivity. The slopes of the *J-V* curves show the resistivity of the MoS_2 samples. The curve of the Er: MoS_2 film has good linearity and a small slope, with the films showing a significant reduction in resistivity when Er ions are doped. According to the equation for calculation of mobility: $\sigma = nq\mu$ (σ is conductivity, n is electron concentration, q is electron charge, μ is mobility), the electron motilities in the MoS_2 and Er: MoS_2 films are $3.996 \times 10^3 cm^2/Vs$ and $5.547 \times 10^3 cm^2/Vs$, respectively. Note that the mobility value for the MoS_2 film is obviously improved by doping Er^{3+}. Furthermore, According to the equation for the Hall coefficients: $\varepsilon_y = R_H J_x B_z$ (ε_y is electric field intensity, R_H is Hall coefficients, J_x is current density, B_z is magnetic induction intensity), the Hall coefficients of the MoS_2 and Er: MoS2 films are $1.905 \times 10^7 cm^3/C$ and $4.581 \times 10^8 cm^3/C$, respectively, showing that the films are p-type se- miconductors. The J-V curves in the MoS_2 film show a significant decrease in resistivity after Er doping. Good conductive properties can reduce the surface heat loss in the photodetector, thereby increasing the lifetime and frequency response of the MoS_2 photovoltaic device.

Figure 4 shows the absorption spectra of the pure MoS_2 and Er: MoS_2 films in the visible light range. Clearly, the absorption of the Er: MoS_2 film is enhanced

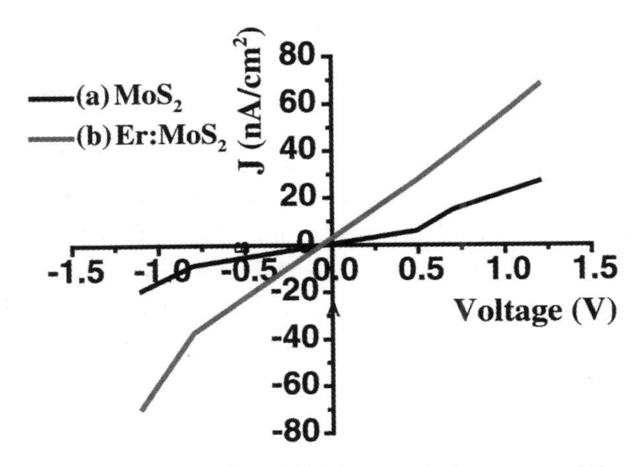

Figure 3. The surface J-V characteristic curves of the MoS_2 film and the Er:MoS_2 film.

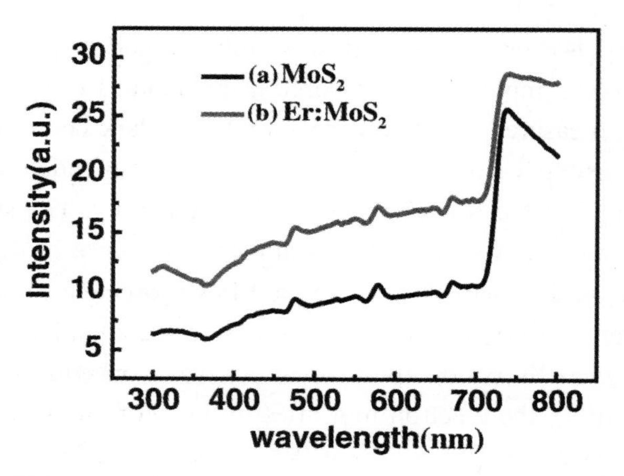

Figure 4. The absorption spectra of the MoS_2 film and the Er:MoS_2 film.

significantly by doping Er, attributing to the absorptions of Er ions and the impurity energy level in the bandgap of MoS_2 by doping Er. Additionally, a few maximum values emerge at 475, 578, 670, and 735nm in the absorption spectra, showing that the film has strong light absorption in these wavebands. The absorption peak at 670nm corresponds to a bandgap width of 1.85eV in the MoS_2 film, close to the energy gap of a monolayer of MoS_2, 1.80eV. MoS_2 films have strong absorption at 735nm, which can be considered as the optical absorption edge, corresponding to a bandgap width of 1.69eV. Therefore, the doping of Er significantly improves the light absorption and does not change the position of the absorption peak. The increase of the light absorption of Er^{3+}-doped MoS_2 film can be improved by the photoelectric transformation and photovoltaic effect of MoS_2 semiconductor devices.

Figure 5 shows the photoluminescence spectra of the MoS_2 and Er: MoS_2 film excited by 360nm light at room temperature. In the pure MoS_2 film, an obvious PL peak is centered at 693nm, coinciding with the intrinsic radiative transition photoluminescence of the single-layer MoS_2. In the Er: MoS_2 film, two significantly enhanced PL peaks are located at 394 and 693nm each. The peak at 394nm is due to the transitions from the $^2H_{11/2}$ energy level to the ground state $^4I_{15/2}$[16]–[18]. The intense peak at 693nm is largely enhanced result from the direct-gap luminescence of MoS_2. It is important to note that the PL intensity of the Er: MoS_2 film is almost twice as strong as that of the undoped film, *i.e.*, the doping of Er in MoS_2 can largely improve the absorption and photoluminescence efficiency of MoS_2, which in turn acts as an exciting active center in the film.

The photocurrent *I-V* behavior of the MoS_2-Si hetero-junction was obtained while irradiating the surface of the films by a standard white light with a power of 100mW/cm^2, as shown in **Figure 6**. For two samples, the current increases exponentially with an increase in the voltage. The short-circuit currents (ISC) of the MoS_2 and Er: MoS_2 film samples are 0.392 and 4.35mA, respectively, and the open-circuit voltage (UOC) is 49.98 and 90.02mV, respectively. Obviously, after Er doping the short-circuit current and open-circuit voltage both increase significantly. This is because the doped Er ions will increase light absorption, resulting in

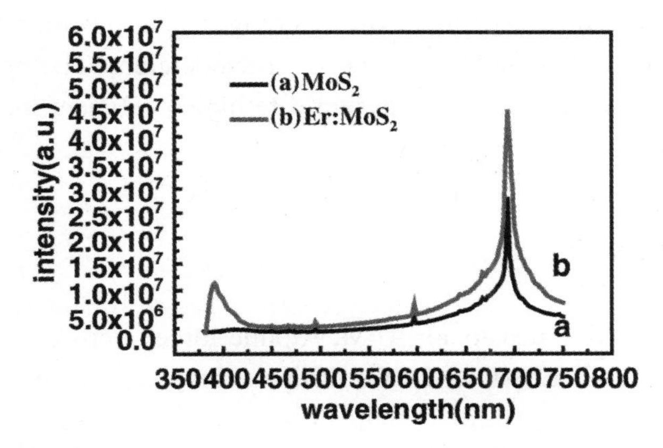

Figure 5. The photoluminescence spectra of the MoS_2 film and the Er:MoS_2 film.

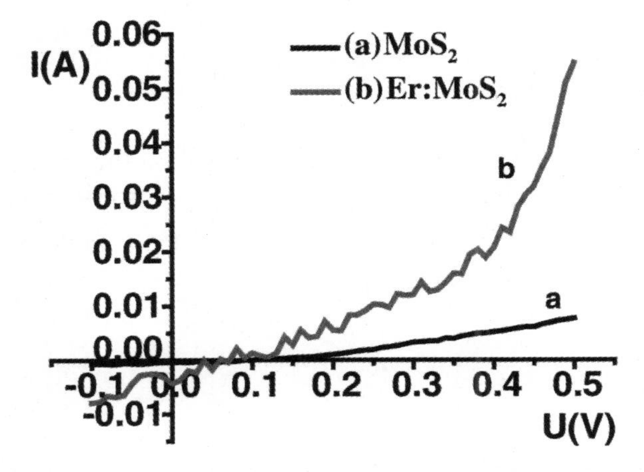

Figure 6. Photocurrent *I-V* curves of the doped and undoped MoS_2/Si heterojunction.

an increase in the number of photo-generated carriers and finally enhancing the photocurrent response.

4. Conclusions

We have studied the effects of Er doping on the surface morphologies, crystalline, optical absorption, PL, and photoelectrical properties of MoS_2 films. We found that the Er^{3+} ions do not change the crystal structure of MoS_2 films but make the crystallinity better. At the same time, Er^{3+} doping improves the carrier mobility and enhances the current-voltage (*I-V*) characteristics of the MoS_2 thin films. Additionally, Er^{3+}-doped MoS_2 films exhibit stronger light absorption and photoluminescence in the visible light range at room temperature. The results show that Er^{3+}-doped MoS_2 film can be used to fabricate highly efficient luminescence and optoelectronic devices.

Abbreviations

2D: Quasi-two dimensional; AFM: Atomic force microscope; CVD: Chemical vapor deposition; $Er(NO_3)_3{\cdot}5H_2O$: Erbium nitrate pentahydrate; Er: Erbium; Er:MoS_2: Erbium-doped MoS_2; H_2SO_4: Sulfuric acid; I_{SC}: Short-circuit currents; *I-V*: Current-voltage; PL: Photoluminescence; REE: Rare earth elements; U_{OC}: Open-circuit voltage; UV-vis: Ultraviolet-visible; XRD: X-ray diffraction.

Acknowledgements

This work was supported in part by the Innovation Program for Postgraduate of Suzhou University of Science and Technology (No. SKCX15_065), the National Natural Science Foundation of China (No. 31570515), and the Scientific Project Program of Suzhou City (No. SYN201511).

Funding

The Innovation Program (No. SKCX15_065) acts as guide to the design of the study and the collection, analysis, and interpretation of the data. The others

(No. 31570515 and No. SYN201511) support the collection of data and the publication of the study.

Authors' Contributions

MM participated in the fabrication of MoS_2, measured and analyzed the data, and wrote the manuscript. XM interpreted the data. Both authors read and approved the final manuscript.

Authors' Information

MM is a graduate student major in the fabrication of semiconductor nanometer materials. XM is a professor and PhD degree holder specializing in semiconductor materials and devices, especially expert in nanoscaled optical-electronic materials and optoelectronic devices.

Competing Interests

The authors declare that they have no competing interests.

Source: Meng M, Ma X. Improving the Photoelectric Characteristics of MoS_2 Thin Films by Doping Rare Earth Element Erbium [J]. 2016, 11.

References

[1] Muller GA, Cook JB, Kim H, Tolbert SH, Dunn B (2015) High performance pseudocapacitor based on 2D layered metal chalcogenide nanocrystals. Nano Lett 15: 1911–1917.

[2] Lembke D, Bertolazzi S, Kis A (2015) Single-layer MoS2 electronics. Accounts Chem Res 48:100–110.

[3] Maitra U, Gupta U, De M (2013) Highly effective visible-light-induced H2 generation by single-layer 1T-MoS2 and a nanocomposite of few-layer 2H-MoS2 with heavily nitrogenated graphene. Angewandte Chemie 52:13057–13061.

[4] Na J, Joo M, Shin M (2014) Low-frequency noise in multilayer MoS2 field-effect transistors: the effect of high-k passivation. Nanoscale 6:433−441.

[5] Kim W, Son JY (2013) Single-layer MoS2 field effect transistor with epitaxially grown SrTiO3 gate dielectric on Nb-doped SrTiO3 substrate. B Korean Chem Soc 34:2563−2564.

[6] Gourmelon E, Lignier O, Hadouda H, Couturier G, Bernede J, Tedd J, Pouzet J, Salardenne J (1997) MS2 (M=W, Mo) photosensitive thin films for solar. Sol Energ Mat Sol C 46:115−121.

[7] Xu X, Hu J, Yin Z (2014) Photoanode current of large-area MoS2 ultrathin nanosheets with vertically mesh-shaped structure on indium tin oxide. Acs Appl Mater Inter 6:5983−5987.

[8] Kin Fai Mak, C.L.J.H.: Atomically thin MoS2 a new direct-gap semiconductor Phys Rev Lett. 2010;136805.

[9] Cappelluti, E., Roldán, R., Silva-Guillén, J.A., Ordejón, P.,Guinea, F.: Tight-binding model and direct-gap/indirect-gap transition in single-layer and multilayer MoS2. Physical review, B. Condensed matter and materials physics. 2013;88:075409-1-075409-18.

[10] Cho B, Kim AR, Park Y, Yoon J, Lee Y, Lee S, Yoo TJ, Kang CG, Lee BH, Ko HC, Kim D, Hahm MG (2015) Bifunctional sensing characteristics of chemical vapor deposition synthesized atomic-layered MoS2. Acs Appl Mater Inter 7:2952−2959.

[11] Tsai DS, Lien DH, Tsai ML, Su SH, Chen KM, Ke JJ, Yu YC, Li LJ, He JH (2014) Trilayered MoS2 metal-semiconductor-metal photodetectors: photogain and radiation resistance. Ieee J Sel Top Quant 20:1−6.

[12] Splendiani A, Sun L, Zhang YB, Li TS, Kim J, Chim CY, Galli G, Wang F (2010) Emerging photoluminescence in monolayer MoS2. Nano Lett 10:1271−1275.

[13] Singha SS, Nandi D, Singha A (2015) Tuning the photoluminescence and ultrasensitive trace detection properties of few-layer MoS2 by decoration with gold nanoparticles. Rsc Adv 5:24188−24193.

[14] Rifai SAA, Ryabtsev SV, Smirnov MS, Domashevskaya EP, Ivanov ON (2014) Synthesis of europium-doped zinc oxide micro- and nanowires. Russian J Physical Chem 88:108−111.

[15] Yourre TA, Rudaya LI, Klimova NV, Shamanin VV (2003) Organic materials for photovoltaic and light-emitting devices. Semiconductor 37:807−815.

[16] Tiwary M, Singh NK, Annapoorni S, Agarwal DC, Avasthi DK, Mishra YK, Mazzoldi P, Mattei G, Sada C, Trave E, Battaglin G (2011) Enhancement of photoluminescence in Er-doped Ag-SiO2 nanocomposite thin films: a post annealing study. Vacuum 85:806−809.

[17] Thomas S, Sajna MS, George R, Rasool SN, Joseph C, Unnikrishnan NV (2015) Investigations on spectroscopic properties of Er3+-doped Li-Zn fluoroborate glass.

Spectrochimica Acta 148A:43–48.

[18] Chen S, Dierre B, Lee W, Sekiguchi T, Tomita S, Kudo H, Akimoto K (2010) Suppression of concentration quenching of Er-related luminescence in Er-doped GaN. Appl Phys Lett 96:181901-1-181901-3.

Chapter 30

Investigation of Optoelectronic Properties of N3 Dye-Sensitized TiO_2 Nano-Crystals by Hybrid Methods: ONIOM (QM/MM) Calculations

Mohsen Oftadeh, Leila Tavakolizadeh

Department of Chemistry, Payame Noor University, Tehran 19395-4697, Islamic Republic of Iran

Abstract: In this article, Ru(4,4′-dicarboxy-2,2′-bipyridine)$_2$(NCS)$_2$ dye (N3) and some derivatives were investigated using Density Functional Theory (DFT) calculations in solution to elucidate the influence of the environment and substituted groups on electronic properties. Full geometry optimization and investigation of electronic properties of N3 dye and some derivatives were performed using DFT and HF calculations. The singlet ground state geometries were fully optimized at the B3LYP/3-21G** level of theory through the Gaussian 98 program. Based on the computed results, the optoelectronic properties are sensitive to chemical solvent environments. Moreover, the properties of anatase cluster (TiO_2) models have been investigated, and N3 dyes have been adsorbed on TiO_2 nano-particle with diprotonated states. The modified N3 dyes highly affected the electronic structure. This leads to significant changes in the adsorption spectra as compared to the N3 dyes. Through hybrid methods, the properties of interfacial electronic coupling of the combined system were estimated. The results of some combined systems showed that the electronic coupling, lowest lowest unoccupied molecular orbitals, and the TiO_2 conduction band resided in the visible region.

Keywords: DFT, Gap Energy, Sensitized Dye, Optical Properties

1. Background

Photo-induced phenomenon, based on photon absorption with enough energy, causes the excitation of electrons and then the current generation. Electron transfer (ET) at organic/inorganic interfaces plays a key role in a variety of applications including photocatalysis, photoelectrolysis, and solar cell and nano-scale electronics[1]-[5]. The dye-sensitized nano-crystalline solar cell, or Grätzel cell, is a popular alternative in comparison to the costly traditional solar cell[3][6]. The attached chromophore dye with tuned desired wavelength in a well-known Grätzel cell should absorb photons and generate electrons within dye-sensitized titanium dioxide nano-particles[7]-[10]. Quantum chemical calculations can provide information about geometrical and electronic structure in dye-sensitized solar cells (DSSCs)[11]. The N3 dye, as shown in **Figure 1**, is one conventional dye which can show optical adsorption spectra[12] related to the size and conjugation of linkers to the surface at semiconductor TiO_2. Surface electron transfer accompanying the initial light absorption and excitation of ruthenium dyes often leads to efficient photo-induced charge separation across the dye and TiO_2 interfaces.

TiO_2 can be formed in several phases such as anatase, rutile and brookite. There are some properties of TiO_2 nano-crystals such as electro-opticality, low cost, chemical stability, non-toxicity, abundance, availability, and lack of erosion and corrosion against light which are common in the literature[13]. The anatase crystal phase of TiO_2 has an octahedral structure with wide gap energy, photo-induced activity, large surface area-to-bulk ratio, and can absorb UV light more than other phases. Due to the lack of anatase stability in its structure and a high degree of dangling bonds, calculation of structural and electronic properties of the large number of under coordinated atom is complicated. However, the size and complexity of TiO_2 nano-particles impose challenges for choosing the methods. Several titanium oxide nanocrystals have been investigated in some studies[14]-[19]. The calculation for anatase $(TiO_2)_5$ and $(TiO_2)_{16}$ clusters has been performed by DFT,

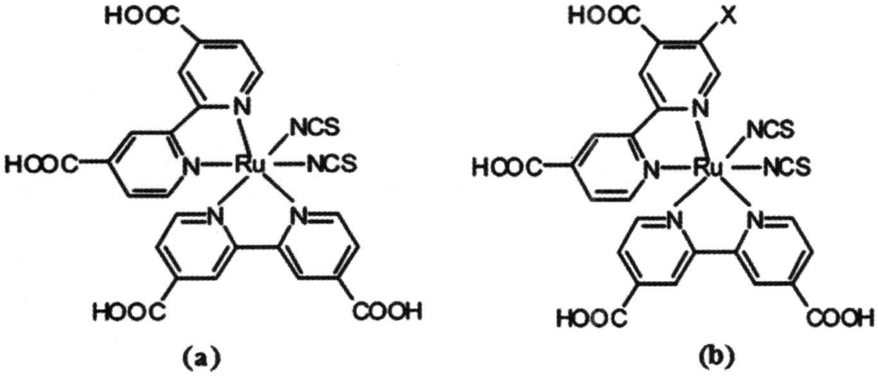

Figure 1. N3 dye (a) and its α substituted derivatives. The structural formulae of N3 dye (a) and its α substituted derivatives (b) with X = CH_3, C_2H_5, NH_2, $N(CH_3)_2$, F, and Cl.

but because of computational demand for large systems, the semi-empirical methods have been used. The $(TiO_2)_{54}$ cluster has large surface and has more tendency toward sphericity, and it is selected as the basis for computing absorption dyes and its derivatives.

The Ru(4,4′-dicarboxy-2,2′-bipyridine)$_2$(NCS)$_2$ or 'N3' dye and its derivatives are probably the most well known sensitized dyes which transfer electron by adsorbing sunlight followed by electron injection to the conduction band of TiO_2, so the gap energy will be increased[20][21]. The electron transfer depends on different parameters, for example, the type of attachment such as bridge, chelating, mono-dentate which are shown in **Figure 2**[22][23]. The dye is adsorbed in a prototype binding of the two carboxylic acid anchor groups of one of the bipyridine ligands in a bridging bi-dentate adsorption mode. **Figure 2** shows the types of attachment anchor groups on the adsorbent layer[4].

The purpose of this paper is to investigate the role of acceptor and donor groups on N3 dyes and to understand the electron transfer phenomena, variation of gap energy, and the geometrical and electronic coupling among the parts in the combined system, $(TiO_2)_{54}$-N_3 derivatives.

2. Methods

Through Density Function Theory (DFT), the full geometry optimization

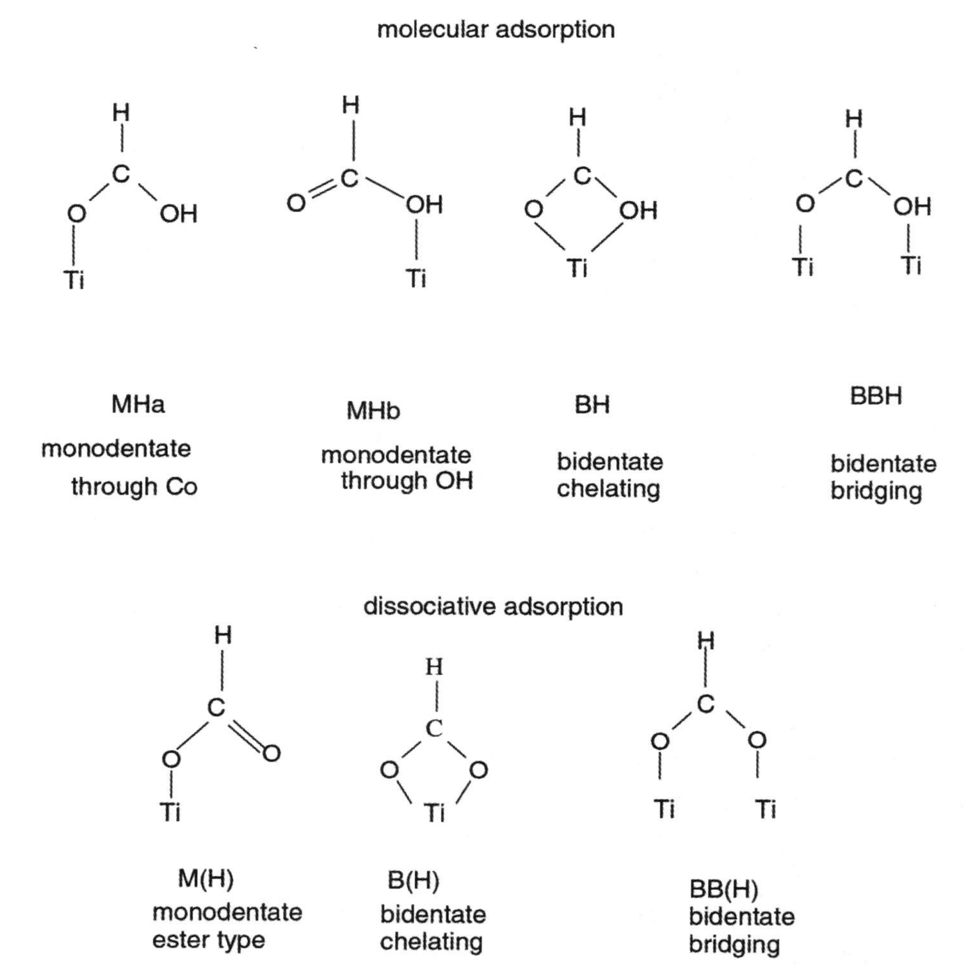

Figure 2. The types of attaching anchor groups on the adsorbent layer[4].

and investigation of the electronic properties of N3 dye and some derivatives were performed[11]. The results were obtained by DFT and supported by Hartree-Fock (HF) calculations. The singlet ground state geometries were fully optimized at the B3LYP/3-21G** level of theory. All calculations were performed with Gaussian 98 program (Gaussian, Inc., Wallingford, CT, USA)[24].

The variation of gap energies was investigated by inserting electron donor groups such as CH_3, C_2H_5, NH_2, $N(CH_3)_2$, or electron accepter groups such as F and Cl on the α position related to carboxylate on N3 dye as shown in **Figure 1**. To compare of the results, all calculations were done first in gas phase and then in solvent phase. In this study, PCM model of ethanol (dielectric constant, $\varepsilon = 25$) and acetonitrile ($\varepsilon = 37$) was used.

Since the DFT calculation of the large clusters is very costly, the $(TiO_2)_{54}$ nano-crystals have been optimized by MSINDO semi-empirical method[3]. Further, we used the integrate B3LYP:ZINDO1 approach for the combined dyes and $(TiO_2)_{54}$ cluster systems and some of the dye derivates.

To evaluate the accuracy of the proposed hybrid method results, the combined system of N3 and $(TiO_2)_{54}$ has been optimized via the DFT method by a 16-core supercomputer. The duration of the each calculation was about 2 months.

3. Results and Discussion

The computational results for the N3 dyes and some derivatives which were performed by B3LYP method and 3-21G** basis set in both gas and solvent phases are presented in **Table 1**. A visual comparison of the highest occupied molecular orbital (HOMO) and the lowest unoccupied molecular orbital (LUMO) of N3 and

Table 1. The calculated HOMO, LUMO (in Hartree), gap energy, and adsorption threshold by DFT.

	Gas phase				Ethanol				Acetonitrile			
-X	HOMO	LUMO	Gap (V)	λ (nm)	HOMO	LUMO	Gap (eV)	λ (nm)	HOMO	LUMO	Gap (eV)	λ (nm)
-H	−0.1063	0.1085	1.494	830	−0.1976	−0.1202	2.105	620	−0.1988	−0.1203	2.136	581
$-CH_3$	−0.1726	−0.1172	1.508	822	−0.1943	−0.1178	2.082	595	−0.1955	−0.1179	2.109	588
$-C_2H_5$	−0.1627	−0.1052	1.565	792	−0.1867	−0.1164	1.913	648	−0.1910	−0.1172	2.008	614
$-NH_2$	−0.1696	−0.1117	1.574	788	−0.1872	−0.1145	1.978	626	−0.1881	−0.1147	1.996	621
$-N(CH_3)_2$	−0.1613	−0.1022	1.609	770	-	-	-	-	−0.1592	−0.1055	1.462	848
-F	−0.1741	−0.1191	1.498	828	−0.1942	−0.1216	1.977	627	−0.1980	−0.1223	2.059	602
-Cl	−0.1741	−0.1197	1.479	838	−0.1928	−0.1232	1.894	654	−0.1981	−0.1243	2.008	618

At B3LYP/3-21G** level of theory for N3-X dye in the gas, ethanol, and acetonitrile phases. X denotes the substituted groups by one of the hydrogen on pyridine besides the carboxylate group.

some of its derivatives in the gas, acetonitrile, and ethanol phases shows an increasing and obvious shifting in absolute energy and reveal a clear difference in the energetic ordering of the frontier orbitals in different environments.

The results show that the substituted groups do not have an important role in transferring the absorption wavelength from UV to visible region in the gas phase. Moreover, the groups have a few influences on the decrease of N3 gap energy.

The energy silently decreases in solvent environment, so the difference between gas and solvent environment is about 0.5eV for all groups. The solvent effect on the gap energy variation occurs through the interaction of the dye and ethanol or acetonitrile molecules. It has a significant influence on the optical properties of these complexes by the environment, so the wavelength of N3 complexes is shifted from 830 to 580 nm by inclusion of the solvent. The range of variation of optoelectronic properties is taken into account under a model for complexes with conjugated properties whether or not the theoretical modeling is capable of providing a physically realistic description of the complexes with extended conjugation.

Using the integrate B3LYP:ZINDO1 approach, the results of the combined systems of N3 dyes and some of the derivatives with $(TiO_2)_{54}$ cluster are summarized in **Table 2**. The table shows that some of the absorption wavelengths of a combined system are located in the visible area. Furthermore, the computed results by the mentioned methods for heterogeneous super molecule are given in **Table 3**. However, the gap energy almost differs between ONIOM and DFT; ONIOM results are in agreement with our expectation.

4. Conclusion

To significantly develop DSSCs and related photo-electrochemical devices, the quantum chemical calculations were performed to provide a better theoretical understanding of the basic physical and chemical processes of dye-sensitized semiconductors. As shown before, the modified N3 dyes have highly affected the electronic structure. This leads to significant changes in the absorption spectra as compared to the N3 dyes. The optical absorption spectra are related to the environment, but not probably related to the type of *X*-substituted groups. The results

Table 2. The calculated gap energy and adsorption threshold by ZINDO1 of theory.

Molecules	Gap(eV)	λ(nm)
$(TiO_2)_{54}$	3.97	313
$(TiO_2)_{54} + N_3$	1.54	800
$(TiO_2)_{54} + N_3 + F$	4.21	295
$(TiO_2)_{54} + N_3 + Cl$	6.6	93
$(TiO_2)_{54} + N_3 + CH_3$	2.7	445
$(TiO_2)_{54} + N_3 + C_2H_5$	2.5	490
$(TiO_2)_{54} + N_3 + NH_2$	1.6	780
$(TiO_2)_{54} + N_3 + N(CH_3)_2$	0.3	4,353

For combined system of N3-*X* dyes and $(TiO_2)_{54}$ cluster in the gas phase. *X* denotes the substituted group by one of the hydrogen on pyridine besides of carboxylate group.

Table 3. The calculated gap energy and adsorption wavelength of combined system of N3 and $(TiO_2)_{54}$ cluster.

Method	Gap(eV)	λ(nm)
ONIOM	1.54	800
B3LYP	3.93	320

In the gas phase by ONIOM and B3LYP methods.

of the combination system showed that the electronic coupling of the lowest dye LUMOs and the TiO_2 conduction band is negligible. Moreover, the lowest LUMO dye and TiO_2 band conduction, using hybrid method and the electronic structure, agree with the experiment evidence. Finally, the absorption wavelength of some substituted dyes of combined system is located in the visible area.

Competing Interests

The authors declare that they have no competing interests.

Authors' Contributions

MO participated in the design and the sequence alignment of the study. LT carried out the jobs by Gaussian and participated in the evaluation of the data and in drafting the manuscript. Both authors read and approved the final manuscript.

Acknowledgements

The authors greatly appreciate Dr. HS Wahab from the Baghdad University of Technology for his instructions.

Source: Oftadeh M, Tavakolizadeh L. Investigation of optoelectronic properties of N3 dye-sensitized TiO_2 nanocrystals by hybrid methods: ONIOM (QM/MM) calculations[J]. International Nano Letters, 2013, 2(1):5–9.

References

[1] Ferrere, S, Gregg, BA: Large increases in photocurrents and solar conversion efficiencies by UV illumination of dye sensitized solar cells. J. Phys. Chem. B 105, 7602–7605 (2001).

[2] Derosa, PA, Seminario, JM: Electron transport through single molecules: scattering treatment using density functional and Green function theories. J. Phys. Chem. B 105, 471–481 (2001).

[3] O'Regan, B, Grätzel, M: A low-cost, high-efficiency solar cell based on dye sensitized colloidal TiO2 films. Nature 353, 737–740 (1991).

[4] Pan, J, Benko, G, Xu, YH, Pascher, T, Sun, LC, Sundstrom, V, Polivka, T: Photoinduced electron transfer between a carotenoid and TiO2 nanoparticle. J. Am. Chem. Soc. 124, 13949–13957 (2002).

[5] Biju, V, Micic, M, Hu, DH, Lu, HP: Intermittent single-molecule interfacial electron transfer dynamics. J. Am. Chem. Soc. 126, 9374–9381 (2004).

[6] McConnell, RD: Assessment of the dye-sensitized solar cell. Renew. Sustain. Energy Rev. 6, 273–295 (2002).

[7] Anderson, NA, Lian, TQ: Ultrafast electron transfer at the moleculesemiconductor nanoparticle interface. Ann. Rev. Phys. Chem. 56, 491-519 (2005).

[8] Rego, LGC, Batista, VS: Quantum dynamics simulations of interfacial electron transfer in sensitized TiO2 semiconductors. J. Am. Chem. Soc. 125, 7989–7997 (2003).

[9] Stier, W, Duncan, WR, Prezhdo, OV: Ab initio molecular dynamics of ultrafast electron injection from molecular donors to the TiO2 acceptor. SPIE Proc. 5223, 132–

146 (2003).

[10] Duncan, WR, Stier, WM, Prezhdo, OV: Ab initio nonadiabatic molecular dynamics of the ultrafast electron injection across the alizarin-TiO2 interface. J. Am. Chem. Soc. 127, 7941–7951 (2005).

[11] Lundqvist, MJ: Quantum chemical modeling of dye-sensitized titanium dioxide. Uppsala University, Sweden (2006). PhD thesis.

[12] Aghtar, A: The making of the dye-sensitized solar cell with TiO2 nanoparticles for use in the photocatalytic oxidation-reduction reactions. Payame Noor University (PNU), Shiraz, Iran (2009). MSc dissertation.

[13] Diebold, U: Structure and properties of TiO2 surfaces: a brief review. Appl. Phys. A 76, 681–687 (2003).

[14] Nilsing, M, Lunell, S, Persson, P, Ojama, L: Phosphonic acid adsorption at the TiO2 anatase (101) surface investigated by periodic hybrid HF-DFT computations. Surf. Sci. 582, 49–60 (2005).

[15] Homann, T, Bredow, T, Jug, K: Adsorption of small molecules on the anatase (100) surface. Surf. Sci. 555, 135–144 (2004).

[16] Vittadini, A, Selloni, A, Rotzinger, FP, Gratzel, M: Formic acid adsorption on dry and hydrated TiO2 anatase (101) surfaces by DFT calculations. J. Phys. Chem. B 104, 1300–1306 (2000).

[17] Nilsing, M, Persson, P, Ojamae, L: Anchor group influence on molecule-metal oxide interfaces: periodic hybrid DFT study of pyridine bound to TiO2 via carboxylic and phosphonic acid. Chem. Phys. Lett. 415, 375–380 (2005).

[18] Rappoport, D, Crawford, NRM, Furche, F, Burke, K: Approximate density functionals: which should I choose? In: Solomon, EI, King, RB, Scott, RA (eds.) Computational Inorganic and Bioinorganic Chemistry, pp. 159–172. Wiley, Chichester (2009).

[19] Nazeeruddin, MK, De Angelis, F, Fantacci, S, Selloni, A, Viscardi, G, Liska, P, Ito, S, Takeru, B, Gratzel, M: Combined experimental and DFT-TDDFT computational study of photoelectrochemical cell ruthenium sensitizers. J. Am. Chem. Soc. 127, 16835–16847 (2005).

[20] Wahab, HS, Bredow, T, Aliwi, SM: Computational modeling of the adsorption and photodegradation of 4-chlorophenol on anatase TiO2 particles. J. Mol. Struct. Theochem 863, 84–90 (2008).

[21] Lundqvist, MJ, Nilsing, M, Persson, P, Lunell, S: DFT Modeling of bare and dye-sensitized TiO2 nanocrystals. Int. J. Quantum Chem. 106, 3214–3234 (2006).

[22] Lundqvist, MJ, Galoppini, E, Meyer, GJ, Persson, P: Calculated optoelectronic properties of ruthenium tris-bipyridine dyes containing oligophenyleneethynylene rigid rod linkers in different chemical environments. J. Phys. Chem. A 111, 1487–1497 (2007).

[23] Iozzi, MF: Theoretical models of the interaction between organic molecules and

semiconductor surfaces. Napoli University, Italy (2006). PhD thesis.

[24] Frisch, MJ, Trucks, GW, Schlegel, HB, Gill, PMW, Johnson, BG, Robb, MA, Cheeseman, JR, Keith, T, Petersson, GA, Montgomery, JA, Raghavachari, K, Al-Laham, MA, Zakrzewski, VG, Ortiz, JV, Foresman, JB, Cioslowski, J, Stefanov, BB, Nanayakkara, A, Challacombe, M, Peng, CY, Ayala, PY, Chen, W, Wong, MW, Andres, JL, Replogle, ES, Gomperts, R, Martin, RL, Fox, DJ, Binkley, JS, Defrees, DJ, et al.: Gaussian 98: revision D.2. Gaussian Inc, Pittsburgh (1999).